W9-AQB-094

ON TYCHO'S ISLAND

TYCHO BRAHE (1546–1601), the premier patron-practitioner of science in sixteenth-century Europe, established a new role of scientist as administrator, active reformer, and natural philosopher. This book explores his wide range of activities, which encompass much more than his reputed role of astronomer. Christianson broadens this singular perspective by portraying him as Platonic philosopher, Paracelsian chemist, Ovidian poet, and devoted family man. From his private island in Denmark, Tycho Brahe used patronage, printing, friendship, and marriage to incorporate men and women skilled in science, technology, and the fine arts into his program of cosmic reform. This pioneering study includes capsule biographies of two dozen individuals, including Johannes Kepler, Willebrord Snel, Willem Blaeu, several artists, two bishops, a rabbi, and various technical specialists, all of whom helped shape the culture of the Scientific Revolution. Under Tycho's leadership, their teamwork achieved breakthroughs in astronomy, scientific method, and research organization that were essential to the birth of modern science.

JOHN ROBERT CHRISTIANSON is a Research Professor of History at Luther College, where he taught history from 1967 to 1997 and served as Chairman of the History Department during 1967–82. In 1995, he was dubbed Knight of the Royal Norwegian Order of Merit by King Harald V. He was also awarded the Bronze Medal of the League of Finnish-American Societies in 1985 and received the Alf Mjøen Prize in 1989. Christianson is a former Fellow of the American Council of Learned Societies and has held postdoctoral grants from the American Philosophical Society, Danish Bicentennial Foundation of 1976, Danish National Bank, George C. Marshall Fund in Denmark, and the U.S. National Endowment of the Humanities, among others. Christianson has written, edited, and/or translated nine books and more than a hundred articles on Scandinavian and Scandinavian-American topics, including twenty publications on Tycho Brahe in such journals as *Scientific American, Isis, Centaurus, Sixteenth Century Journal, Fund og Forskning i Det Kongelige Biblioteks Samlinger,* and *Personalhistorisk Tidsskrift.*

"PORTRAIT OF TYCHO BRAHE OTTESEN THE DANE AT THE AGE OF FIFTY, when after a long period of exile from his native land through divine providence he regained the liberty he had sought after." He is dressed in the black attire of a courtier, with a white lace collar and cuffs, gold jewelry including a signet with the Brahe arms on his index finger, and King Frederick II's Order of the Elephant on two golden chains. The prosthesis is evident on the bridge of his nose. The emblem at top left shows a cone (which Tycho called a "pyramid") on a pedestal, protected from the surrounding wind, water, and lightning by a canopy suspended from a hand emerging from the sun. The legend on the cone reads: "Standing on firm ground, I am protected though the wind, fire, and waves do rage." Honnens de Lichtenberg (1989, 362–3) interpreted this emblem as symbolizing that the changing fortunes of the world could not overthrow Tycho because the four elements of air, fire, water, and earth combined to give him strength as he rose to immortality under the protection of Apollo–Jupiter–Jehovah. (Courtesy Skokloster Castle)

ON TYCHO'S ISLAND

—

TYCHO BRAHE, SCIENCE, AND CULTURE IN THE SIXTEENTH CENTURY

Abridged Paperback Edition

JOHN ROBERT CHRISTIANSON

Research Professor of History, Luther College

CAMBRIDGE
UNIVERSITY PRESS

Publisher's Note: This paperback edition is slightly abridged from the hardcover edition, *On Tycho's Island: Tycho Brahe and His Assistants, 1570–1601* (2000) in that Part II, the Biographical Directory, has been shortened. No illustrations have been omitted, and the final one has been improved.

CAMBRIDGE
UNIVERSITY PRESS

32 Avenue of the Americas, New York NY 10013-2473, USA

Cambridge University Press is part of the University of Cambridge.

It furthers the University's mission by disseminating knowledge in the pursuit of education, learning and research at the highest international levels of excellence.

www.cambridge.org
Information on this title: www.cambridge.org/9780521008846

© John Robert Christianson 2000, 2002

This publication is in copyright. Subject to statutory exception and to the provisions of relevant collective licensing agreements, no reproduction of any part may take place without the written permission of Cambridge University Press.

First published 2000
First paperback edition 2002

A catalogue record for this publication is available from the British Library

Library of Congress Cataloguing in Publication data
Christianson, J. R. (John Robert)
On Tycho's island : Tycho Brahe, science, and culture in the sixteenth century / John Robert Christianson.
p.   cm.
Includes bibliographical references and index.
ISBN 0-521-65081-X (hc.) – ISBN 0-521-00884-0 (pb.)
1. Brahe, Tycho, 1546–1601.   2. Science – Denmark – History – 16th century.
3. Astronomers – Denmark Biography.   I. Title.
QB36.B8C54   1999
520´.92 – dc21
[B]   99-33118
CIP

ISBN 978-0-521-00884-6 Paperback

Cambridge University Press has no responsibility for the persistence or accuracy of URLs for external or third-party internet websites referred to in this publication, and does not guarantee that any content on such websites is, or will remain, accurate or appropriate.

To
Birgitte

How Apollo, the leader of the Muses, overcame the god Pan in a music competition:

"[Apollo's] golden head was wreathed with laurel of Parnasus, and his mantle, dipped in Tyrian dye, swept the ground. His lyre, inlaid with gems and Indian ivory, he held in his left hand, while his right hand held the plectrum. His very pose was that of an artist. Then with trained thumb he plucked the strings and, charmed by those sweet strains, Tmolus ordered Pan to lower his reeds before the lyre."

Ovid, *Metamorphoses,* translated by Frank Justus Miller

The goddess, Pallas Athena, visited the home of the Muses on Mount Helicon to see the spot where the hoof of Pegasus, the winged horse, had struck the ground, causing Hippocrene, the sacred spring, to flow. She was greeted by Urania, the Muse of Astronomy. Athena told why she had come, and Urania replied,

"'Whatever cause has brought thee to see our home, O goddess, thou art most welcome to our hearts. But the tale is true, and Pegasus did indeed produce our spring.' And she led Pallas aside to the sacred waters. She long admired the spring made by the stroke of the horse's hoof; then looked round on the ancient woods, the grottoes, and the grass, spangled with countless flowers. She declared the daughters of Mnemosyne to be happy alike in their favourite pursuits and in their home. And thus one of the sisters answered her: 'O thou, [Athena], who wouldst so fitly join our band, had not thy merits raised thee to far greater tasks, thou sayest truth and dost justly praise our arts and our home. We have indeed a happy lot – were we but safe in it.'"

The other daughters of Mnemosyne and Zeus, sisters of Urania, Muse of Astronomy, were these: Calliope, Muse of Epic and Heroic Poetry; Clio, Muse of History; Erato, Muse of Love Poetry; Euterpe, Muse of Lyrics and Music; Melpomene, Muse of Tragedy; Polyhymnia, Muse of Sacred Hymns; Terpsichore, Muse of Dance; and Thalia, Muse of Comedy.

Ovid, *Metamorphoses,* translated by Frank Justus Miller

CONTENTS

ILLUSTRATIONS

PREFACE

T HIS BOOK BEGAN IN A SHOEBOX. In the course of my research, whenever I ran across the name of a coworker of Tycho Brahe, I made a notecard and put it in the shoebox. Then I tried to find out who they all were and what they did. As years passed, the shoebox filled up, its contents became a data base, and I decided it was time to write a book about Tycho Brahe and his associates. So here we are.

Some scholar, deep into the lore of the late sixteenth century, may ask why I did not include this or that figure, a Petrus Severinus, Duncan Liddel, Bartolomæus Scultetus, or any number of others. The answer is that I had spent more than twenty-five years compiling names, trying to track down each one of them and make sense of the cultural, intellectual, and material networks that connected them, and the task could easily have gone on for another quarter-century, but the time had come to publish what I had found. I encourage others to carry on the work and can only advise, by way of incitement: Seek and you will find!

I owe profound thanks to those who have supported my research on Tycho Brahe and Tycho's island. At the beginning of my academic career, a Danish government grant (42.Dan.5/3.b) in 1962, together with two United States Office of Education Title VI Foreign Language Fellowships in 1962–3 and 1963–4, allowed me to research and write my dissertation on Tycho Brahe. A research associateship and travel grant from the University of South Dakota in 1966 let me continue my Tychonic research and writing, as did summer stipends from the National Endowment for the Humanities in 1968 and 1980. A summer grant from the Penrose Fund (no. 5865) of the American Philosophical Society in 1971 allowed me to focus specifically on Tycho Brahe and the patronage of science. In 1973–4, my work on Tycho Brahe was supported by a fellowship of the American Council of Learned Societies and a grant from the George C. Marshall

Fund in Denmark. In 1989, the National Bank of Denmark fostered my research by arranging the use of an apartment at Nyhavn 18 in Copenhagen. Luther College has generously encouraged my Tychonic research through sabbatical leaves in 1973–4, 1980, and 1988–9, a Paideia Endowment Sabbatical Support grant in 1988–9, and smaller grants from various research funds.

Václav Babicka, Bartlett R. Butler, Richard G. Cole, Lesley B. Cormack, Jesper Düring Jørgensen, Noah Efron, Elizabeth Eisenstein, Hanne Honnens de Lichtenberg, W. A. Huijsmans, Erik Iversen, Kjell Lundquist, Bent Kæmpe, Peter Kristiansen, Kristian Peder Moesgaard, the late Wilhelm Norlind, Sarah Tyacke, Dieter Veldtrup, Alex Wittendorff, Peter Zeeberg, and many other scholars, archivists, and librarians have kindly responded to my inquiries over the years, and I thank them for it. The late Victor E. Thoren, a sensitive and generous scholar, discovered many things about the astronomy of Tycho Brahe that had evaded his great predecessors, Gassendi and Dreyer, and I benefited from discussing Tycho with him. J. C. Baron Bille Brahe did a great deal to encourage my work. Four individuals read the manuscript of this book in its entirety in one version or another, and I value their comments: Paul Christianson, Owen Gingerich, Jole Shackelford, and James R. Voelkel. Michael Gnat sensitively polished the manuscript and "packaged" the book. Alex Holzman was the acquisitions editor for Cambridge University Press. Finally, I owe more thanks than I can express here to my wife, Birgitte Christianson, who read the manuscript, discussed it frequently, suggested the title, and supported my work and well-being in untold ways from 1962 until the present day. Any errors remaining in the book are my own.

PART ONE

ON TYCHO'S ISLAND

INTRODUCTION

THIS BOOK IS ABOUT POWER. It shows how one man, Tycho Brahe, used his powerful position to bend the lives of hundreds of others toward a goal that he deemed important: a new understanding of the cosmos. It shows how he established a new role for the astronomer as large-scale organizer, active reformer, and natural philosopher.

Patronage was his means to create structures of power and incorporate others into his program of reform. This book therefore examines the hierarchies of patronage and clientage that extended from the highest ranks of human society to the broad laboring masses, hierarchies that incorporated men and women skilled in science, technology, learning, and the fine arts to work with Tycho Brahe and take him as their role model.

He used his power and patronage to build teams of people working together to carry out his ends, so this book is also about teamwork, which it aims to show was essential to the birth of modern science. Newton once remarked that he could see far because he stood on the shoulders of giants. This book shows how much of seventeenth-century European culture rested on the shoulders of a late sixteenth-century giant, Tycho Brahe: scientist, natural philosopher, technical expert, and poet; connoisseur of music, courtly grace, and the fine arts; and one of the most innovative organizers known to history.

In sixteenth-century Europe, Tycho Brahe was the patron of science par excellence. Because his use of patronage goes to the heart of the transition from Scientific Renaissance to Scientific Revolution, this book is a case study of patronage during that key period. The first chapter deals with Tycho Brahe as client of the king of Denmark and differentiates the two types of royal patronage he received. Chapter 2 shows how he reorganized the island of Hven and brought all of its inhabitants into his system, linking their labors to those of his dependents in Norway and two provinces of Denmark. Chapter 3 focuses on theories of friendship that established reciprocal links with learned individuals throughout Europe and created an international support network for his work. Chapter 4 examines the structures of household and family as they were used by Tycho Brahe to shape his working staff. Chapter 5 shows how scientific expeditions, staff organization, the technology of printing, the fine arts, and the manipulation of objects and information in "museums" all served to reinforce reciprocal relationships of friendship and patronage while strengthening the pursuit of science and disseminating its results. Chapter 6 examines how the social linkage of patronage to family could shape the elements of continuity and discontinuity in sixteenth-century science. Chapter 7 looks at the problem of cultural conflict when scholars of various nationalities brought different mentalities and assumptions to their common work.

Chapter 8 shows how marriage strategies aimed to reinforce bonds of friendship and patronage, and how they could fail: The betrothal of Tycho Brahe's daughter, Magdalene Tygesdatter, was an attempt to bridge a gap between social classes, and its failure revealed much about the dynamics and the limits of late sixteenth-century social bonding. Chapters 8 and 9 show the importance of honor as a personal attribute, and how reputations could be destroyed and patronage lost in fierce battles between rival individuals, institutions, and ideologies. Chapter 10 deals with the difficult but not insuperable task of establishing new patronage relationships on the international level.

The scientific, research, and familial legacies of Tycho Brahe are discussed in Chapter 11, which brings Part One of this volume to a close. Part Two is a Biographical Directory, which sketches the careers of individuals who were drawn into Tycho Brahe's network of patronage or, in a few cases, of friendship.

Tycho Brahe is the richest example of scientific patronage and friendship in sixteenth-century Europe, but his structures of power and support interlocked within a broader cultural ambience that also needs to be examined. In his Latin poetry, Tycho Brahe described his island of Hven as *Insula Venusinus,* a magical place where gods and goddesses dwelt on earth. The island's inhabitants lived and worked together in the Temple of Urania, devoting themselves to eternal matters and ignoring all lowly, earthly ones. In pursuing this elevated life, he asserted, they became like demigods. He said that visitors to the island had the privilege of experiencing the realm of the divine and left as better people. Such was Tycho Brahe's own description of a place that modern historians have seen as a crucible of the Scientific Revolution. The disparity between the way he understood what he was doing and the twentieth-century assessment of his scientific achievement is quite astounding.

An important task of this book is therefore to focus upon the lives of Tycho Brahe's coworkers in order to see who these "demigods" really were, how they lived and worked together, what they made of their lives, and how they related to the mainstream of European science and culture. To bridge the gap between their mental world and ours, it is necessary to examine how these men and women of the sixteenth and seventeenth centuries saw the world, and how their world view changed. On this basis, we can move to a new understanding of Tycho Brahe's role in the Scientific Revolution and of his innovative achievement in organizing large-scale scientific research.

Tycho Brahe's idea of a large-scale, multifaceted scientific-research institution was inspired by the Neoplatonic tradition of the Renaissance, reinforced by the memory of the ancient Museum of Alexandria, and given focus through some of the places Tycho had visited in Denmark, Germany, Switzerland, and Italy. Once he had the idea for such an institution, financing it depended upon the fact that Tycho Brahe was born into the high nobility and had close connections to the Danish crown. Tycho recruited his staff by means of a European network of learned men and courtiers who felt connected by ties of Platonic friendship. Together, Tycho, his students, coworkers, and colleagues searched for the mathematical unities of the cosmos by means of huge new instruments, standardized procedures, innovative methods of observation and experimentation, and an epistemology that insisted upon quantifying and verifying the reliability of data. The result was a paradigm of research that gradually became a general European phenomenon.

The life and scientific achievement of Tycho Brahe were described at length in the late Victor E. Thoren's *The Lord of Uraniborg: A Biography of Tycho Brahe* (New York: Cambridge University Press, 1990), to which I contributed, but the focus of this book is quite different. Here, the staff, structure, and culture of Tycho's island move to the forefront, including the networks of patronage and clientage that converged on the island. What life and scientific work was like on Hven (also called Hveen or Ven) in the years 1576–97 is considered in some detail, followed by a look at how he conducted his search for a new patron in the years 1597–1601.

Scores of assistants, poets, scholars, scientists, and technicians scattered throughout Europe when they left Tycho Brahe's service, working in many fields and frequently patterning their lives on his. Some of them tried to establish new institutions on the model of Hven. In the seventeenth century, they infused the scientific culture of Tycho's island into the mainstream of European life. This is their story.

I

IN KING FREDERICK'S SERVICE

———

1575–1576

THE RELUCTANT LORD

Toward the end of the year 1575, Tycho Brahe returned to Denmark from an extended trip abroad and immediately presented himself to King Frederick II at Sorø Abbey, where the court had arrived to celebrate the Yuletide.[1] Tycho Brahe was a consummate courtier, twenty-nine years of age, barrel-chested and noble of bearing, with pale eyes, reddish-blond hair and beard, immense moustaches, and a flesh-colored prosthesis on his nose. He wore the flowing cape, plumed bonnet, and sword of a nobleman. King Frederick was forty-two, regal, magnanimous, tall, and athletic, with a Van Dyke beard and close-cropped, curly hair.[2] He was known as a patron of learning and the fine arts, and he also loved rich banquets with mead and rhenish wine and hearty draughts of ale, as well as the witty, learned talk around a table like that of the Abbot of Sorø.

During his travels abroad, Tycho Brahe had acted as a royal agent to recruit artists and artisans for King Frederick, who was building a great new castle at Elsinore. Tycho had previously brought Venetian glassmakers to Denmark, as well as German papermakers, and he reported to the king on his recent contacts with painters, sculptors, hydraulic engineers, and other skilled artisans in Hesse-Kassel, Augsburg, Nuremberg, and Venice. He also described his visits to various courts and told about the coronation of the king of the Romans, which he had attended in Regensburg.

King Frederick beamed at the success of the mission and assumed that Tycho Brahe was now ready to serve the Danish crown in the tradition of his ancestors. Tycho had grown up in castles commanded by his father and foster father, and he was well aware that the opportunity to wield such authority was reserved to a privileged few. He was also aware of the burdens of command. The king offered him a choice of fiefs. Hammershus and

Visborg Castles were on distant Baltic islands of considerable strategic importance; Helsingborg and Landskrona Castles warded over the Sound (Øresund), the gateway to the Baltic.[3] In all of these fiefs, the Lord Lieutenant held power over hundreds of peasants and villagers, while knights, soldiers, servants, petitioners, tenants, and couriers swarmed around him. King Frederick offered a choice of fiefs, but Tycho Brahe was noncommittal. He replied with the grace of a polished courtier, yet he demurred. He needed time to think it over. The king was puzzled, but he accepted Tycho's tactful demurral.

Soon after, Tycho Brahe wrote more frankly to a friend, Johannes Pratensis, in Copenhagen: "I did not want to take possession of any of the castles our benevolent king so graciously offered me. I am displeased with society here, customary forms and the whole rubbish. . . . Among people of my own class . . . I waste much time." What Tycho really wanted, he wrote to his friend, was to live in a location favorable to a "student interested in learned subjects, or one who loves Apollonian tranquility and the Muses."[4] Beleaguered Baltic islands and great castles were not in that category. He longed to exchange his noble garb for the starry robe of a magus and natural philosopher, but he realized that he needed to do so in a way that would not arouse the ire of his liege lord, the king.

A NOBLE UPBRINGING

Despite his unusual preferences, Tycho Brahe's career to the age of twenty-nine had generally been typical of his generation of Danish aristocrats. He was born at his family's ancestral seat of Knutstorp Castle [Fig. 1] on 14 December 1546, the son of Otte Brahe and his wife, Beate Bille, who could both blazon their eight quarterings of nobility. Like many in his class, Tycho was raised by foster parents; in his case, they were a childless aunt and paternal uncle, Inger Oxe and Jørgen Brahe, who effectively kidnapped him. Like other young Danish aristocrats, he learned Latin from tutors, entered the University of Copenhagen at an early age, and then traveled abroad with a preceptor to continue his studies in the famous Lutheran universities of Germany. Tycho's studies took him to Leipzig, Wittenberg, Rostock, Basel, and Augsburg during eight years abroad.[5] Like many aristocratic students, he once fought a duel: During the Christmas season of 1566, after a fair amount of drinking, Tycho Brahe and Manderup Parsberg, a fellow Danish nobleman, sparred with broadswords in the dark of a Rostock night. Tycho took a hard blow that gashed his forehead and hacked off the bridge of his nose. For the rest of his life, after the

Fig. 1. Knutstorp Castle, built ca. 1551 by Otte Brahe. (Photograph by the author)

wounds were healed, he wore a nasal prosthesis, rubbing it frequently with ointment or adhesive from a small jar.[6]

In his early twenties, Tycho Brahe started to deviate from what his father considered to be the normal aristocratic pattern: He began to consider a scholarly career. His father was commander of great castles and a state councillor, but Otte Brahe was no scholar, and he saw no reason why his eldest son should be one. These scholarly interests came from his mother's side and from his foster mother: There had been many learned bishops, archbishops, royal secretaries, and ambassadors among the Billes, Ruds, and Oxes, whereas the Brahes had always been soldiers, lords lieutenant, and companions of kings.

Royal patronage encouraged young Tycho Brahe in his scholarly interests, despite his father's disapproval. The year after his duel, Tycho came home from Germany for the Christmas holidays of 1567, twenty-one years of age. After the holidays, he returned to the University of Rostock, where he received a promissory charter from Denmark in the spring of 1568, granting him the next vacant canonry of Roskilde Cathedral. Some influential person or persons at court had moved the king to issue this charter.[7] The broker of this patronage was probably Lord Steward Peter Oxe [Fig. 2], Tycho's foster uncle and a leading figure in the Danish administration, but Tycho's maternal uncle, Steen Bille, was probably also involved. During the holidays in 1567, Tycho must have discussed his scholarly interests with these sympathetic kinsmen, moving them in turn to act on his behalf. The promissory charter became the first link in a patron–client relationship between King Frederick II and young Tycho Brahe. Canonries in cathedral chapters of the Lutheran state church were the only type of learned space available to both nobles and commoners.[8] The appointments were for life, and there was a great deal of competition for them, so powerful patronage brokers were essential. Even Tycho Brahe, for all his influential relatives at court, would have to wait several years for the canonry promised to him in 1568.

Tycho Brahe's father died in the spring of 1571, and Tycho inherited considerable landed wealth, but it did not come to him immediately because it took several years to settle the estate. According to Danish law, a lifetime widow's jointure was set aside and the remainder of the estate was divided among all the sons and daughters. Each daughter received an equal share and each son received twice the daughter's share, with sons given preference with regard to manors.[9] Tycho Brahe was the eldest son, but Danish law and custom gave him no more than each of his brothers. In the matter of inheritance, as in so much else, his life was governed by the normal practices of the day.

LOVE AND MORGANATIC MARRIAGE

Tycho Brahe had returned home to Denmark in the year of his father's death. He had studied abroad for eight years, but now his student days were over. In the tradition of his family, he entered the service of the Danish court and soon established close personal ties with King Frederick II, twelve years his senior. Their relationship was based on shared aristocratic values within a courtly context that ran on the principle of reciprocity. Tycho Brahe brought family honor and personal service into the relationship, and the king granted honors while requiring deference and counsel. Tycho's ancestors had served the crown for many generations in the high-

Fig. 2. Peter Oxe. Portrait by unknown artist, 1574. (Courtesy Det National-historiske Museum på Frederiksborg, Hillerød)

est offices of the kingdom, and his foster uncle was Lord Steward and commander of Copenhagen Castle, his foster mother was head of the queen's court, and his foster father, Jørgen Brahe, had died in 1565 from saving the life of King Frederick. These facts were part of Tycho's family honor, and his bearing, striking intelligence, and ability to give a rare kind of learned counsel contributed to his personal honor.[10] Royal fiefs formed a structure of shared power that cemented ties between the crown and high aristocracy. King Frederick knew that this talented, courtly, and headstrong young man deserved an honor comparable to a major fief. Tycho's powerful advocates at court, working as brokers on his behalf, helped to define what it should be.

When not at court, Tycho resided mainly with his mother at Knutstorp Castle or at Herrevad Abbey, the seat of his maternal uncle, Steen Bille. Working one day to settle the affairs of the estate, his mother must have told him the circumstances of his own birth: He had been a twin, but his brother was stillborn. This moved Tycho deeply, and he composed a Latin ode to his dead twin, wondering who had gotten the better fate, the one who went directly to heaven or the one who stayed on earth. In 1572, he took the ode to Mads Vingaard in Copenhagen and had it printed as his first publication.[11]

It was also at Knutstorp that he fell in love. Perhaps they met in the village church of Kågeröd on a Sunday. He customarily sat with his mother and younger siblings in the patron's pew under the coats of arms of their ancestors. Among all the blue-eyed and hazel-eyed village maids, she alone caught his gaze.

Who was she?

Kirsten Jørgensdatter was her name.[12] Tycho's mother could have told him that, and also that she was not of the nobility. Her father was Pastor Jørgen Hansen, the Lutheran clergyman of Knutstorp's village church since 1546, who stood before them in liturgical vestments every church day to conduct the service.[13] Perhaps the somber elegance of Kirsten's attire, white lace collar and cuffs, set off her hands and face and flaxen hair, in contrast to the peasant girls with their bright costumes and ruddy cheeks.[14]

Tycho knew the old ballads of love, of knights and maidens fair, of kidnapped brides, rune spells and potions, boldness and grace, and of fair maidens who were as gentle and good as they were beautiful.[15] In real life, Tycho lived at Knutstorp Castle, with its towers and pinnacles and cobbled court, its broad moats and drawbridge, while Kirsten Jørgensdatter spent her weekdays in the low, half-timbered and thatched parsonage with her family, wearing her apron and going about her everyday tasks. Only a bold stroke, like those of knights of yore, could bridge the social gap between them. Those olden times were really not so far removed: Tycho's brothers still rode to battle in armor; Tycho himself wore a sword, fought a duel, rode spirited horses, delighted in courtly pastimes. We can only imagine his thoughts at that important time in his life. Perhaps he dreamt of enticing her with rune spells as in the ballads. Perhaps he recalled the ballad of King Valdemar and Little Tova, his morganatic wife, so beautiful and light of foot, whom his other wife, the queen, murdered in a fit of jealousy. Tycho did know that some noblemen of his own day still took morganatic wives, sealing the relationship with a morning gift in place of a betrothal and wedding, and he knew that the common law of Denmark still recognized the validity of such marriages.

A morganatic match was certainly not what Tycho's mother, Beate Bille of Knutstorp, wanted for her eldest son: It would deprive her and her kinswomen of their right to play the role of matchmaker, and it would doom the family line of her eldest son. Nonetheless, Tycho Brahe loved Kirsten Jørgensdatter, and Kirsten loved Tycho, and love finds a way. They lived together, despite his mother's disapproval. On 12 October 1573, Kirsten Jørgensdatter gave birth to their daughter, Kirstine.[16] With Beate Bille making her dower seat at Knutstorp, Tycho now spent much of his time at Herrevad Abbey. Kirsten Jørgensdatter may have lived with him at Herrevad, but she probably stayed at his town house in Helsingborg. Perhaps it was in Helsingborg that he gave her the keys to the larder, cabinets, and doors of the household.

Under Danish law, when a nobleman and a commoner woman lived together openly as husband and wife, and she wore the keys to the household at her belt like any true wife, their alliance became a binding morganatic marriage after three years. The husband retained his noble status and privileges; the wife remained a commoner. Their children were legitimate in the eyes of the law, but they were commoners like their mother and could not inherit their father's name, coat of arms, or landed property.[17]

Tycho and Kirsten made no attempt to conceal their relationship, and the word quickly spread. Soon everybody at court and all the neighboring castles and manors knew about it. Some were sympathetic, and a few of Tycho's kinsmen may even have been pleased because the relationship would ultimately make them his heirs; but some proud aristocrats and legalistic clergymen disapproved in principle of all marriages that crossed class lines.

One thing was certain, and this was that Tycho's relationship with Kirsten Jørgensdatter did not put him at odds with King Frederick II. If anyone in that era could have understood Tycho's feelings, it would have been the king, for he too had fallen in love with a woman beneath his station. Her name was Anne Hardenberg. She had come as maiden-in-waiting to the court of Frederick's mother, Queen Dorothea, around 1557, when the prince was twenty-three.[18] The queen became extremely fond of her. Anne was lively, charming, tactful, well bred, and deeply religious, like the queen herself. She became her majesty's intimate confidant and often played music for her. Young Prince Frederick fell head over heels in love.

His father, King Christian III, old and ill, vociferously opposed the relationship. Anne Hardenberg was of distinguished noble birth, but she was not royalty, and in the king's mind, that made a world of difference. Royalty ruled over states; nobles were merely privileged subjects. Kings of Denmark married royalty, and their marriages were carefully arranged. The old king held the common view that falling in love was utter foolishness.

King Christian III died in 1559, and Frederick became king. Anne Hardenberg remained at Queen Mother Dorothea's court and continued, year after year, to live within the royal family circle. King Frederick remained loyal to her, despite constant negotiations to match him with this or that foreign princess: Duchess Renée of Lorraine, Queen Elizabeth of England, Mary Queen of Scots, a Habsburg princess.

Around 1569, King Frederick finally took charge of his own nuptial arrangements and announced that he would enter into a morganatic marriage with Anne Hardenberg, though he knew that their children could never succeed to the throne of Denmark and Norway. Then the real pressure began. All his advisors, all his royal relatives at home, and all his princely kin at foreign courts opposed the king's plans. The Danish court was thrown into a crisis that persisted for two long years, until even Anne herself let the king know that she opposed his plan.[19] Left without a choice, King Frederick II, in his thirty-eighth year, reluctantly consented to give up the love of his life and enter into an arranged alliance with his fourteen-year-old cousin, Princess Sophie of Mecklenburg [Fig. 3]. They were betrothed in 1571 and married in Copenhagen on 20 July 1572.

At that very same time, Tycho Brahe and Kirsten Jørgensdatter were finding each other. When King Frederick heard of their love affair, he understood full well what was involved.[20] His fondness for this scholarly young nobleman was not diminished. Secrets of the heart went even deeper than principles of reciprocity and shared aristocratic values to strengthen the ties between the king and his subject.

THE PATRONAGE OF LEARNING

King Frederick II had a great love of learning, but he was no scholar, and for a long time historians believed that he was not very intelligent. Recently, however, it has been shown that he had a learning disability: King Frederick was dyslexic.[21] His teachers never understood this, so dyslexia hampered his formal schooling, but it could not quench his love of learning. The king shared chemical, astrological, technical, and esthetic interests with Queen Sophie, her parents in Mecklenburg, his sister Electress Anna in Saxony, and others in his family and court circle, including young Tycho Brahe. The king was not conservative in his intellectual interests but was drawn to the ideas of innovative thinkers like Luther, Copernicus, Paracelsus, Ramus, Giordano Bruno, and the later Italian humanists. His personal interest in Paracelsian medicine led him to bring Petrus Severinus to court in 1572 as his personal physician and Johannes Pratensis to the University of Copenhagen as professor of medicine.[22]

Fig. 3. Queen Sophie of Denmark. Oil on paper, artist unknown. (Courtesy Det Nationalhistoriske Museum på Frederiksborg, Hillerød)

King Frederick wanted his kingdom to become famous for its learning,
and he was willing to do what it took to achieve that result. With the guid-
ance of Lord Steward Peter Oxe, the king built a solid infrastructure of
higher education. This infrastructure and adequate sources of sustained
funding were essential preconditions of large-scale scientific activity. Lutheran universities were regarded among the best in Europe during the late
sixteenth century, and they had a curriculum that fostered mathematics
and science. The Philippist curriculum, as it was called after its formula-
tor and Luther's friend, Philipp Melanchthon [Fig. 4], was based on the
humanist ideal of a well-rounded education and consisted of a reciprocal
"world of learning" or *orbis litterarum,* in which comprehension of each
part depended upon mastery of the whole.[23] The Master of Arts program
included arithmetic, geometry, and astronomy. This Philippist curriculum
had been in use at the University of Copenhagen since 1539.[24] The result

(b)

(a)

Fig. 4. Influential friends: (a) Philipp Melanchthon. Engraving by Albrecht Dürer, 1526. (From K. A. Knappe, *Dürer: The Complete Engravings, Etchings, and Woodcuts,* 1965) (b) Martin Luther. Woodcut 1530 by Hans Brosamer. (From Geisberg 1974; courtesy Hacker Art Books)

was that students in Copenhagen and other Philippist universities learned to consider mathematics and astronomy as mainstream subjects that every educated person should master, rather than as tedious disciplines accessible only to specialists.

King Christian III had begun to establish a solid financial basis for the reorganized university in the years around 1539. Three decades later in 1571, King Frederick II enlarged the university endowment to provide a steady, inflation-adjusted income from a large landed estate, tithes, and church incomes.[25] This income was partly in kind and included, among much else, the right to every eighth swine grazing on mast in the forests of the university, which provided the professors with plenty of hams, pork roasts, sausages, and boars' heads at Christmas time.[26] The newly endowed chairs attracted a distinguished faculty and put the university on a solid footing.

Every bit as important were the students, some of whom were gallant young nobles but many of whom were poor peasants' and pastors' sons.

Since 1537–9, a dozen university students (and later twenty) had received free board from an endowed institution called the *Communitatis Regiæ*. In 1569, King Frederick enlarged the endowment to provide one hundred students – a good half of the total student body – with two square meals and ale daily for up to five years, in a new refectory called "the Cloister" along Nørregade on the university grounds.[27] In 1569, the crown established four *stipendium regium* scholarships for advanced study abroad. Candidates needed outstanding recommendations and a master's degree in the curriculum of the *orbis litterarum*. The stipend provided 100 dalers a year, as much as the salary of a junior professor, and one scholarship was specifically designated for a medical student.[28] Previously, almost the only way a poor Danish student could study abroad was to serve as preceptor for a nobleman's son, as Anders Sørensen Vedel and Hans Aalborg had done for Tycho Brahe.

With the educational infrastructure in place, King Frederick II and his advisors began to look for world-class scholars. They already had Niels Hemmingsen, a protégé of Philipp Melanchthon whose theological works were read throughout Protestant Europe. In their search for a natural philosopher of equal renown, it began to appear that young Tycho Brahe might be the man.

King Frederick II followed with interest the developments at Herrevad Abbey, where Tycho Brahe and his uncle, Steen Bille, were establishing a series of interrelated technological and scientific facilities: a papermill, glassworks, instrument manufactory, chemical laboratory, and astronomical observatory [Fig. 5], all in a single, self-contained institution.[29] Just as those facilities were taking shape, an opportunity arose to use them in a manner that proved their worth. A great new star, a supernova, burst forth in the constellation Cassiopeia in 1572. Tycho observed it from Herrevad Abbey. His observations are famous in the history of science because he was able to prove for the first time, on the basis of empirical evidence, that change could and did occur in the celestial regions, just as on earth. This disproved a fundamental axiom of the Aristotelian world view, celestial immutability, and became an important first step in resolving the post-Copernican crisis. In a short Latin treatise on the new star, Tycho stated quite unequivocally that "this new star is not located in the upper regions of the air just under the lunar orb, nor in any place closer to earth . . . but far above the sphere of the moon in the very heavens . . . in the eighth sphere [of the fixed stars], or not far from there, in the upper orbs of the three superior planets," and he explicitly emphasized the revolutionary cosmological implications of this discovery.[30]

Lord Steward Peter Oxe encouraged Tycho to publish his treatise on the supernova as part of a patronage strategy emphasizing the *political* importance of astronomy. The publication was not dedicated to the king because it would have been an inappropriate gift from a magnate of Tycho's lineage. Instead, Tycho presented it as a casual humanist bauble, dashed off for the amusement of friends and published with reluctance. Nevertheless, everybody knew that the supernova would induce powerful astrological effects and that it was a serious omen. Tycho Brahe devoted a whole section of his treatise to these effects, speaking in guarded language of political changes unprecedented since the rise of the Roman Empire, influencing rulers born under the sign of Taurus, and working especially in Russia, Livonia, Finland, Sweden, southern Norway, and adjacent regions. Tycho said that he knew a good deal more about these coming events than he was willing to reveal in writing.[31] His whole line of thought made it clear that the interpretation of stellar events was a matter of national security. National security was one of the most powerful motives for state support of science, and it now came into play in the life of Tycho Brahe: He was summoned to court to discuss with King Frederick the political implications of the new star. The king wanted to know what the heavens foretold in order to plan his foreign policy accordingly.

These conversations about the new star of 1572 strengthened the ties between patron and client by demonstrating the value of service that only a nobleman of Tycho Brahe's learning could render. King Frederick saw the need to build this crucial knowledge into the structure of learning in his kingdom, and he encouraged Tycho to lecture on astronomy at the University of Copenhagen. Normally, noblemen did not become lecturers, but the king's urging allowed Tycho to do so in a way that conferred aristocratic prestige upon the university, rather than reducing the noble astronomer to the status of a middle-class academician. Tycho consented. Early in September of 1574, he delivered an inaugural Latin oration on the mathematical disciplines in the presence of the French envoy and virtually the whole university community.[32] Following his oration, he lectured on planetary theory and gave private instruction to a small group of students during the winter semester of 1574–5. Tycho Brahe became one of the first instructors at any university to base his lectures on the theories of Copernicus.[33] **Peter Jacobsen Flemløse** from the Danish island of Fyn, who later became Tycho's associate on Hven, was among his students.

Tycho Brahe had already shown that a courtier and aristocrat could be a natural philosopher; now he demonstrated that an aristocrat and courtier could also be an academic scholar.[34] As he moved from court to univer-

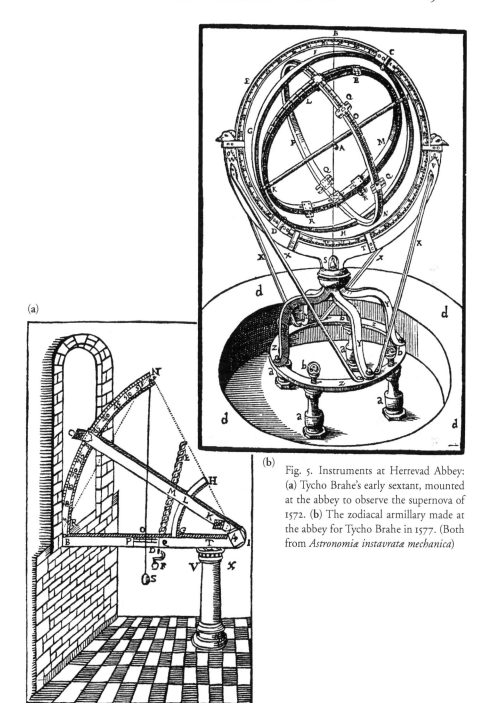

Fig. 5. Instruments at Herrevad Abbey:
(a) Tycho Brahe's early sextant, mounted
at the abbey to observe the supernova of
1572. (b) The zodiacal armillary made at
the abbey for Tycho Brahe in 1577. (Both
from *Astronomiæ instavratæ mechanica*)

sity, propelled by the innovative patronage of King Frederick and the mediation of his influential relatives, he began to reshape the mainstream of
Danish cultural life. His powerful kinsmen at court limited their roles to
those of patron and patronage broker, but Tycho actually moved from one
cultural space to the other, blurring the lines between court and academy
as he immersed himself in the neo-Latin culture of the late Renaissance.

In the spring of 1577, the professors would elect him Rector Magnificus, head of the university community. By 1577, however, Tycho's situation had changed, and he would decline the office.[35] By then, as we shall
see, Tycho Brahe had created still another innovative type of cultural space,
separate from court and university but tied inseparably to both, and he
would have no desire to abandon that new space in order to become an
academic administrator.

In the spring of 1575, however, Tycho Brahe was still lecturing at the university on Copernican astronomy when the legacy of his father was settled and he came into an income worth around 650 dalers per annum. A
senior professor at the University of Copenhagen received 300 dalers a year
plus a residence, the incomes of a prebend, and various perquisites.[36] Tycho's inheritance provided an income nearly twice that large, and it gave
him the independence to chart his own path in life. Despite his devotion
to learning, he was not tempted to cast aside the aristocratic life for that
of a professor. Instead, he abandoned his university teaching and set off
on another trip abroad.

He did not do so without the approval of his royal patron. Indeed, the
main motivation for this journey seems to have been the desire of King
Frederick II to send a cultural broker abroad. The king was planning to
embark on some ambitious building projects and was in need of technical
experts in many fields: architects, hydraulic engineers, decorative smiths
and bronze founders, sculptors, painters, tapestry weavers, and many others. Tycho Brahe knew how to contact such people. Kirsten Jørgensdatter
stayed home with their daughter, Kirstine, but Flemløse may have accompanied Tycho on his travels.

Tycho Brahe traveled as a courtier, and his route went mainly to courtly centers of learning. He visited the learned Landgrave William IV of
Hesse in Kassel, saw the new villas of Palladio in the Veneto, attended aristocratic academies in the Venetian Republic, and was present at the coronation of the future Emperor Rudolf II as king of the Romans. He visited the great technical workshops of Augsburg and Nuremberg. Tycho had
a private agenda as well: In Basel, he found a place that appealed to his
search for peace, wisdom, and universal harmony, close to Italy, France,

and Germany, and a veritable bridge between diverse creeds and cultures. On his return to Denmark, he harbored plans to emigrate and live as an astronomer and patron-practitioner of natural philosophy in Basel.

Now it is clear that we have returned to where we began. This was the background that Tycho Brahe brought to his audience with King Frederick II at Sorø Abbey during the Christmas season of 1575. He told the king of his attendance upon Landgrave William in Kassel, who surrounded himself with artists, astronomers, and philosophers. Of those who had served the landgrave, Tycho recommended the hydraulic artist **Georg Labenwolf**, whom he had contacted in Nuremberg.[37] Tycho told the king of the portrait painter **Tobias Gemperle** of Augsburg and the Italian-trained sculptor **Johan Gregor van der Schardt**, both of whom were willing to accept commissions in Denmark. He told about the great celestial globe that he had personally commissioned from **Christopher Schissler** of Augsburg. Tycho reported in full detail on his attendance at the coronation and many other matters; but he did not mention his plans to emigrate.

King Frederick listened to this report and searched for a way to reward Tycho Brahe for the success of his mission. The king was determined to retain him in his service. He offered a choice of castles and fiefs, but Tycho demurred and refrained from accepting any of them. To the king, it was a matter of honor to recognize the value of this member of one of Denmark's most illustrious families. His royal reputation demanded that he bestow gracious favor upon a courtier with Tycho's talent and connections. There were plenty of influential brokers to help the king define those favors, but the process was complicated by Tycho's refusal to accept a major royal fief. Moreover, what if King Frederick learned of Tycho's secret plans?

ENTERING ROYAL SERVICE

Two months after the audience at Sorø Abbey, everything suddenly changed: Tycho's plans, his interests, and the whole course of his life. It happened as the result of a long private audience at the royal hunting lodge of Ibstrup, where King Frederick made an offer that Tycho Brahe could not refuse. Three days after the event, Tycho himself described it in great detail.[38] He had been lying awake at Knutstorp Castle, wondering how he could escape from Denmark without arousing the attention of his relatives, when he was told that a royal courier, a nobleman and relative of his, had ridden all night, arriving at Knutstorp about two hours before dawn, and requested to see him without delay. The courier delivered a let-

ter from the king and departed posthaste.[39] Tycho took the letter, broke
the seal and unfolded the heavy, hand-laid paper. It had been dictated by
King Frederick on the previous day, 10 February 1576, at Ibstrup in the
forests north of Copenhagen. The letter commanded Tycho Brahe to come
to him immediately.

Tycho obeyed the summons. He arrived at Ibstrup that same day, 11
February, in the afternoon before sunset. A court chamberlain, Niels Parsberg – brother to Manderup Parsberg, Tycho's erstwhile dueling opponent
– brought him into the king's chamber for a private audience.[40] Three
days later, Tycho reported the king's words as he remembered them. He
wrote that when they were alone, King Frederick said:

Tycho, you know the offer I made to you when you last visited me at Sorø. I asked
you to return to me soon and tell me what you desire of such things, or else send
your request by letter to my chancellor, Niels Kaas, or my secretary, Hans Skov-
gaard. But because you have not even asked for the things that others covet and
fight to get, I do not know what you are thinking or why you do not hurry to
make up your mind. One of my courtiers reported that your uncle, Steen Bille,
told him in secret that you are planning to return to Germany. Therefore, I wanted to summon you to me, so I could hear from you yourself what you think of
my proposal and why you have not accepted it. I suspect that you do not want
to accept a great castle as a sign of royal favor because the studies you enjoy so
much would be disturbed by the external affairs and concerns required to maintain and command a castle, and because you do not want to neglect your learned
investigations.[41]

"I was staying recently at the castle I am building in Elsinore," the king
continued.[42] This was Kronborg Castle, later famous as the setting of
Shakespeare's *Hamlet*.

I looked out one of the windows, and I saw the little island of Hven, lying in the
Sound in the direction of Landskrona. No nobleman possesses it. As I recall, your
uncle, Steen Bille, told me once – and this was before your last trip to Germany
– that you liked its location. It occurred to me that it would be very well suited
to your investigations of astronomy, as well as chemistry, because it is high and
has an isolated location. Of course, there is no suitable residence, and the necessary incomes are lacking as well, but I can provide those things. So if you want
to settle down on the island, I would be glad to give it to you as a fief. There you
can live peacefully and carry out the studies that interest you, without anyone
disturbing you. And when I have built the castle which I have now begun in Elsinore, I will sail over to the island from time to time and see your work in astronomy and chemistry, and gladly support your investigations, not because I have
any understanding of astronomical matters or know what is involved in all of it,

but *because I am your king* and you my subject, you who belong to a family which has always been dear to me, and you who are said to have considerable insight into such matters. I see it as my duty to support and promote something like this. What good would it do for you to return to Germany and be a stranger there, when you can accomplish every bit as much in your native land? We should see to it that Germans and people of other nations who want to know about such things should come here to see and learn that which they could hardly acquire knowlege of in any other place.[43]

In conclusion, Tycho reported that the king said, "Think it over for a few days, then come and tell me what you have decided to do, as soon as you hear that I have returned to my residence at Frederiksborg Castle. I do not want you to feel compelled to do anything, but rather, I want to give you free choice to decide whether you think the support I am offering will serve your interests."[44]

When King Frederick was finished, Tycho reported that he thanked him sincerely. He was stunned by the proposal. Tycho knew the island of Hven. Groping for words, he said that the island was surrounded by the sea on all sides, with difficult, frequently dangerous landing places. He promised to give serious consideration to the king's proposal and bring his humble reply to Frederiksborg Castle in the near future.[45]

The next day, Tycho returned to his seat of Knutstorp Castle to contemplate King Frederick's proposal. It seemed remarkable to him, almost like an omen, that the royal courier had arrived as he lay wrestling with his plan to leave for Germany.[46] He had been deeply moved by the king's words because they revealed such penetrating insight into his own state of mind, and because he was convinced that the king himself had personally decided to make this proposal. Advised by courtiers who were close relatives of Tycho Brahe, King Frederick had conceived a plan to harness Tycho's unique talents in order to bring fame to Denmark and crucial scientific, technical, and political advice to the Danish court.

Tycho wrote the long letter to his close friend Johannes Pratensis at the University of Copenhagen, describing the audience in detail and reiterating the king's words. He rode to Herrevad Abbey and discussed the audience with Steen Bille, who had acted as a broker of patronage in shaping the king's proposal. He then went to Copenhagen and talked it over with Pratensis and the French envoy, Charles de Dançay.[47] Undoubtedly he also discussed it with Kirsten Jørgensdatter, with his mother, Beate Bille, and with his brother, Steen Brahe, who also made his seat at Knutstorp Castle.

Then he made his decision: He would accept. This offer was his destiny, given shape and form by the productive, innovative patronage of King

Frederick.[48] When Tycho heard that the king had arrived at Frederiksborg Castle, he hurried to meet him. There he renewed his service to King Frederick II of Denmark and Norway on new terms on 18 February 1576. As a token of his grace, the king immediately granted Tycho Brahe an annual pension of 500 dalers, which nearly doubled his yearly income.[49] For his part, Tycho promised to be loyal and true and ready to serve the king according to his ability whenever the king had need of him and summoned him to court.[50] These were the terms of his contract of service.

Four days after his royal audience at Frederiksborg Castle, Tycho Brahe was on the island of Hven. On the evening of 22 February 1576, he observed a conjunction of the moon and Mars near the head of Hydra. This was his first recorded observation from Hven.[51]

Three months later, on 23 May 1576, King Frederick II granted to Tycho Brahe the whole island of Hven in fee, with all its resident peasants and servants of the crown and all royal incomes and rights, quit and free, "for the rest of his life and as long as he lives and desires to continue and pursue his mathematical studies," subject only to the laws of Denmark and Tycho's remaining loyal and true to his liege lord, the king.[52] Since there was no royal manor on Hven, the king granted Tycho the sum of 400 dalers in cash to construct a suitable residence.[53] At the age of twenty-nine, Tycho Brahe had finally become a royal Danish vassal and crown governor of Hven. The infrastructure for his life's work was in place.

As suzerain and as patron, King Frederick II had the power to bestow considerable benefits upon a favorite, especially if he or she came from one of Denmark's great aristocratic families, because that allowed the king to draw on resources of the royal domain as well as his cash income from tolls and duties. In 1577, the fief of Kullen in Skåne, just north of Hven, was granted to Tycho Brahe, and in the spring of 1579, eleven farms in Skåne.[54] Extensive forests in these areas made up for the lack of fuel on Hven. That same spring, Tycho Brahe also received in fee the whole district of Nordfjord in Norway until the promised canonry became available.[55] This Norwegian fief, which Tycho never visited but administered through a bailiff, provided a substantial income in kind, which the bailiff could convert to cash. During the summer of 1579, Tycho finally received his canonry in Roskilde, the Chapel of the Three Holy Kings, but he persuaded King Frederick to let him keep Nordfjord as well.[56] The canonry fief included fifty-three tenant farms, the manor of Gundsøgaard with nine serfs, and the tithes and advowson of Uby and Store Heddinge parishes.[57] Besides the pastors of those two calls and vicars performing religious obligations in Roskilde, this fief added two studious choirboys from the cathedral school and two students at the University of Copenhagen to the learned clients whose livings depended on Tycho Brahe. As a client of the king, Tycho was also becoming a substantial patron in his own right, and he held his fiefs on unusually favorable terms.

PATRONAGE

Sixteenth-century practices of patronage lie at the heart of this book. The patronage system was based on an unequal, reciprocal relationship between a patron and a client for the mutual benefit of both, and it was shaped by the social concepts of reciprocity, gift exchange, service, and loyalty.[58] Although voluntary in theory, patronage by the late sixteenth century had become an essential and pervasive instrument of social binding throughout the European world. When Tycho Brahe received a fief from the king, or when he granted a stipend to a chorister or university student, the conventions of patronage came into play as power and influence circulated from one level of society to another.[59]

A reciprocal exchange of "gifts" was at the core of the patronage process. The value of such a gift was determined in terms of honor and prestige, not by monetary value, because patronage operated as an "economy of honor" and not a market system. Patronage gifts might have no monetary value or might be worth a princely sum.[60] In Tycho's life, these "gifts" would range from intangible forms of royal favor, grace, and access, to invitations, privileges, monopolies, royal visits, hospitality, dedications, chemical secrets, medicaments, learned insights, noble and exotic animals, golden chains, medallions bearing the patron's image, fiefs, islands, pensions, and gifts in kind or cash. The gift from a patron always had a social value commensurate with the patron's lofty status and was granted with "heroic" public display, so that part of the patron's prestige accompanied the gift, bestowing worth upon the recipient. The patron's gift was far more valuable in the hierarchy of honor than favors brought to the patron by a client, and this created an enduring sense of obligation on the part of the client. As a recipient of royal grace, Tycho Brahe felt obligated to transform the island of Hven and all its achievements into a monument to his royal patron, King Frederick II. Moreover, when Tycho was absent from court, he counted on friends and kinsmen to keep him informed of the changing desires of the king, so he could arrive when needed with the appropriate favor from his observatory, laboratory, or museum. The royal physician, Petrus Severinus, was one of Tycho Brahe's court "informers" in the early days on Hven.[61]

The binary relationship of patronage was unequal, binding, and reciprocal for as long as it endured. The patron provided prestige, material benefits, advancement, and protection. The client served as a loyal subordinate, helping the patron to carry out his duties and requirements of honor; appearing in his following at court; riding in his tournaments and progresses; offering pleasure, information, and advice; dedicating his works to the patron; and in some cases, even lending money and fighting to preserve the patron's honor.[62] Undefined and unlimited reciprocity between unequals was at the heart of the relationship.

King Frederick II had substantial sources of income for his grants of patronage. Cash grants came mainly from the Sound Tolls, which were collected at Elsinore from every ship sailing to and from the Baltic Sea.[63] Sound Toll revenues flowed directly into the royal coffers, and King Frederick could dispense them with a free hand. This was ideal because patronage by its very nature was informal, undefined, and personal, and an unhampered source of income allowed the patron to express his grace as he chose.

Grants in fee, like the island of Hven, Kullen, Nordfjord, and the Chapel of the Three Kings, came from the Danish royal domain. The domain was also administered as an "economy of honor" and used to establish dynamic, binary, reciprocal relationships that distributed power, prestige, and wealth. Domain "gifts" were fiefs of substantial value, however, and the crown was not allowed to bestow them without consultation. According to the charter of the reign, the king could bestow grants in fee only on individuals of noble rank, and he was obliged to grant major fiefs to state councillors, who saw to it that their near kinsmen received fiefs as well. The State Council was dominated by a handful of interrelated magnate families, among them Tycho's close relatives of the Brahe, Bille, Oxe, Rud, Rosenkrantz, and Ulfstand families.[64] Tycho's father, both of his grandfathers, all four of his great-grandfathers, seven of his eight great-great-grandfathers, and many of his other kinsmen had been state councillors and officers of the realm. The women of his kinship were also courtiers: Tycho's foster mother, Inger Oxe, and his mother, Beate Bille, served in turn as Lady Stewardess at the head of Queen Sophie's court. All of Tycho's brothers and brothers-in-law would pursue careers in the State Council or as lords lieutenant holding major fiefs. These ties made Tycho Brahe one of the consummate insiders in the power elite of Denmark and made it possible for him to hold the island of Hven and other grants in fee.

Patronage was productive, as Mario Biagioli pointed out, when it propelled clients into action, promoted and gave structure to their activities, rewarded novelties like Tycho's treatise on the new star of 1572 and his contacts as a cultural broker, and legitimized knowledge that might otherwise have been unacceptable, such as Tycho's views on astrology and Copernican astronomy as expressed in his university lectures.[65] Patronage could also be innovative, as when King Frederick II chose to legitimize Tycho Brahe's new "socioprofessional identity" of courtier-philosopher by grants that allowed Tycho to develop a new type of learned space.[66] King Frederick did not intend to micromanage Tycho's activities but transferred complete control to the new lord of Hven, reserving the right to make specific requests from time to time. The king knew that science brought national prestige and strengthened national security, and that was what he wanted.

Using the formulas of late feudalism, the wealth of the domain state, and a Renaissance prince's cultural patronage, King Frederick II of Denmark and his vassal-client, Tycho Brahe, were poised in 1576 on the verge of creating a new type of scientific space: neither court, cloister, cathedral chapter, university, nor academy, but rather, a state-supported, autonomous research center with an extensive staff, large-scale equipment, and projects and goals of its own. Now the work on the island of Hven could begin.

JUNKER AND PEASANTS

1576–1581

THE ISLAND OF HVEN

The island of Hven already had an organizational structure when Tycho Brahe arrived, but its goals were those of an agricultural community, not those of science. Nearly fifty households of villagers lived in simple isolation on the island – until the day Tycho Brahe arrived with his royal grant of lordship. This grant gave him the seigneurial right to govern the island's inhabitants. Tycho's first problem was to reorganize the economic and human resources of the island to serve his own needs over those of local farmers.

The island was like a natural fortress, surrounded by cliffs a hundred feet high. Prehistoric carvings marked the cliffs, and ruins of Viking fortresses guarded the island's four headlands. Above the cliffs, the island was an undulating plain of glacial drift, rising gently toward the center. Open fields of the village plowland, six to seven hundred acres in all, covered the northern part and was farmed in the old two-field system.[1] Each of the two fields was divided into furlongs, and each furlong into long, narrow strips or acres. One field was fenced and sown with rye in the fall, while the other lay fallow and open for grazing. Oats were raised on marginal land.[2] The plowland must have lain under snow when Tycho Brahe had first come sailing to Hven in the winter of 1576.[3] St. Ibb's Church, built in the thirteenth century, and the windmill, high on Möllebacken Hill [Fig. 6], were both clearly visible as his boat bobbed toward the island.

Amid the fields was the village of Tuna, the only inhabited place on the island (the name simply means "town"). Here the fifty households of islanders were clustered.[4] Each farming family had a "toft" or lot in the village, with a cobblestone courtyard surrounded on two, three, or four sides by low, half-timbered buildings under thatched roofs. The larger farms, like those of the pastor and the hundred bailiff (a crown official), were "full-farms" and claimed thirty to forty acres of arable land in scattered strips throughout the village fields. Most of the farms were "half-farms," however, with smaller buildings and ten to twenty acres of arable land besides a share in the open village commons to the south.

The typical farmhouse of the day had a hall kitchen with an open hearth under blackened rafters. Hams hung from the ceiling, and bowls of milk stood on high shelves to curdle. Here the peasant family took its meals by the hearth. Broad sleeping benches ran along the walls as in Viking times. There were undoubtedly also a few modern dwellings with a paneled chamber in Renaissance style, containing alcove beds along the inner wall and warmed by a stove instead of an open hearth. Here they sat on benches at a long table under a row of windows with small leaded panes, with the patriarch on the high seat at the end of the table. Both types of farmhouse also had unheated rooms for sleeping, storage, weaving, and other household activities.[5] Some farms had separate servants' quarters, a dower house, perhaps even a cottage or two on the toft. This meant that there could be two or three households in a single farm.

The other wings of the farm complex contained stables, stalls, pens, threshing floors, haymows, and a wagon shed. Ducks, geese, pigs, cats, and dogs wandered about. There were also cattle, calves, and horses, with a manure pile near the stable and stalls. Behind or near the dwelling was an enclosed croft of an acre or more, with a garden laid out in small beds, an orchard, and sometimes a fold.[6] Women went to the village well for water.

There were nearly fifty households of tithe-paying farmers in the village, living on full- or half-farms, plus cottages with no plowland. Some of the cottage folk were fairly prosperous, like the miller and the smith; others were the likes of a tailor, cobbler, carpenter, thatcher, boatwright, and village schoolmaster; still others were poor laborers, fishermen, ferrymen, and widows. In short, the two hundred villagers of Tuna were a varied lot, ranging from fairly well-to-do to very poor. The farms were smaller on average than farms on the Danish mainland. The Tuna folk turned to the sea to supplement their livelihood, fishing for herring, cod, and plaice, hauling cargo in their small skiffs, and engaging in coastal trade.

There were no forests on the island, only a grove of alders in a marshy meadow, north of the village. On a hill at the southern edge of the village was the windmill, a post mill, belonging to the church. At the center of the island was an ancient circle of boulders. This was the village assembly and site of the Hundred Thing, the local court of law. When the village "grands," or heads of farming families, met as the Hundred Thing, the

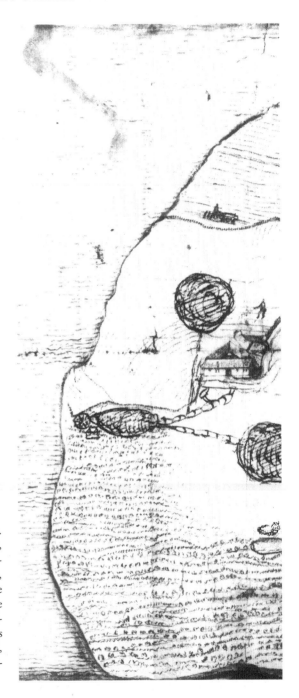

Fig. 6. Bird's-eye view of Hven ca. 1580. St. Ibb's Church is in the middle at top, Uraniborg dead center, and the windmill on Möllebacken hill between them, to the right. Faintly visible between the windmill and the manor house, near the latter, is the stone-ringed site of the village assembly. The three outbuildings of the grange, with triangular courtyard, are left of center. (Courtesy Österreichische Nationalbibliothek, Vienna)

presiding officer was the hundred bailiff, a local farmer appointed by the crown. When they met instead to decide matters of communal life, however, an alderman chosen from their midst presided. At this assembly, the patriarchs ruled the communal life of the island.[7]

The southern two-thirds of the island was commons: open land, dotted with thorn bushes and crabtrees, and valuable as pasture and meadow. A hazel thicket filled a ravine near the southeastern shore. There were no rats or snakes on the island, and peasants in Skåne had a saying that rats could not live where Hven soil was scattered.[8]

When Tycho Brahe arrived on Hven, the island had never been surveyed or mapped. Today we know it is roughly pear-shaped, seven miles in circumference, almost three miles long from northwest to southeast, a mile and a half wide, with a total area around 1,850 acres (750 hectares).[9] Tycho Brahe had visited the island in February of 1576. Three months later, he had received it in lifetime fee, and sometime in the spring of 1576, when snows were gone and rye was greening, he came to proclaim his possession. He and his party struck sail, pulled their skiff on shore, and made their way up the hill to the village assembly. Tycho Brahe had likely already announced his arrival to the hundred bailiff and the pastor of St. Ibb's.

At the assembly, Tycho Brahe and his party addressed the bailiff and grands, many of whom knew Tycho from previous visits. These village worthies wore beards in peasant fashion and dressed in long trousers, belted smocks with ruff collars, and high hats with narrow brims. Tycho was attired like a gentleman, in knee breeches, sword, buttoned vest of rich fabric, linen shirt, lace ruff and cuffs, short cape with high collar, gloves, and plumed beret. The whole village of Tuna turned out because they knew that something important was afoot. Housewives wore long-sleeved blouses, long woolen dresses in various colors and weaves, and long aprons. Married women covered their hair with linen caps, but maidens showed their long, flaxen braids. Children were dressed like small adults. They all thronged around the assembly circle and studied the strangers.

A parchment document, lettered in a bold gothic chancery hand and bearing the seal of King Frederick II, was displayed to the assemblage and read aloud by Tycho's clerk. The document proclaimed that the island of Hven had been granted in lifetime fee to the honorable and wellborn Tycho Brahe of Knutstorp. Eyes turned to the young nobleman with the blond beard and the long moustaches who did not doff his hat at the assembly. The clerk read on: "But he shall observe the law and due right towards the peasants living there, and do them no injustice against the law, nor burden them with any new dues or other uncustomary innovations."[10]

The bailiff and grands exchanged glances. They listened carefully as the parchment was read aloud, pondering every word.

Here were two worlds moving inexorably toward conflict: the world of these independent island farmers, and that of this rich young nobleman.

COMMUNAL VERSUS MANORIAL ORGANIZATION

Tycho Brahe came to the island of Hven intending to be its seigneur and lord of the manor. People called him "the Junker." He was born to rule, and he already exercised seigneurial rights over hundreds of tenant families on his private estates around Knutstorp. Tycho Brahe marched onto the commons and stood at the very center of the island, declaring that here he would build his manor house. He asked no man's leave: As royal vassal, he could do as he pleased on the island of Hven. At least, he thought he could.

The hundred bailiff, village alderman, and grands also thought that they had a right to rule the island. They and their predecessors had run its affairs since time immemorial. The grands of Hven did not consider themselves to be the tenants of a young nobleman from Skåne: They were freehold farmers like their ancestors, not villeins, and they were subject to no man's beck and call. Like Tycho Brahe himself, they recognized the king alone as their overlord. These sturdy farmers saw no reason to give up their traditional freedoms and communal institutions simply because the honorable and wellborn Tycho Brahe and his commoner wife were moving to the island. This was a serious disagreement. It would take several years to reach a settlement.

The farmers of Hven were living in the past, whereas Tycho Brahe was in step with the times. They wanted to continue living in medieval self-sufficiency, but this was the late sixteenth century, and times were changing. New estates and manor houses were springing up all over Denmark. The laws of the land favored large-scale estates, not peasant villages.

Nevertheless, the agrarian tradition of the Hven villagers was deeply rooted in Denmark. It was the tradition of self-governing peasant farming communities that scholars call *Grundherrschaft*.[11] Such communities controlled their own land and village life. If the farmers were freeholders like those on Hven, they simply paid dues and taxes to the crown and tithes to the church; if they were tenants, they also paid annual rent and dues to their landlord. Otherwise, they ran their own affairs through the traditional structures of the village assembly, patriarchal household, and a subsistence and barter economy.

Landlords benefited from the system of *Grundherrschaft* because it was relatively easy to administer. Dues were paid mainly in kind, and the tenants were obligated to transport their dues to the lord's manor or the nearest market town. A bailiff and a few assistants could collect the dues, market some goods for cash, and supervise the storage and consumption of the rest. A small demesne at the seat of the lord might be cultivated by serfs or boon workers under the direction of a bailiff and overseer.[12]

Tycho Brahe's patrimony was of this kind. From his father, he had inherited half the estate of Knutstorp Castle. His brother, Steen Brahe, had inherited the other half, and their mother, Beate Bille, continued to claim part of the estate as her widow's dower. Most of the Knutstorp lands lay in the province of Skåne. The total estate comprised 322 copyhold farms, 29 cottages, 7 mills, and the patronage of Kågeröd Church, including the right to collect the tithe.[13] Since Tycho did not control the whole estate, it would have been difficult for him to alter the system of *Grundherrschaft* at Knutstorp, but there was no reason for him to do so: Incomes collected by the bailiff were divided among the three owners, and Tycho's share was sent wherever he wanted it. Tycho had also inherited a town house in Copenhagen and additional property in Helsingborg and other towns.

The island of Hven was another matter. Tycho Brahe intended to reside there, and he planned to develop it into a more highly organized estate of the type scholars call *Gutswirtschaft*.[14] The central element was the estate itself, rather than control of rents and dues from peasant villagers. Production was organized to bring forth a marketable surplus on a large demesne surrounding the manor. In the Danish system of *Gutswirtschaft*, labor was furnished not by serfs or hired hands but by the corvée or boon work requisitioned from peasants in nearby villages. These peasants continued to practice traditional communal agriculture at home and paid the customary dues to their lord, but in addition, they were required to work a day or two each week on the lord's demesne.

The system of *Gutswirtschaft* required strong, day-to-day management by the lord's bailiff and a staff including an overseer, summoner, and clerk. Despite the added administrative expense, the system was profitable for several reasons. First of all, there was a steadily growing market for Danish grain, cattle, and dairy products, primarily around Lübeck, Hamburg, and in the Netherlands.[15] Estate production could be oriented toward that market. Second, the lord paid nothing for boon work, except the cost of a midday meal on the job. The peasants were compensated by exemption from extraordinary tax levies, but this cost the landlord nothing, since taxes went to the crown. Moreover, in an era when the link between power and status was quite direct, the system of *Gutswirtschaft* became desirable simply because it gave such obvious and complete social control over the villages surrounding the manor, and because only the richest and most powerful aristocrats could even think of implementing it. Tycho Brahe was not interested in supplying the Dutch market with beef, but he knew the value of status, and he wanted complete control in order to transform the island of Hven into a center for large-scale scientific research. He believed, in other words, that a modern, highly centralized estate could serve science as well as commerce.

The first step in *Gutswirtschaft,* after acquiring control of the land, was to establish a demesne. The method used by Tycho Brahe was to expropriate part of the village commons. The ecology of village husbandry was distorted when common pastures and meadows were taken for the lord's personal use, but the basic rhythms and patterns of communal village life, the rotations of plowland, the traditional landholding systems of fields, furlongs, and acres, and the village patterns of toft and croft could persist without fundamental change.[16] Therefore, the basic peasant economy of the island was not seriously disturbed when Tycho Brahe decided to build his manor, with its mansion, garden, grounds, barns, and fields, on the common lands that formerly covered two-thirds of the island. Around the same time, there was a shift from the two-field to a three-field system on the village plowlands. This was accomplished by changing rotations on the furlongs, and it increased grain production, but at the price of even less grazing land.

A dynamic, late-sixteenth-century landlord like Tycho Brahe also wanted to control the advowson of the local parish church. Tycho gained it very soon after arriving on Hven. The windmill and two-thirds of the tithes went with the advowson of St. Ibb's, which provided a good subsidiary income.[17] Also included was the right to call the pastor of St. Ibb's Church, and the pastor played a key role in the indoctrination and social control of the island population. Tycho built a patron's pew of oak in Renaissance style, bearing the arms of Brahe and Bille and the date 1576. He and his brother, Steen Brahe, already shared the advowson of Knutstorp's parish church in Kågeröd, and Tycho acquired control of two advowsons on Sjælland when he became a canon of Roskilde Cathedral in 1579.

Finally, the progressive landlord wanted to control justice. On the island of Hven, the authority to appoint the hundred bailiff devolved to Tycho Brahe with the fief. The matter of boon labor soon became a sticking point of law and custom. There had never previously been a resident lord on Hven, and the villagers of Tuna had never done boon work for any man. Now Tycho's bailiff took down their names, noting the size of their farms and cottage holdings so that their new lord could demand his rightful dues of unpaid labor.

There were two main types of labor dues.[18] The first was the aforementioned boon work (Danish: *hoveri*): The bailiff's summoner called a man from each farm and cottage to work one or two days a week without pay. Farmers could usually send a hired hand or a grown son, but cottagers generally had to answer the call in person. The boon workers brought their own equipment, including plows and harrows, sickles and pitchforks, spades and hoes, as directed by the summoner. A certain number of days each year were span-days, when boon workers had to bring a span of horses for heavy work. The second main type of labor dues was cartage (Danish: *ægt*). For a set number of days each year, the bailiff's man summoned boon workers to bring horses and wagons to haul the lord's goods. Part of this obligation was an annual "long haul" to local markets, which was carried by skiff from Hven. The burden of span-days and cartage rested on farmers who owned horses, wagons, and skiffs. Cottagers and poor farmers ran errands, delivered letters and messages, drove animals, and carried light goods instead.

Tycho Brahe's bailiff quickly introduced these labor dues to the island of Hven. There was no particular malice nor any tyrannical inclination on his or Tycho's part in doing so. They simply introduced a new, more productive, and centralized agrarian system that was becoming increasingly widespread in the market-oriented parts of northern Europe. Similar labor dues were helping to increase estate production throughout Denmark and the Baltic region. On the island of Hven, however, the villagers had known nothing of boon work and cartage, *hoveri* and *ægt*. No wonder they objected when Tycho's bailiff began to summon a dozen or more men, every day except Sunday, to labor under the overseer of the nascent manor.

As the labor and production of Hven were modernized by the application of *Gutswirtschaft,* the island's economy became integrated into wider systems, and the goal of production shifted from local self-sufficiency to the support of scientific research. As production intensified, some aspects of the island economy became overburdened. The main problem was a chronic shortage of forest resources for fuel and lumber. Tycho mentioned it to his royal patron in 1577 and received two fiefs on the forested peninsula of Kullen in Skåne, an easy sail up the coast of Skåne.[19] From then on, the villagers of Hven were required to haul Tycho's timber and firewood in their own boats as part of their cartage dues. Swine production also demanded forest resources beyond the island because swine were

Fig. 7. Fragment of the Uraniborg cornerstone. (From Christensen & Beckett 1921, 1)

grazed for part of the year on the mast of beech and oak forests. The Hven farmers customarily brought their animals across the water to forests in Skåne. Through his bailiff, Tycho Brahe directed them to use his forests around Knutstorp, so he could be the one to collect the fee of *oldengæld*, every sixth swine, paid for the consumption of mast. Some did as he directed, and some did not.

In the years following 1576, the way of life on Hven changed rapidly. The islanders did not like it, but Tycho Brahe and his staff forced it upon them. Painful changes in traditional ways of life were occurring in many other parts of Denmark by the transition to estate production, just as on Hven. Tycho Brahe differed from most landlords, however, because his thoughts were on neither the Amsterdam grain exchange nor the Wedel livestock market. He aimed to restructure the production of Hven in order to support a superstructure of scholars, scientists, and servants centered on his new manor.[20] From this base, he intended to grasp the mysteries of the starry heavens and unlock the secrets hidden in earth and sea. For those lofty purposes, the life of the island's peasants had to change.

FOUNDING URANIBORG

The work of construction went fast. By early August of 1576, a group of Tycho Brahe's learned and aristocratic friends was able to gather on Hven for a festive ceremony to dedicate the cornerstone of his new manor house. The stone was brought to the island by the French envoy to Denmark, Charles de Dançay, who had commissioned it and composed its inscription [Fig. 7]. "On the 8th of August in the morning, when the rising Sun together with Jupiter was in the heart of Leo, while the Moon was in the western heavens in Aquarius," wrote Tycho Brahe, "he laid this stone in the presence of all of us, having first consecrated it with wine of various kinds and praying for good fortune in every respect, in which he was joined by the surrounding friends."[21] The foundation and cellars must already have been constructed when the stone was laid in the southeast corner. The building was dedicated to Urania, the Muse of Astronomy, and named Uraniborg, "The Castle of Urania." Tycho Brahe referred to it as his *museum*, using the word in the literal sense of a "temple of the Muses."

Meanwhile, Tycho and his household needed a place to live on the island because Uraniborg would not be ready for several years. Henrik Jern thought Tycho probably constructed the outbuildings of the grange first of all.[22] A map from the 1580s [see Fig. 6] shows three barnlike buildings enclosing a triangular courtyard, with a portal leading through one building into the courtyard.[23] These may have been thatched, half-timbered, wattle-and-daub structures on fieldstone sills, like farmhouses and barns, and they could have been built quickly. A paneled chamber in the portal building, which had a tower and chimneys, could have served as Tycho's temporary residence, and later, the bailiff could live there.

While boon workers dug foundations, ditched, built dikes, hauled timber and bricks, cleared boulders and brush, made stone fences, and transformed rough commons into cultivated demesne fields, the construction of the castle was carried on by hired masons, stonecutters, carpenters, glaziers, and other skilled craftsmen. Some of the cottager craftsmen of Tuna may have been hired, but most of the master craftsmen came from the mainland, especially from Elsinore.

VILLEINAGE AND VILLAGE CHARTER

The farmers of Tuna may not have been familiar with humanist culture and large-scale estates, but they were no fools, and their vision was not bounded by the cliffs of Hven. They sailed regularly to cities within sight of the island: Landskrona, Elsinore, Helsingborg, Malmö, and Copenhagen. They saw ships of many lands drop anchor before the Tollhouse of Elsinore to pay the tolls established by command of the King of Denmark. They knew that much about the realities of power, and their elders were practiced in the procedures of customary law, if not in the great written codifications. Their culture was essentially traditional, aural and tactile, not based on books, but they were experienced and shrewd. They knew when to bend, when to stand fast, and when to sail away. At least, they thought they did.

The years following Tycho Brahe's arrival struck many as a good time to sail away. Every morning at daybreak, a dozen men from the village had to walk the path to Uraniborg and spend the long day in labor. Landowning farmers were not very mobile, but young cottagers, farmhands, and farmer's sons were, as long as they were not married. Their ranks soon thinned. Tycho's bailiff, overseer, and summoner knew every villager by name, and they soon noticed who was missing. When Tycho Brahe became aware of the flight, he went to court and reported to the king that his tenants were violating the law of villeinage (*vornedskab*).[24]

On 10 April 1578, King Frederick II addressed a stern missive to the inhabitants of Hven.[25] The king had learned that some of them had dared to move away from the island since Tycho Brahe came there and began to build, because the burdens placed upon them were heavier than before. This served only to make the burden even greater on those who remained. Therefore, King Frederick strictly forbade any inhabitants to leave the island to live elsewhere, unless they had Tycho Brahe's permission. In a strictly legal sense, this missive did not introduce a new restriction because the peasants of Sjælland were already bound to their home manor by the law of villeinage: The king's missive simply made it clear that this law also applied to the island of Hven. Tycho must have been pleased. His supply of tenants and boon labor was ensured.

The villagers of Tuna did not give up. Two years later, in the summer of 1580, they themselves appealed to the crown. Their supplication to King Frederick II complained of harmful, uncustomary burdens of boon work, cartage, and the like demanded by Tycho Brahe. In that year, Tycho and his household moved into Uraniborg, and there was plenty of work to do.[26] Craftsmen were finishing the mansion and gatehouses, while boon workers did the heavy lifting, built the huge earthen-and-stone ramparts, and helped to plant elaborate gardens and Renaissance orchards within the ramparts: three hundred trees with many varieties of apple, pear, plum, and cherry, as well as exotics like apricot, fig, walnut, and quince.[27] Moreover, the new demesne was under cultivation, and that meant plowing, harrowing, planting, harvesting, and threshing for Tycho's land as well as their own, while wood had to be freighted continually from Kullen for the stoves and chemical ovens of Uraniborg. When one task was done, another appeared. It was clear that the burden would never end in Tycho Brahe's lifetime, and the villagers of Tuna appealed to the king for relief.

King Frederick received their supplication and asked Tycho Brahe for an explanation. The king soon got one, both orally and in writing. Tycho knew the law, and he could also consult with experts like his uncle, Steen

Bille, the former chief justice of Skåne. Tycho was confident that he was in the right and requested a full hearing. He said that if he were in the wrong, he was willing to accept any punishment the law prescribed, adding ominously that the complaining villagers should also be willing to take the consequences if it turned out that their complaints were unjustified. In fact, similar complaints were reaching the crown from peasants throughout the land, especially from places where new manors were under construction, and Tycho's reaction was typical. Moreover, many aggressive, progressive landlords had the same ties to the inner circles of government that Tycho enjoyed.

The king and his administration eventually developed a way to deal with such matters. What they did was to issue a charter. Old customs became outdated because they did not recognize the role of a seigneur in the community, but new relationships needed to be defined impartially and in writing. The island of Hven was one of the first places where this was done. On 20 August 1580, King Frederick appointed a commission consisting of the provincial chief justice of Skåne and the Lord Lieutenant of Helsingborg.[28] The commissioners were instructed to go to Hven, take depositions from Tycho Brahe and the inhabitants of the island, and determine whether there were grounds for complaint. If not, they were to identify the authors of the unjust complaints so they could be brought to trial. Furthermore, the commissioners were instructed to draft a charter for the island, prescribing the proper amount of dues in cartage, labor, and boon work as was customary elsewhere in Denmark, how fences and dykes were to be maintained, how to protect the trees of the island, how the island assembly was to be held, and so on. The commissioners were instructed to submit a clear written report and seal it with their signets. Their work was done by the end of the year, and the charter of Hven was issued over the royal seal on 8 January 1581.[29] Its provisions were as follows:

1. *Labor dues.* Each farm of the normal size (a "half-farm," Danish: *halvgård*) had to give Tycho Brahe two days of boon work every week on land or sea, from sunrise to sunset. If a worker did not show up until ten, eleven, or twelve, it counted as a half-day and he had to make it up the next day if requested and pay a fine of three marks, unless he had a very good excuse. In general, labor dues would follow the customs of Sjælland and Skåne. This provided Tycho's bailiff with some fifteen boon workers daily, six days a week.

2. *Fences and dykes.* Any farmer who did not finish the enclosures assigned to him by St. Valborg's day (the first of May) and maintain them in good repair would be required to pay a fine of three marks to the lord

and a good barrel of ale to the village men, and he would also be liable for any resulting damage. This was the normal provision in villages that sowed spring barley. It probably needed to be spelled out on Hven because the new three-field rotation of spring barley–winter rye–fallow, with oats on the margins of the sown fields, was replacing the old two-field rotation.

3. *Trees and thickets.* The inhabitants of Hven could not tamper with hazel, apple, alder, thorn, or other trees or thickets on the island, nor with their fruit, unless Tycho Brahe or his bailiff showed them where to go. Any person who plucked or chopped without permission would be punished according to the laws protecting forests in Sjælland and elsewhere in Denmark. By this provision, Tycho managed to have the scrub and thickets of Hven classified as forest land under the lord's direct control.

4. *Assembly.* The village assembly was to meet every other Wednesday, and the hundred bailiff was obligated to help all persons achieve justice. Hven had always been under the jurisdiction of Sjælland, but the farmers complained of the great distance to the provincial court, so the island was hereby transferred to the jurisdiction of Skåne. These changes had their basis in the seigneurial right to police, incarcerate, initiate criminal proceedings, collect fines, and execute court sentences.[30] The village assembly remained a local hundred court, but it came under the guiding influence of Uraniborg because Tycho Brahe now held the power to appoint the hundred bailiff who functioned as judge. Disputes between lord and villager led to more frequent appeals to the provincial court, and the fact that the Skåne court in Lund was much closer than the Sjælland court in Ringsted became important to the islanders.

5. *Swine on mast.* As long as Tycho Brahe held the island in fee, the people of Hven had to pay him the customary fee of *oldengæld* if they foraged their swine in his forests in Skåne. If not, they still had to give him the next-best swine when they returned to the island, as was customary elsewhere. This provision moved in the direction of transforming *oldengæld* from a mast fee into a seigneurial privilege.

6. *Rents and dues.* In the past, the farmers of Hven had been negligent in paying their annual dues for land (*landgilde*) because they regarded themselves as freeholders and not tenants. Henceforth, they were to pay *landgilde* as the law provided, that is, dues in butter on St. John's Midsummer (24 June), dues in grain and cash at Martinmas (11 November). Those who failed to do so would be punished according to the law. These dues were paid at Tycho Brahe's grange south of Uraniborg.

7. *Freehold.* Although the farmers of Hven previously claimed their farms as freehold property, when Tycho Brahe summoned them to court

it was discovered that they had no charters, patents, or deeds to document their claims. Therefore, from the day of the charter, when a farmer died or retired, his farm would be granted and leased like any other crown farm and no longer be regarded as freehold property. Rents and dues would not be increased but would remain as in the past, and no additional burdens would be placed upon the farmers except as the king himself might ordain and decree. This method of seizing freehold farms was being used by sharp landlords throughout Denmark. Peasant freeholders generally held their land on customary tenure because their ancestors had owned it since time immemorial, and not because they or their ancestors had purchased it or received it in hereditary fee. Consequently, they had no written documentation of ownership; possession itself, supported by the consensus of tradition, was the basis of their freehold. That was good enough for the village assembly, but not for higher courts, where charters, deeds, and documentation were required. By taking the matter to a higher court, Tycho Brahe was able to invalidate their freehold claims and reduce the Hven farmers to the status of crown tenants. This allowed the lord, in consultation with his bailiff and overseer, to put in solid, docile tenants, and it also meant that the lord would collect an entry fee from every new tenant. Entry fees were an important source of income for the landlord because they were flexible, and when farms were in demand, they could be quite high.

8. *Wares.* There was some import of wares to the village of Tuna, but the villagers also exported goods from the island. Tycho Brahe had requested that these goods be offered first to him, so that he might have the opportunity to buy them for what they would be worth in the nearest market town. In the presence of Tycho and the aforenamed commissioners, the farmers had willingly agreed to this. They were often delayed in sailing to town by adverse winds, which hampered their commerce and meant that some of their wares spoiled. Therefore, before they put to sea from the island on their own account, the farmers of Hven would henceforth be required to offer their wares to Tycho Brahe, who in turn would be required to pay promptly as much as they could get in the nearby towns, and not force anybody to give up wares for less than they were worth. This provision conflicted with an ordinance of 1536, which gave Danish farmers the right to sell to whomever they chose. Therefore, the charter emphasized that the villagers had freely agreed to it, and that they saw it to be to their own benefit.

If so, it was the only positive outcome they got from the whole charter: All the other tangible benefits went to the royal vassal, Tycho Brahe. Perhaps it was beneficial to all parties, however, to have these rights and obligations clearly spelled out in a document with the authority of a royal charter. At least the long trips to the higher courts would be less frequent, because most of the disputed points were resolved by the charter. When this charter was signed on 8 January 1581, the long battle between Tycho Brahe and the villagers of Tuna seemed to come to an end. The resolution left Tycho in charge. His base was secure.

In one sense, Tycho Brahe was transforming the villagers of Tuna from freehold farmers into tenants and villeins, and they naturally resisted with all the means at their disposal. In another sense, Tycho was making them part of one of the great achievements of the sixteenth century. Two hundred peasants on the island of Hven were being harnessed to the service of science, along with hundreds of others on Tycho's fiefs in Skåne, Sjælland, and Nordfjord, and on his patrimonial lands under Knutstorp. Tycho was confident that his achievement would be a great one, but he rather oddly conceived of Uraniborg and his island of Hven as a place that embodied *amor,* fond and passionate love, which he believed to be the force that drove the natural universe. We need next to examine how a sixteenth-century learned mind could come to such an extraordinary conclusion.

3

AMONG FRIENDS

—

1570–1576

THE LANGUAGE AND BENEFITS OF FRIENDSHIP

Tycho Brahe had learned at an early age that life in a great aristocratic house was not always tranquil and supportive, regardless of how many doors could be opened by family ties. As an infant, Tycho had experienced violence when he was abducted by his uncle, Jørgen Brahe. Although this brought his childhood into the cultural orbit of his foster mother's learned and courtly family, it also weakened his emotional bonds to the chivalrous values of the house of Brahe. Later, Tycho's marriage outside the nobility delivered a huge blow to family and class solidarity. He remained a courtier, a great noble landowner, and a holder of fiefs, but Tycho Brahe by the age of twenty-nine had moved to the margins of aristocratic solidarity through his own action.[1]

Solid ties of friendship replaced his weakened ties of class and kinship. Tycho made many friends, including those who gathered on Hven for the dedication of Uraniborg in 1576. These learned aristocrats and middle-class scholars shared a concept of friendship that had been shaped by the culture of the late Renaissance. They could point out that the ancient Greeks put family love in the category of φιλια (*filia*) or filial love, whereas love between friends was εροσ (*eros*), more intense and full of desire.[2] Tycho and his friends referred to friendship in Latin as *amicitia* and saw it as a reflection of *amor*, the force that drove the natural universe. Therefore, in order to be in touch with nature, they believed that it was necessary to pursue *amicitia*.

Tycho revealed something of this line of thought in a poem of 1572 expressing the theme of "friendship communicating through nature." Written at Herrevad Abbey and dedicated to his friend Johannes Pratensis in Copenhagen, this poem showed how the contemplation of nature could unite friends and bring spiritual release:

We cannot join our beaming eyes because
We live so far apart, and yet the beams
Of radiant Olympus join our eyes at last.
For every time the starry flock does graze
The heavens at night, then every star I see
With eyes turned heavenward, that very star
Your eyes will see. So heaven itself unites
Our eyes, and earth releases, too, our heavy flesh.[3]

Such a poem was an expression of *amicitia,* and such friendship typically led to an exchange of "benefits": gifts and acts of service or kindness
among friends.[4] An elegy or epigram dedicated to a friend, a metrical letter in the style of Ovid, a shared secret, advice, counsel, a banquet or symposium with music and learned conversation, the solemn celebration of
an event like the dedication of Uraniborg, the gift of a publication or portrait, and on rare occasions even a crown of laurel were among the benefits that Tycho and his friends shared with one another. Such benefits were
meritorious and deserved the office of gratitude, so friends naturally re-
paid them: Once established, friendship became reciprocal and bestowed
mutual benefits without end. Friendship was essentially a relationship between equals, but sometimes poems of friendship were written to draw
superiors, usually patrons or respected former teachers, into the circle of
amicitia; such poems were modeled on those of Virgil and Horace to their
patron, Mæcenas.

In general, Latin poetry was a preferred expression of *amicitia* among
Tycho and his friends.[5] As schoolboys, they had learned to write, speak,
and think in Latin, and at the university, all of their lectures had been in
Latin. In time, they even Latinized their names: Hans Frandsen from Ribe
became Johannes Franciscus Ripensis, and Rasmus Glad achieved fame as
Erasmus Lætus; likewise, Tyge Brade Ottessøn became Tycho Brahe Ot-
tonides, Anders Sørensen Vedel became Andreas Severini Velleius, Peder
Sørensen became Petrus Severinus, and Jehan du Pré, the son of a French
immigrant to Denmark, styled himself Johannes Pratensis. Among these
Danish humanists and their friends throughout Europe, Latin verse reflected the author's erudition, values, and character. It expressed bonds of
gratitude and *amicitia,* reinforced genealogies of learning and friendship,
and bestowed prestige upon poet and recipient alike.[6]

The Latin literature of *amicitia* marked great events in the life of Tycho
Brahe. One such event was his discovery of the new star of 1572, when
the spirit of friendship formed an elegant part of his first published book,
commemorating the discovery. Two friends, Johannes Franciscus and An-

dreas Velleius, graced the book with Latin poems dedicated to Tycho, while letters exchanged between Tycho and Johannes Pratensis introduced the subject of the volume.[7] The heart of the book was Tycho's analysis of the supernova in three parts. First came his astronomical assessment of the celestial position of the supernova, then his astrological analysis of the supernova as an omen of the future, and finally, his explication of the intellectual or mental significance of the supernova. This last part took the form of a Latin poem entitled "Elegy to Urania."

Two Danish scholars have shown that Tycho based this elegy on a poem by Ovid that used the theme of "epiphany," the appearance of a divine being.[8] Ovid's poem described the author wandering in an ancient forest by a spring, where he suddenly meets the Muses of Elegy and Tragedy, who ask him to dedicate his life to them. Tycho's poem began with a description of Herrevad Abbey and the marvels produced there under Steen Bille, from brilliant crystalline glass to smooth paper. Tycho had helped his uncle to establish a glassworks and papermill and to transform the old Cistercian monastery into a center of technical and scientific innovation.[9] The poem moved on to a picture of Tycho himself, wandering one evening at sunset in the forests of Herrevad along Rönne Brook. Suddenly, the Muse Urania comes down from the heavens, identifies herself, and tells him not to be afraid. She then describes her attributes at some length:

> And boldly I soar to heaven, beyond the clouds.
> With Jove, I taste ambrosia divine
> And search the sevenfold heaven's ceaseless orbs
> Among the stars, wandering well-worn paths:
> The stars that reveal the skill of godly will
> That built a sparkling roof above the earth;
> The stars that far in advance know our fate,
> Fate – often good, often bad;
> The stars that silently exercise justice among you.
> Much do they grant with grace. They hinder too,
> But do not force the soul that has a mind,
> For he does all according to his will,
> But few will take the way of the mind on earth,
> So, very few can bend the heavenly force.

In words inspired by Italian Renaissance philosophers like Ficino and Pico della Mirandola, Tycho's Muse was saying that the person who used his or her intellect could break free from the influence of fate and astrological forces and follow his or her own free will. Such a person stood above ordinary people in the great chain of being. Like the angels, these people of the mind lived in the world of ideas, whereas most people lived in the material world of sensory experience. Urania urged Tycho Brahe to be a person of the intellect.

THE ASTRAL PHILOSOPHY OF LOVE

Tycho Brahe and his friends did see themselves as superior beings who lived in the mental world of ideas, and they were not unique in this attitude. It was part of a cult of *amicitia* that had spread throughout Europe, and its roots were in the Renaissance revival of Platonic thought with Pythagorean and Hermetic undertones. Part of this pattern of thought was the idea that poetry could induce physical effects by spiritual means. Petrus Severinus would write a few years later,

Plato asserts in one of his dialogues that there is a divine power in the words of poets. . . . The very bonds which hold body and soul together seem to be loosened when the senses are overwhelmed by a beauty and delight which they have never before known, when they are blinded by an unexpected brilliance, like one with diseased eyes by the noonday sun. These invisible ideas possess such great power that their force can color, sustain, and enliven the physical bodies of nature and endow them with a multitude of allurements.[10]

To Severinus, Tycho, and their circle, there was truly something magical about poetry, and poetry had to be considered as a part of science because it could literally alter the physical universe.

Two millenia earlier, Plato had asserted that human beings were caught between two worlds: the world of Ideas and the world of the senses. Plato had taught that the human soul was supernatural and divine in origin: Before birth, it dwelt in the world of Ideas, and it retained a memory of that world and continued to be attracted to it. This upward attraction of the soul, away from the material world, was called εροσ (*eros*), later translated into Latin as *amor*.[11] The god Eros, who personified this attractive force, was linked to the goddess, Aphrodite (Venus), and she in her heavenly form was Tycho's Muse, Urania.[12] Beauty, the brightest and most radiant of all Ideas, had the power to awaken Eros, and when the soul was awakened, it "gains wings and is able to ascend to the super-sensible."[13] Plato described the "heavenly ladder" by which the soul ascended, step by step, until it reached the goal of Absolute Beauty. Plato's Heavenly Eros was an acquisitive form of love that fled the world in order to mount to the divine and achieve individual happiness.[14]

Aristotle transformed this Platonic concept of Eros into a concept of cosmic force. He taught that all matter was impelled by Eros to seek its natural place in the universe, and that Eros was an inner desire kindled in matter by Pure Form, the unmoved mover of the cosmos.[15] Five hundred years later, Alexandrian philosophers further redefined the Platonic concept of Eros in the light of the mystery religions of late antiquity. They described a sharp dualism between God and matter, but they believed that divine emanations could reach the material universe through intermediaries. Plotinus spoke of emanations that flowed downward and upward through the three worlds of the cosmos: Eros, the Divine One, illuminated the intelligible world, which was the highest of the three worlds and transmitted light to the world-soul in the second world, while the world-soul in turn radiated the beauty and goodness it had beheld, attempting to reproduce in the empirical world that which it had received from above. Bodily things derived their beauty from the soul, the soul from reason, and reason from the Divine One, Eros. Emanations of divine light and love thus descended by stages, and they were reflected upward again from the lower levels. When the human soul turned from the world toward contemplation of Eros, it became stronger and more beautiful, ultimately achieving ecstatic union with the Divine One, "having become, or rather being, God."[16] To Plotinus, Eros was God, the ultimate source and goal of all things, absolutely eudaemonistic and self-sufficient.

Syncretic thinkers of the following centuries took this Neoplatonic philosophy in different directions. Hermetic and Gnostic authors embedded Neoplatonism in a web of astrological, magical, talismanic, and mystical lore, emphasizing in various ways the emanations of Eros throughout the cosmos, as well as the divinity of the human soul and the divine power over the natural world of the contemplative Magus or god-man.[17] Early Christian theologians strove to reconcile two Greek concepts of love, the Neoplatonic Eros and the Christian αγαπη (agapē), and they ultimately achieved synthesis in a new concept of love called *caritas*.[18] The most influential of these theologians was Augustine of Hippo. He asserted that God imparted divine love (*agapē*) to humans, and that this was the elemental force in human life. In humans, however, love became an acquisitive force, striving to possess its object. When directed upward toward God, he called it *caritas*, and when directed toward temporal objects, *cupiditas*. He used the Latin *amor* as the equivalent of *eros* and treated it as another name for *caritas*. After Augustine, as the ancient world gradually metamorphosed into the Middle Ages, the Hermetic tradition faded into obscurity and the *caritas* view of love became a dominant Christian motif.

By Tycho Brahe's day, the *caritas* synthesis itself was in the process of falling apart, while classical traditions of *eros* and *agapē* reappeared in new forms and contexts as humanist scholars strove to recapture the heritage of antiquity. Earlier in the sixteenth century, Martin Luther had founded Protestant theology around a restatement of the *agapē* concept of love.[19] Luther followed St. Paul in describing *agapē* as a spontaneous, powerful, and unconditional love flowing from God, far different than the acquisitive *amor* of an intellectual human being. Luther's view of love was theocentric, not egocentric: He described divine love (*agapē*) as a vital force that permeates the whole universe and infuses it with life. God's love is irresistible, and it flows to the worthy and unworthy alike. This powerful, redeeming, and life-giving love moves human beings to love God in return, and also to love and serve their fellow humans, making them "lords of all and servants of all."

Tycho Brahe and other Lutheran natural philosophers in the Philippist tradition reflected Luther's theology of *agapē* in their ceaseless search for manifestations of a divine spiritual immanence activating the material universe. However, more than one intellectual tradition exerted its force upon them, and they were even more strongly influenced by the humanist philosophies of Eros or *amor,* coming from Marsilio Ficino and the Florentine Academy.[20]

Giovanni Pico della Mirandola had described the Neoplatonic cosmology as a hierarchy of three worlds emanating from the One: the angelic or intellectual world of the mind or spirit, the celestial world, and the material or sublunar world. From God, angelic intelligences received the ideas or types of all things, which they infused into the celestial world as virtues. The celestial world comprised nine spheres ruled by nine angelic orders, receiving light from the angelic world and diffusing it through all the spheres while moving in perfect, circular, changeless paths. From the virtues infused into it, the celestial world endowed the dark, corrupt matter of the sublunar world with forms and qualities. Moreover, Pico asserted that God was immanent, active, and manifest throughout the cosmos, though not identical with it, and that the mind of man could apprehend Him and become like Him through contemplation of His creation.[21]

Marsilio Ficino's theory of *amor* also placed humans at the center of a Neoplatonic universe. He saw the individual as a fourth world or Microcosm, reflecting the other three. Endowed with the seeds of every form of spiritual and material life, the individual chose his or her own destiny by the exercise of free will. He (or she) could live the vegetative life of the plants, the sensual life of animals, the rational life of a human soul, devote himself to intelligible truth and become like one of the angels, or even "withdraw into the solitude of his own soul, and there be made one with the spirit of God."[22] At the highest level, the individual was the lord of the earth, the link between matter and spirit, created by God to "contemplate and understand the rest of creation and to unify and recreate the universe in himself."[23] In Ficino's view, *amor* created a godlike, Olympian fellowship among friends who aspired to live this intellectual life of the mind.[24]

Such thoughts were in Tycho's mind when he wrote to his friend Pratensis, "So heaven itself unites / Our eyes, and earth releases, too, our heavy flesh." Tycho Brahe was strongly influenced by the tradition coming from Ficino and Pico through figures like Cornelius Agrippa and Paracelsus, and he implied in his treatise on the supernova of 1572 that this school of thought could explain the new star, whereas other schools of his day, including the Aristotelian tradition of the scholastics, could not.[25]

Although the Platonic philosophy of the Renaissance claimed to be a revival of ancient wisdom, it was different in one fundamental aspect. When Ficino's theory of immanence replaced the transcendental dualism of Plato and Plotinus, the realms of spirit and matter were no longer seen as sharply separated. Following Ficino, Renaissance Platonists believed that divinity was immanent within matter, energizing it, giving it form, quality, and life. To Tycho Brahe and his friends, *amor* permeated the cosmos, charging everything with life, and the individual was a Microcosm precisely because all the elements of the cosmos, divine, intelligible, rational, and material, came together within a single human being.

This line of thought led to a radically new view of human potential.[26] The Renaissance individual did not strive to realize his or her divinity by fleeing from Plato's shadowy cave of the world to a nonmaterial world of eternal Ideas. Tycho saw himself as living in a cosmos where a divine spirit was immanent within matter, not separate and far above it, and he took as his task to probe and comprehend that cosmos, not to escape from it. Contemplation to him could not be a passive, introspective process of withdrawal; rather, it was a bold, outgoing, creative probing of the immanent cosmos, using innate powers of the cosmic individual to penetrate the divine secrets within nature. The aim of such active contemplation was to understand God the Creator and even to become like a god, grasping the whole plan of creation and re-creating it within oneself, on one's own island, within a circle of friends. This process would lead to a cosmic reform that could eventually bring about the instauration of the Golden Age, when men lived like gods upon the earth. For these reasons, Tycho

Brahe established his research institute on the island of Hven and assembled his friends to help him dedicate it. In Tycho's mind, ties of *amicitia* to his learned friends formed an essential part of his ambitious plan of cosmic reform.

Tycho intended to create on the island of Hven a magical talisman of the universe, a garden of the gods, peopled by demigods who could penetrate all the secrets of the heavens and earth, move through the spheres, change the world, and recreate the Golden Age when the world was young and *amor* was manifest. This is what he had in mind when he put the following words in the mouth of the divine Muse, Urania, speaking to him in the forests of Herrevad along Rönne Brook:

> But I recall an ancient, worthy time
> When I was worshiped, honored here on earth,
> And I recall when, in the halls of kings,
> Proudly I went forth in glory. Then
> No men but kings and those of royal blood
> Would dare approach my sacred temple site.
> But you do not neglect to show me honor,
> For you have strewn your incense on my altar
> And often stand at night and watch the stars.
> Then spoke Apollo: "He belongs to you!"
> His very words! And both of us did hope
> That you would dedicate yourself to me,
> And serve me under Ursa Major's sign
> And spread abroad your northern homeland's fame.

Tycho Brahe was of noble blood, like the astronomer-kings of old. His friends, Johannes Franciscus and Anders Vedel, emphasized his noble birth and noble learning in their poems of tribute, whereas Tycho contrasted his own pursuit of celestial insight with the mundane interests of most noblemen, who worshiped Bacchus and Venus instead of Urania and Apollo. Because of his noble mind, Tycho was worthy to revive the worship of Urania. At first, however, he had not persisted. In his years at Herrevad, he had turned from astronomy to chemistry, and that had caused Urania's fame to dim, just when it had seemed on the verge of revival.

At this point, Tycho's elegy took up the relationship between chemistry and astronomy. Tycho explained it in terms of Renaissance Neoplatonism, including references to the Paracelsian theory of "stars within" and to the Hermetic view that the lower spheres of creation reflect the upper spheres. Urania continued to address him:

But soon you fell to Vulcan's secret arts
And labored many hours with sacred fire.
My glory dimmed, for no one worshiped me
And I was stripped of honor as before.
But I could bear it, since the earth has stars
And they are not at enmity with mine;
The earth has suns, and it has moons as well,
It takes its hosts of stars in broad embrace.
For that which is above is also hid below,
And these two regions have a common nature.
But earthly stars are treated just like matter:
The powers they have appear in Vulcan's art.
Our stars in heaven's sphere the eye can see,
But mind alone, not eyes, can see their force.

Tycho's elegy went on to contrast ordinary humans with the superior person who used his or her mental powers to seek the divine immanence within nature. Urania concluded:

Like blind moles, lethargic mobs see
No more than earthly, perishable things.
So very few Apollo grants to see
The riches which Olympus hides away,
For they must show contempt to earthly gain
And lift their eyes unto the heavenly beams.
And Venus cannot lure them, nor the glass
Of wanton Bacchus, riches, power, fame.
More beautiful by far the goal they seek,
For it is not a goal unknown to gods:
Through mental force control the heaven's stars,
Subject the ether to his conquering spirit.

THE ALLEGORY OF NATURE AND TYCHO'S FRIENDS

How much of this did Tycho and his friends really believe? Did he actually think that he could control the very stars of heaven through mental power alone, and master the hidden powers of earthly minerals and jewels by means of chemical manipulation, or was this simply poetic allegory? Did Tycho and his friends truly believe that they were demigods with semidivine powers?

When Tycho Brahe told in poetic language about the epiphany of Urania at Herrevad Abbey, he was using allegory to describe the appearance of the supernova of 1572. He did not literally mean that a goddess in human shape had come down from the skies to chat beside Rönne Brook. On the other hand, he certainly did believe that the stars exerted influence upon the lives of human beings, and that the supernova did indeed play the role of divine messenger in his own life, reshaping its whole direction. Tycho did also truly believe that there was a spark of divine power in the human intellect, and that the person who used his or her mind could penetrate to a knowledge of the hidden, nonmaterial, forces within nature. He truly believed that friendship, *amicitia,* among people who used their minds could bring them into spiritual union with one another and with the natural world. In his poetry, Tycho Brahe used the language, imagery, and conventions of Augustan Rome, but this poetry was not mere fantasy or play. His vivid, allegorical language expressed what Tycho considered to be the truth about man and nature.[27] His friends agreed; their poetry expressed the same things.

On 8 August 1576, they all gathered on Tycho's island for the dedication of Uraniborg, these "demigods" and friends, men of the mind, beloved of Apollo and the Muses, who felt capable of flying through the spheres and able to control the ethereal stars by mental power alone. All around them were the peasant men and women who rowed the boats, drove the wagons, dug the foundation, fired the bricks and made the fences, baked the bread, brewed the ale, caught the fish, and smoked the hams. Those were the tasks of molelike men and women who lived in the vegetative, sensual, or at best the rational world. Tycho ruled the island. "He is the undisputed lord of the earth," as Ficino had said of the man of intellect, "linking it to heaven by his activities and so rendering the very soil he cultivates almost divine."[28] Yet *amor* animated every living being, and even such "demigods" enjoyed a splendid symposium together, with many wines, rich meats, piquant sauces, all the splendors and gifts of the earth, prepared and served by earthbound servants.

Who were they, these friends who attended the dedication of Uraniborg? We know only one by name: Dançay. He and Tycho's best friend, Johannes Pratensis, had planned to donate the cornerstone of Uraniborg as a service of *amicitia.* Dançay had procured a block of dark porphyry stone flecked with feldspar crystals [see Fig. 7], and Pratensis had reserved the right to embellish it with an incription and "emblems mystical and sacred." Then Pratensis had died suddenly in May of 1576, only thirty-two years of age, and the inscription had to be composed instead by Dançay and Tycho.[29] Charles de Dançay, nearly seventy, had been the French envoy to the Danish court for almost thirty years. He had been a Protestant since youth, when he had studied with Johan Sturm and John Calvin. His verve, charm, experience, and erudition made him a central figure in the circle of Tycho's friends.[30]

Tycho Brahe's maternal uncle, Steen Bille, may also have attended the dedication. An ex-diplomat with broad experience throughout Europe, Steen Bille now resided at Herrevad Abbey, where he maintained a community of aging monks and a Lutheran Latin school for boys, as well as a force of soldiers and knights. He shared many of Tycho's learned interests and had often acted as his advocate at court. A gentle, charitable person with curly hair and a huge moustache, he may not have allowed declining health to prevent his attendance.[31]

Anders Sørensen Vedel [Fig. 8] must have been there. He was four years older than Tycho and the grandson of a burgomaster of Vejle. When Tycho left for the University of Leipzig in 1562, fifteen years of age, Vedel had accompanied him as his preceptor in charge of the purse and plan of study. Their three years in Leipzig had cemented a lifetime friendship. Vedel was now the royal chaplain at Copenhagen Castle and cultivated a strong interest in history. He had powerful patrons at court, and his star was definitely on the rise.[32]

Hans Aalborg may also have been there. Since serving as Tycho Brahe's preceptor at the University of Rostock, he had also remained a friend. Aalborg continued his studies abroad and was about to leave for Leipzig in 1576 as preceptor to Tycho Brahe's kinsman, Oluf Rosensparre, but they may not have left before the Uraniborg dedication in early August.[33]

One friend apparently unable to attend the dedication of Uraniborg was Petrus Severinus, who was two years older than Vedel and Aalborg. As a student, he had wandered the continent for eight years with his inseparable friend Johannes Pratensis. His book *Idea medicinæ philosophicæ* (Basel 1571) made him famous and did much to make Paracelsian thought respectable in European learned circles.[34] In 1571, King Frederick II had appointed Severinus his personal physician, and from then on, he rarely left court;[35] but both he and Tycho Brahe were Paracelsians, and they corresponded regularly.

Johannes Franciscus Ripensis [Fig. 9] undoubtedly was there. Forty-four and married to a rich widow, he was a professor of medicine in Copenhagen, lecturing on mathematics, astronomy, philosophy, and Galenic medicine. He also basked in far-flung fame as poet in Latin and Greek.[36]

Erasmus Lætus may also have attended the dedication of Uraniborg. Almost sixty, he had retired from the university to live as poet laureate of Denmark. His European renown was even greater than that of Johannes Franciscus, and one of his famous works had described Tycho Brahe, dis-

Fig. 8. Anders Sørensen Vedel. Oil painting by Tobias Gemperle, 1578.
(Courtesy Det Nationalhistoriske Museum på Frederiksborg, Hillerød)

coverer of the supernova of 1572, as a jewel in the crown of Danish learning.[37]

Certain noblemen among Tycho's friends also attended the dedication
of Uraniborg. Besides Steen Bille, the most likely possibilities were Niels
Kaas, Falk Gøye, Erik Lange, and Axel Gyldenstierne. All except Axel Gyldenstierne had spent several years studying abroad. Niels Kaas was one of
the great intellects of his generation. He was appointed Royal Chancellor
in 1570, and after the death of Peter Oxe in 1575, he became the foremost
patron of learning in the realm, save only the king. Though Kaas was a
strong supporter of Tycho, Vedel, and their circle, he was probably too
busy to attend the dedication ceremonies on Hven.[38] Falk Gøye and Erik
Lange were younger than Tycho and had traveled abroad together, studying in Strasbourg, Tübingen, Wittenberg, and Paris. Both were highly

Fig. 9. *Typus amicitiæ* ("The picture of friendship"). Johannes Franciscus Ripensis designed this woodcut emblem to express the idea of friendship and dedicated it to Philipp Melanchthon. A young man, barefoot and in rags, holds forth one hand and points with the other to his heart. This emblem shows that friendship is eternally young, unaffected by material circumstances, always faithful and true. The legends, *mors et vita* ("life and death"), *æsta hiems* ("summer, winter"), *longe prope* ("far, near"), suggest that neither death, time, nor distance can diminish true friendship. (Courtesy Det Kongelige Bibliotek, Copenhagen)

cultivated men, obsessed with alchemy, and both became close friends of Tycho Brahe; one or both them may have attended the dedication of Uraniborg.[39] Axel Gyldenstierne was Tycho Brahe's cousin, four years older, and a soldier and administrator, not a man of learning.[40] In May of 1576, he was appointed Lord Lieutenant of Landskrona, which made him a close neighbor to Hven. He was a personal friend of Tycho Brahe but might have been a bit uncomfortable in the group that came to dedicate

Uraniborg. The same was true of Tycho's brothers, Steen, Axel, Jørgen, and Knud Brahe.[41]

Latin learning was a man's world in those days, but Tycho Brahe did find a woman with whom he could share his love of natural philosophy. She was his own sister, **Sophie Brahe**. On 8 December 1573, when she was fourteen and still living at Knutstorp with their mother, Sophie Brahe had helped Tycho to observe a lunar eclipse.[42] Tycho called her Urania – his Muse. She loved her big brother, and in later years, nobody visited the island of Hven more frequently than she. Perhaps Sophie Brahe was there for the dedication ceremonies in 1576. At seventeen, she could have played the role of muse among courtly elder gentlemen like Charles de Dançay, Uncle Steen Bille, and the poet laureate, Erasmus Lætus; learned scholars like Johannes Franciscus and Anders Vedel; and dashing noblemen like Erik Lange and Falk Gøye; all friends of her devoted brother, Tycho Brahe.

We cannot be certain who attended the dedication of Uraniborg: The veil of time falls over their faces and hides them. We know only that Dançay attended, "accompanied by several noblemen and also by some men of learning among our common friends." Tycho's closest friends in those years were the people named above, and they were the ones most likely to have been on Tycho's island at dawn on 8 August 1576, when the great porphyry stone was laid by Dançay, having first been consecrated with wine of various sorts and prayers for good fortune in every respect, expressed in Latin verse that has not survived the centuries.

The learned cult of *amicitia* had created strong bonds between Tycho Brahe and his influential friends at the Danish court and university. Bonds of friendship cast splendor over the founding of Uraniborg. In years to come, these bonds would play an important role in sustaining Tycho's activities on Hven by linking them to the international community of learning. Tycho Brahe needed the reciprocal benefits and convivial support of friends, but he also needed a complement of skilled and learned servants, coworkers, and collaborators, on the spot, day in and day out, at Uraniborg on Hven. His circle of friends could help him to build and sustain what he called his *familia*.

4

FOUNDING THE FAMILIA

———

1576–1584

ONE BIG FAMILIA

Craftsmen, artists, bailiffs, and boon workers worked to prepare the physical facilities of the Uraniborg complex as Tycho Brahe directed their labors. At the same time, Tycho began to build his scientific, technical, and support staffs and organize them into coordinated research teams of under his central leadership. The model he used to build his staff was that of the *familia*.

Humanist scholars like Tycho Brahe used the word *familia* in the ancient Roman sense, though in a sixteenth-century context. To them, the *familia* was not what we would call a "family"; rather, it meant all those who lived under the authority of a *paterfamilias,* or patriarch.[1] Tycho used the words *famulus* and *famula* (male and female members of a *familia*) to refer to servants or staff members of Uraniborg. In the broadest sense, he might have considered all the inhabitants of his patrimony and fiefs on Hven, in Skåne, Sjælland, and Norway – hundreds and hundreds of people – as part of his *familia,* but the core of the *familia* was the coresidential unit: the household where the paterfamilias resided. The household was a unit of fundamental importance at all social levels. It formed the normal, everyday link between the individual and the world at large.[2] On the island of Hven, Tycho experimented with new uses for the household and *familia* as he worked to make them into the structural foundation of his scientific research.

PEASANT HOUSEHOLDS

Most of the fifty-odd households on the island of Hven were not devoted to scientific research. They were farming households, and they were organized around a nuclear family. The particulars on Hven are not known, but the main lines have been sketched out in studies of other Danish peasant communities.[3] The farmer was the patriarch who held the farm in a legal sense; sat as a grand in the village assembly; paid rents and dues; and led the work in fields, commons, and at sea. The household was a working unit, however, with work patterns differentiated by age and gender, and it required a matriarch as well as a patriarch. The wife was matriarch of the women's domain of house, croft, cattle and smaller animals, food preparation, herbal family medicine, and textile production.[4] In addition to the partnership of husband and wife, the household typically included children; if the children were small, a farmhand or milkmaid might be hired to help with the work. Some households also contained older relatives, an impaired or disabled person, or crofters living in a cottage on the farm.[5] All were under the authority of the peasant husbandman and goodwife.

Mortality rates were high, and many children died at an early age. Danish peasant communities like Tuna had a high birth rate of around thirty-five per thousand and swarmed with children, but on average, only around half the crofter children and two-thirds of the farmer children lived to the age of ten, though three-quarters of Danish noble children did.[6] Life expectancy at birth was only twenty to thirty years for villagers because epidemics and high infant mortality counterbalanced the high birth rate. The living standard on Hven was precarious because the population level was close to the island's carrying capacity. Fortunately, the sea was abundant, and many islanders found supplementary livelihoods as fishermen, sailors, boatwrights, skippers, and coastal traders. In general, the peasant population of Tycho Brahe's day was more mobile than it would be in later centuries.[7]

Some village farms had an excess of mouths to feed and more labor than was needed. This normally occurred when children had grown to adulthood but had not yet married. Some of these young people managed to leave the island. Others came to Uraniborg as boon workers. Still others knocked on Tycho's portal and sought regular service. The great household of Uraniborg always needed people, and it drew some of its servants from the households of farmers and crofters in Tuna village.

THE HOUSEHOLDS OF TYCHO BRAHE

Tycho Brahe's own household was divided when he came to Hven in 1576. It was not unusual for a great nobleman to have several residences, nor for

noble spouses to maintain separate households. Tycho's seat remained at
Knutstorp Castle, but he lived on Hven because that was where he needed to be. Initially, he may simply have commandeered the parsonage or
moved into one of the best Tuna farmhouses. By the time his friends arrived in August of 1576, he may have built his own temporary residence
within that strange triangular courtyard of the Uraniborg grange.

His *familia* on Hven began without a family.[8] Tycho Brahe and Kirsten
Jørgensdatter now had two daughters, Kirstine, born 1573, and Magdalene,
born 1574, and they were expecting a third child early in 1577.[9] Kirsten
Jørgensdatter continued to live in Skåne with her daughters, probably in
the town of Helsingborg. Little Kirstine died there of the plague on 24
September 1576, three weeks short of her third birthday and less than two
months after the dedication of Uraniborg. She was interred in St. Mary's
Church, Helsingborg. The survivors took refuge from the plague in a
country residence on Kullen peninsula, where Kirsten Jørgensdatter gave
birth to a son, Claus, on 2 January 1577. The infant lived only six days and
was buried in Väsby church. His father never saw him.[10]

Meanwhile, Tycho Brahe began to build his own *familia* on the island
of Hven. A basic staff of housekeeper, maid, secretary, and footmen could
provide for his personal needs. He also needed a bailiff to direct the work
of demesne and manor site, an overseer to supervise boon workers, a summoner to call them up, and one or more clerks to keep records. Some of
these staff members, such as the overseer and summoner, may have been
villagers who entered Tycho's service but continued to reside in Tuna like
the boon workers. Tycho also employed skilled craftsmen on the construction sites of Uraniborg and the grange. Although he was thoroughly familiar with classical and Renaissance architectural theory, and worked out the
theoretical aspects of the building himself, the ratios of the elevations and
plan, Tycho needed a master builder with practical experience, in addition
to masons, stonecutters, carpenters, glaziers, painters, blacksmiths, and
other craftsmen. Some of them may have commuted irregularly by skiff
from Elsinore, while others undoubtedly boarded with villagers in Tuna.

Tycho Brahe's household, or rather, his households, were certainly "por-
ous" in the early years on Hven.[11] He treated the whole island community like one big *familia* under his patriarchal authority. Subunits within
this community – the village, the demesne, the building site, the Junker's
personal staff, the resident scholars – were more or less self-governing, but
always under the watchful eye of Tycho Brahe, who claimed the right of
paterfamilias to run everything on the island. Tycho also made trips to Kirsten Jørgensdatter in Skåne, and he supervised her household, as well as
his share of the Knutstorp estate and his far-flung fiefs. In addition, of

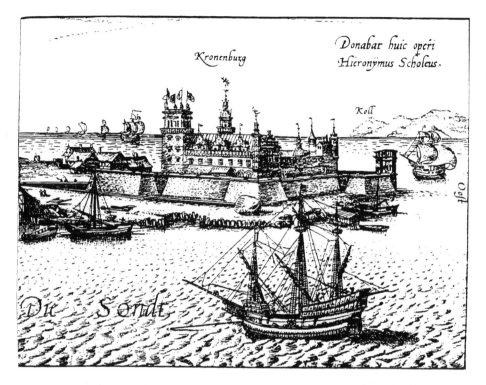

Fig. 10. Kronborg Castle viewed from the direction of Hven. Engraving based on a drawing by Hans Knieper, 1582. "Koll" in the background at right is Kullen, site of Tycho's lighthouse. (From Braunius & Hogenberg 1572–1617 [1588])

course, he made his appearances at court and elsewhere. It was a busy life, but not unusually so for a man of his time and station.

KRONBORG AND URANIBORG

The households of Tycho Brahe in the late 1570s, large and complex though they were, were dwarfed by the households of the crown. King Frederick II and Queen Sophie moved in more or less regular, seasonal progresses among the various castles, monasteries, and royal seats of the realm.

The king was a great builder. In 1574, two years before Tycho came to Hven, King Frederick II had begun a major renovation of his castle in Elsinore. The dark medieval fortress of Krogen was to be transformed into a Renaissance castle with the new name of Kronborg ("Crown Castle") [Fig. 10]. Frederick wanted it to be the pride of the kingdom. His first ar-

chitect was a master stonecutter named Hans van Paeschen, whom some
have also identified as the designer of Uraniborg, but Hanne Honnens de
Lichtenberg has shown that **Johan Gregor van der Schardt** was a much
more likely candidate, always working closely with Tycho himself. Tycho
Brahe had met Schardt during his travels abroad the previous year and had
recruited him to serve the Danish crown. Schardt had worked at the imperial court and probably also on villas designed by Palladio. Tycho Brahe
invited him to work on Uraniborg, and Schardt was on Tycho's island by
November of 1576, if not earlier.[12] Tycho himself had established an elaborate, geometrical plan that aimed to make Uraniborg a microcosm of the
universe according to Neoplatonic cosmic theories. He needed a skilled
artist to carry out this theoretical program in the elevations and decora-
tions, and he also had ideas for the dynamic use of hydraulics, wind, and
sound to bring the building "alive." Schardt understood exactly what Tycho was talking about and was just the man to help him realize these intri-
cate, powerful plans.

Tycho also recruited skilled craftsmen from Kronborg to help with the
construction of Uraniborg. Indeed, Uraniborg was a royal project because
Hven belonged to the crown and the work was largely financed from the
royal coffers, although Tycho Brahe designed and built with a free hand
and held Hven in lifetime fee.[13] By establishing his scientific center on a
remote island, King Frederick II departed from the pattern of patronage
followed by princes like Emperor Rudolf II and Landgrave William IV of
Hesse-Kassel, who kept their savants and collections at court.[14]

Tycho soon had occasion to travel to court and render learned service
to the crown. In April of 1577, Queen Sophie bore a son and heir to the
throne of Denmark and Norway. Inger Oxe was present at the birthing.
It was a great event in the life of the realm, and by July, Tycho had consulted his Muse, Urania, and prepared the infant prince's nativity. This was
precisely the kind of "gift" expected of him in the reciprocal system of
scientific patronage.[15] Tycho Brahe delivered the nativity personally, and
it was a political document of great significance because it foretold (with
uncanny accuracy) the character, strengths, weaknesses, and all major
events in the life of the future king. Tycho Brahe asserted that the stars
affected but did not predetermine the lives of humans because God could
countermand their influences at any time, and because sound education
and an exertion of individual free will could also counteract adverse astral
influences. Tycho's status in the eyes of his royal patrons soared as King
Frederick and Queen Sophie pored over every diagram, table, and sentence
in this precious nativity.

Meanwhile, construction continued in Elsinore and on the island of Hven. In the summer of 1577, Hans van Paeschen was assigned to another royal project, and Antonis van Obbergen took over as master builder of Kronborg. During the winter of 1577–8, Obbergen traveled to his native Flanders to recruit skilled craftsmen.[16] The Dutch wars were forcing Protestants to flee the Spanish rule, and Obbergen found many Flemish and Dutch artisans willing to come to Denmark.

That same winter, Urania once again beckoned Tycho Brahe and led him away from worldly affairs. Around sunset on 13 November 1577, Tycho was by one of the ponds near Uraniborg, catching fish for supper. It was two days after the harvest feast of Martinmas, and the skies were clear. As his workers dragged in the nets, Tycho scanned the skies and noticed a very bright star in the west. He knew it could not be Venus, and it was too bright to be Saturn. The star held its position as dusk turned to darkness, and a long, ruddy tail gradually appeared, stretching away from the sunset: This was a comet, the first Tycho had ever seen.[17] He left the fishing to others and summoned his learned staff.

Tycho Brahe, inventor of the sextant, had a considerable collection of instruments: He had a cross-staff (radius) designed 1564 in Leipzig and another made by Walter Arscenius of Louvain, a large (155 cm) wooden sextant from Herrevad, a portable steel sextant, a couple of rather small quadrants, and a fair-sized (117 cm) zodiacal armillary from Herrevad [see Fig. 5b], among others.[18] Tycho and his assistants immediately brought out one of the cross-staffs, later also the steel sextant [see Fig. 17c] and quadrants, and began to observe the comet. Tycho took a quill pen and some foolscap sheets, folded the paper into a notebook in quarto format, and began to jot down the observations [Fig. 11a]. The next day, he wrote to Petrus Severinus at court, asking him to arrange things with King Frederick so that he would not be disturbed unnecessarily in the days to come.[19] He knew that the king would be eager to learn how the astrological effects of the comet impinged upon national security, but Tycho wanted to study the phenomenon carefully before venturing any prognostication.

For the next two and a half months, whenever the sky was clear, Tycho Brahe and his assistants observed the comet, and he entered the observations in his little notebook. It became tattered by weather and use, but the notebook has survived to the present and reveals each step in the process of Tycho's observations and thought. His studies of this comet became another milestone in the history of science. First of all, Tycho was able to establish that the comet was located above the sphere of the moon, confirming his conclusion from the supernova of 1572 that changes could and did occur in the heavens, as well as on earth. Second, Tycho demonstrated that the path of the comet cut across the orbits of several planets, proving they were not borne on solid crystalline spheres. This disproved another supposed tenet of Aristotelian cosmology and strengthened the evidence for a more dynamic, Neoplatonic cosmology. Third, Tycho began to analyze the position of the comet within the planetary system from a Copernican perspective but incorporated his own data and a geostatic (i.e., fixed-earth) assumption. On this basis, he gradually began to develop a new understanding of the whole planetary system. His work in astronomy took great leaps forward with these cometary studies during the winter of 1577–8. The comet of 1577 also provided the basis for Tycho Brahe's next major "gift" to his liege lord and patron: After studying the astrological effects of the comet in considerable detail, Tycho determined that they had important implications for Danish national security; these he spelled out clearly in a manuscript written, in German, for the eyes of King Frederick and Queen Sophie alone.[20]

That spring of 1578, Antonis van Obbergen returned to Elsinore with a number of Flemish and Dutch craftsmen and their families. The town of Elsinore was already a cosmopolitan little entrepôt with resident colonies of native Danes, Scots, Holsteiners, Hanseatic Germans, sailors of many lands, and now these nascent communities of Flemings and Dutchmen. Among the new arrivals was a native of Antwerp named **Hans van Steenwinckel**, who was the same age as Tycho Brahe, and whom Tycho immediately enticed away from Kronborg to take residence in the *familia* of Uraniborg on Hven.

Steenwinckel had learned his trade from his father, a distinguished master builder, and his training had been in practical construction and design. He knew all the latest styles of the Netherlands, but he did not have a classical education. Tycho tutored him in geometry, astronomical theory, observational methods, classical Vitruvian and Renaissance architectural theory, and perhaps also in the theory of perspective. Schardt probably lent a hand. Steenwinckel needed all of this to understand their highly theoretical approach to the design of Uraniborg; but even as he was learning, he took over as Uraniborg's master builder. Meanwhile, Schardt, who was thoroughly familiar with the theoretical dimensions of construction, continued to work on sculpture and decoration. They were on the cutting edge of mannerist architecture and worked well together, soon becoming good friends.

Kirsten Jørgensdatter and her household remained in Skåne during these years because there was still no appropriate residence for them on

(a)

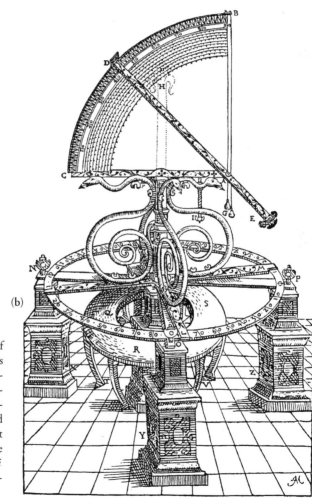

(b)

Fig. 11. The comet of
1577: (a) Tycho Brahe's
first sketch of the com-
et. (Courtesy Det Kon-
gelige Bibliotek, Copen-
hagen) (b) The midsized
brass azimuth quadrant
he used to observe the
comet. (From *De mvndi
ætherei recentioribvs phænomenis*)

the island. A daughter, Sophie, was born on 4 August 1578. At the end of the year, the queen gave birth to a second son, christened Ulrik, and Tycho Brahe duly prepared a detailed nativity as a gift and benefit to his royal patrons.[21]

Tycho's staff on Hven now included a bailiff, a Dutch master builder, an Italian-trained sculptor and architect, a German portrait artist, and two or three young Danish university graduates, besides servants of various kinds. Several languages were heard, and Tycho himself was fluent in Danish, German, and Latin, knew Greek and a little Hebrew, and may have spoken Dutch. Except for the servants, they all gathered each afternoon at Tycho's table for supper and long discussions, according to the custom of the day. The staff of the Uraniborg household was starting to take shape.

STAFF RECRUITMENT

The staff came to Tycho's island along three main axes of recruitment [Fig. 12]. The primary axis for recruiting craftsmen ran from the Netherlands to Kronborg Castle and on to the island of Hven. Flemish Protestants, driven into exile by the wars, formed the main stream along this axis. A second axis for recruiting artists and artisans of exceptional skill ran from northern Italy through Augsburg and Nuremberg to Hven.[22] Tycho Brahe had traveled this route in 1575 and visited Kassel, Augsburg, Nuremberg, Regensberg, Basel, and Venice. He expressed particular admiration for the craftsmen of Augsburg, where his first truly innovative instrument, the *Quadrans maximus,* had been built in 1569 and where Tycho had commissioned his great brass globe with **Christopher Schissler** in 1570.[23] He recruited the Augsburg portrait artist, **Tobias Gemperle**, in 1575, met **Georg Labenwolf** in Nuremberg, and **Johan Gregor van der Schardt** in Nuremberg or Regensberg. Tycho could offer the privileged status of court artist, which gave tax exemptions as well as profitable commissions, and that helped his recruiting efforts. He had previously brought Venetian glassmakers to Denmark on similar conditions. Gemperle and Schardt both made trips back and forth between southern Germany and Denmark, and when matters were arranged to their satisfaction, they joined Tycho's *familia* on the island of Hven in the autumn of 1577. They worked on commissions for Tycho Brahe, the Danish crown, and some of Tycho's friends.

Labenwolf was a bronze founder who could construct fountains with numerous water jets and revolving mythological and allegorical figures, mounted on basins of metal, marble, or decorative stone. These fountains expressed the classical precepts of Vitruvius and gave "life" to architectural complexes in a manner that fascinated Tycho Brahe and his contempo-

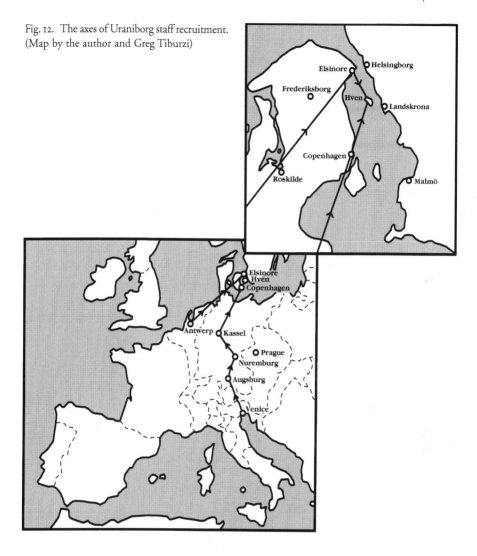

Fig. 12. The axes of Uraniborg staff recruitment. (Map by the author and Greg Tiburzi)

raries. Even more important, they were like the animated statues described by Asclepius, which Ficino had reconciled to Christianity: holy idols full of divine spirit, sensations, and understanding, which could accomplish wonders, foretell the future, cause and cure illness.[24] Tycho arranged for Labenwolf to come to Denmark. In 1577, Labenwolf contracted to deliver a monumental fountain for the courtyard of Kronborg Castle, one that would eventually become his masterpiece. He also agreed to construct the hydraulic system of Uraniborg, including a fountain in the central hall and running water throughout the building.

With his direct ties to the centers of artistic and technological skills in northern Italy and southern Germany, Tycho Brahe provided the valuable service of cultural broker to his royal patron. At the same time, Tycho benefited from the skills brought to Elsinore by the crown. Recruitment and exchange of artists and craftsmen between Kronborg and Uraniborg became another expression of the reciprocal obligations of patron and client.

A third axis of recruitment in building Tycho Brahe's staff ran to Uraniborg from Copenhagen. Along this axis, the flow was primarily university students and graduates. When Tycho needed an assistant in his laboratory or observatory, he called upon friends in Copenhagen to recommend someone. The earliest example was in the spring of 1572, when Tycho needed help with his chemical investigations at Herrevad Abbey and wrote to Pratensis, who found Niels Mikkelsen but gave him a lukewarm recommendation; whether Niels Mikkelsen actually came to the abbey is unknown. Tycho's next assistant came from his family circle: In December of 1573, his fourteen-year-old sister, **Sophie Brahe**, helped him at Knutstorp Castle to observe a lunar eclipse.[25] During the autumn and winter of 1574, Tycho Brahe lectured at the University of Copenhagen and came into direct contact with students. One of them, **Peter Jacobsen Flemløse**, commended himself to the master with an elegant bucolic poem in the style of Virgil.[26] Flemløse attended Tycho's oration and lectures, entered his service, and may have accompanied him abroad in 1575. Two years later, Flemløse was a part of the *familia* on Hven, where he became the principal recorder of astronomical observations. A second university student from Copenhagen, Jens Pedersen Plow, joined Tycho Brahe's *familia* around 1577 but left in 1578 to become pastor of Egebjerg. After the death of Pratensis, Tycho Brahe still had many friends in Copenhagen, including Dançay, Johannes Franciscus, Lætus, Vedel, and the university chancellor, Niels Kaas; Tycho's ertwhile preceptor, Hans Aalborg, was in and out of town.[27] That Tycho was highly respected in the university community was illustrated in the spring of 1577, when the eminent theologian Niels Hemmingsen went so far as to nominate him to the position of Rector Magnificus or head of the university faculty.[28] Tycho was elected in defiance of an unbroken tradition that the rector must always be a professor, but he declined the honor because of his obligations on Hven.

Tycho Brahe's network of *amicitia* also reached out to learned friends and correspondents throughout Europe and served as a channel for recruiting new *famuli*. **Paul Wittich**, a native of Silesia, arrived on Hven in July of 1580 with a letter of recommendation from Thaddeus Hagecius ab Hayck (Tadeás Hájek z Hájku), personal physician to Emperor Rudolf II and Tycho Brahe's friend and correspondent at the imperial court (they had met in 1575 at Rudolf's coronation as king of the Romans). Wittich was the same age as Tycho; they had first met as students in Wittenberg and had several mutual friends; both were great admirers of Copernicus. Wittich owned meticulously annotated copies of Copernicus's *De revolutionibus orbium coelestium,* which he undoubtedly brought to Uraniborg.[29] Two contained extensive annotations copied from Erasmus Reinhold of Wittenberg, as well as Wittich's own notes, and in one, Wittich added numerous drawings of planetary, solar, and lunar motion in his attempts to transpose Copernican astronomy to geostatic models.[30] Tycho had tried the same thing when he had prepared his 1574 lectures and later when he tried to locate the comet of 1577 in planetary space.[31] The two of them may actually have come up independently with some identical hypotheses.[32] Around Tycho's table, they shared their ideas with great excitement and enthusiasm.

Wittich had poor eyesight and was not an experienced observer, but he was a tremendously talented mathematician. He had invented a method of simplifying trigonometric calculations by converting multiplication and division into addition and subtraction. He called it *prosthaphæresis,* and he had worked long and hard to perfect it, teaching it to others including a Scots student, John Craig, who later taught it to John Napier.[33] Tycho put Wittich's method into regular use in processing all of his astronomical data. Eventually, he made handbooks of *prosthaphæresis* and had them on every worktable for his staff to use.[34]

Wittich taught Tycho his mathematical innovations, and in return, he learned the observational methods used on Hven. Tycho explained the features of his most innovative instruments, including the sextant built of steel, rather than the more usual wood, for stability despite changing temperature and humidity [see Fig. 17c]. Its calibration included Tycho's system of transversal dots [Fig. 13a], replacing the original Nonnian calibration that had made the instrument heavier but not more accurate. A new type of sighting, invented by Tycho, allowed four parallel views of a star and was much more precise than pinhole sights. When Wittich saw this sighting device [Fig. 13b], "he uttered a cry of joy," said Tycho, "and assured me that he had now come to know something that he had sighed for in vain for many years."[35] Perhaps Wittich also saw the first model of another instrument Tycho invented: the equatorial armillary [Fig. 13c].[36]

In October of 1580, a comet appeared. Wittich and Flemløse observed it with a quadrant from Uraniborg while Tycho observed it with a radius in Helsingborg. Soon after, Wittich was called home by a death in the family. Tycho gave him a rare and costly edition of Apianus as a parting gift. Wittich never returned to Tycho's island, and Tycho missed him.[37]

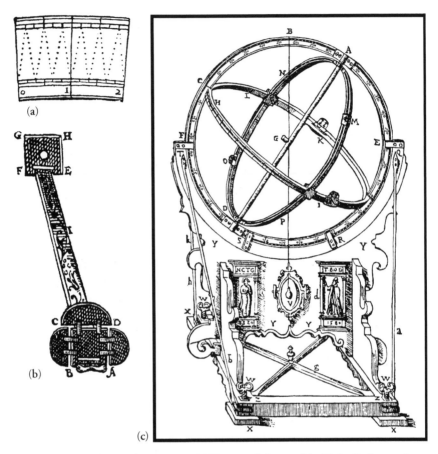

Fig. 13. Innovations and inventions: (a) Transversal dots used by Tycho Brahe to sub-divide calibration on his instruments. (b) Tycho Brahe's sighting device to provide parallel lines of sight along all four sides of the pinnules (sights). (c) Equatorial armillary built ca. 1580, with armillæ of brass-covered plywood, bearing portraits of Ptolemy, al-Battani, Copernicus, and Tycho. (All from *Astronomiæ instavratæ mechanica*)

THE LEARNED HOUSEHOLD

Shortly after Wittich's departure, Tycho Brahe and the whole household finally moved into Uraniborg in November of 1580 [Fig. 14]. At long last, Kirsten Jørgensdatter and her daughters were able to move to the island, and Tycho's nuclear family joined the household of Uraniborg at this time. Boon workers sailed them over with all their trunks, furnishings, servants, and gear, and the same boon workers moved everything into the new man-

Fig. 14. Uraniborg ca. 1580, in red brick with limestone trim. Shown is the east facade, with Schardt's emblem of Astronomy above the entrance and the Pegasus (V) far above. The Winter Room (B, *left of entrance*) was in the southeast corner of the main floor, and beyond it, the museum (M) in the south rotunda, with the chemical laboratory (L) below, the south observatory (O, N) above, and the aviary in front. (From Brahe 1913–29, 6: following 348)

or house. Besides Magdalene, born 1574, and Sophie, born 1578, there was now Elisabeth, born around 1579, and Cecilie, born around 1580. This meant that the household staff included a governess, wetnurses, and maids, and it would soon be in need of a tutor as well. Meanwhile, Uraniborg was buzzing. Craftsmen were still working to finish all the buildings of the complex. Gemperle, Flemløse, and Steenwinckel were there, working under Tycho's direction in their respective spheres of art, science, and architecture.

Once Tycho Brahe was settled in Uraniborg, he began to expand his learned staff by drawing new *famuli* from students at the University of Copenhagen. **Niels Lauridsen Arctander**, son of the archdeacon of Trondheim Cathedral in Norway, spent time in Tycho's service around 1580. Arctander had entered the University of Copenhagen in 1578 at the age of seventeen. He was probably at Uraniborg when Tycho Brahe's son, Tyge, was born in 1581, but he left Denmark to study abroad in 1584. By the end of 1581, Anders Viborg had arrived at Uraniborg; he stayed for three years and later became a clergyman in his native city of Viborg. In February of 1582, Johannes Stephanius, twenty years of age, arrived from the university to spend three months in Tycho's *familia*. He returned to his studies in Copenhagen but came back to Uraniborg in December of 1583 and remained for most of the following year, acquiring a lifetime admiration for Tycho Brahe. Gellius Sascerides, son of a distinguished professor of Hebrew in Copenhagen, joined Tycho's household in 1582. He was twenty years of age, had entered the university at thirteen, and spent three years at Wittenberg before coming to Hven. Gellius was already a Paracelsian and assisted Tycho in the chemical laboratories, located in the well-lit cellars of Uraniborg. He also worked with Tycho and Flemløse in the observatories, and Tycho found him to be a valuable assistant. In 1584, Gellius was awarded the prestigious *stipendium regium* for study abroad, but Tycho arranged for King Frederick to defer the scholarship until Gellius had finished his studies on Hven, and he remained on Tycho's island until the early summer of 1588. In the spring of 1583, another student named **Elias Olsen Morsing** entered the household of Tycho Brahe at Uraniborg. He came from Jutland and was over thirty years old, so he was probably a peasant's son who had had to work long and hard for a university education. Tycho assigned him to keep a daily meteorological journal, while Flemløse and Tycho taught him the observational theory and methods of Hven. Morsing soon became a trusted coworker and stayed in Tycho's service for six years. A student named Severinus was also at Uraniborg around 1583 and stayed just long enough to acquire a reputation as a troublemaker with a wild imagination.

New instruments for the two observatories of Uraniborg were under construction in these same years. First came a large (155 cm radius) quadrant mounted on a ball-and-socket swivel to measure distances between any two points in the sky up to 90° apart.[38] Next came an immense azimuth quadrant of steel [Fig. 15a], with a brass arc over six feet (194 cm) in radius, designed to occupy the south rotunda of Uraniborg, above the room that Tycho called his museum.[39] Then came in rapid succession in the years 1581-3 a large (155 cm radius) but rather unsuccessful bifurcated sextant for two simultaneous observations, followed by an excellent trigonal sextant [see Fig. 17a,b] of the same size, a bipartite arc [Fig. 15b] that proved accurate for simultaneous observations of close objects, a huge mural quadrant (194 cm radius) installed in the southwest room of the main floor of Uraniborg [see Fig. 31], and an even more immense triquetrum (330 cm long) [Fig. 15c] on the north rotunda, as well as a portable azimuth quadrant (58 cm radius) for mapping [see Fig. 34].[40] Tycho Brahe had invented the sextant per se, and the trigonal sextant and mural quadrant were unlike any earlier instruments because they were designed with the arc next to, instead of away from, the observer. This made it much easier to read the scale and kept large instruments from becoming top-heavy.

Tycho and his enlarged staff tested the new instruments and put them to use as soon as they were finished. Gingerich and Voelkel argued that the distance to Mars became "a central, driving theme" of Tycho's observational program, beginning in 1582, the reason being that Mars in the Copernican system comes closer to the earth than does the sun, but it does not in the Ptolemaic system.[41] If Tycho could measure the diurnal parallax of Mars and determine its distance from the earth, he would have observational evidence for choosing between the two systems. The problem was that this parallax was very small, and excruciating precision was needed to measure it. Two of his expensive new instruments, the azimuth quadrant and triquetrum, proved disappointing, but the mural quadrant quickly achieved unprecedented accuracy. This instrument was oriented due south, and Tycho began using it for the preliminary task of determining the declinations of a select catalog of reference stars. Longitudes or right ascensions of these same stars were required to complete the catalog, and those needed to be determined with respect to the sun's position. Ptolemy had linked solar and stellar observations by means of lunar eclipses, but these events occurred too infrequently for Tycho, who experimented with the idea of using Venus instead, since it could be observed both night and day. The trigonal sextant proved to be excellent for these observations, made during the early months of 1582, but the cross-staff and one of the small azimuth quadrants were also used. As the observations progressed, Tycho gradually came to realize that Mars would be a better link than Venus because it was visible throughout the night. By the end of 1582 and early 1583, he and his crews were concentrating much of their observational activity on Mars.[42]

As these observations unfolded, Tycho Brahe's son, Tyge or Tycho, was born on 28 August 1581, and another son, Jørgen, was born in 1583. Tycho and Kirsten Jørgensdatter now had six living children: four daughters and two sons, ranging in age from nine to infancy. They and their governess all joined the household at the supper table each evening, except for the youngest with their nursemaids. Infants were swaddled in those days, and that kept them fairly passive.

In the autumn of 1583, King Frederick II provided Tycho Brahe with a new, fully equipped pilot-boat of fifty or sixty tons displacement, capable of sailing in almost any weather between Hven and all the cities of the

(a)

Fig. 15 *(caption on facing page).*

Sound: Copenhagen, Elsinore, Helsingborg, Landskrona, and Malmö.[43]
Elias Morsing recorded the weather every day in his journal, and he also
began to note down various events of island life. His records showed a
steady flow of visitors, messengers, and prospective students, to and from
Hven. Tycho's island was becoming a very busy place.

Tycho Brahe did not invent the tradition of a learned household swarming with young scholars and their mentors. It was well-known in his day
and went back to various origins. One was the living tradition of medieval bishops and abbots who gathered learned men around their table.[44]

(b)

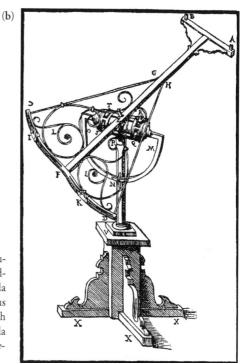

Fig. 15 *(here and facing)*. Observatory instruments at Uraniborg: (**a**) Greatest azimuth quadrant of steel, mounted 1581 in the south-rotunda observatory. (**b**) Bipartite arc for simultaneous observations of two close objects. (**c**) Azimuth triquetrum, installed 1583 in the north-rotunda observatory. (All from *Astronomiæ instavratæ mechanica*)

(c)

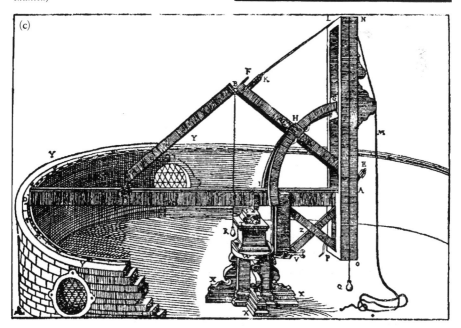

Another was the practice of Italian humanists to assemble in "academies" and emulate the musical and philosophical banquets of Plato's *Symposium*.[45] Still another was the custom of certain princely courts of Renaissance Italy, such as Castiglione's Urbino, to involve women as well as men in philosophical discussions.[46] Finally, the Reformation brought the model of Martin Luther's learned household. Luther lived in a former monastery in Wittenberg, and his household gathered student boarders, relatives, colleagues, and visitors each day for supper at the family table. Luther's wife, Katherine von Bora, understood conversations in Latin, and she brought the nuclear family into the table talk.[47]

Tycho Brahe was familiar with all aspects of this tradition through personal experience. Tycho's maternal uncle, Steen Bille, had spent five years in the household of his own uncle, Archbishop Torbern Bille of Lund, and Tycho had lived in Steen Bille's secularized monastic household at Herrevad Abbey. Tycho had also supped with the king at the table of Abbot Iver Bertelsen in Sorø. During his student days, Tycho had lived in the households of Lutheran professors in Copenhagen, Leipzig, Wittenberg, and Rostock, where women and children joined the colloquies around the table, and where many older scholars could tell stories of their own days as students in the *familiæ* of Martin Luther and Philipp Melanchthon. In addition, he had participated in Platonic symposia during his travels in the Veneto and in Augsburg, as well as among his friends in Denmark.

Tycho had even experienced *familiæ* that combined science and technology in various ways. Symposia in the households of friends in Augsburg, for example, had led to the construction of Tycho's *Quadrans maximus,* and in Kassel, Tycho Brahe had visited the learned staff and seen the collections at the court of Landgrave William IV.[48] Finally, he had lived for several years at Herrevad Abbey. This old Cisterian monastery had come under Danish royal jurisdiction in 1537 and received a lay governor, Steen Bille, in 1565. When Tycho Brahe stayed there around 1570–4, the extensive *familia* included the personal household staff of Steen Bille and Kirsten Lindenow; Tycho and his helpers in the observatory and chemical laboratory; a team of Venetian glassmakers; some German papermakers; the smiths and iron founders who manned a large smithy with a power triphammer; a Lutheran clergyman and the boys of his Latin school (Knud Brahe, Arild Urup, and Claus Lyschander studied here); half a dozen elderly monks with their servants; two or three gentlemen pensioners; a contingent of men-at-arms with squires and servants; a miller family; a bailiff, herdsmen, and milkmaids of the nearby grange; besides hundreds of peasants in the surrounding villages of the Rönne valley. Herrevad was a large *familia* combining learned and technological activities: It was the most direct model for Tycho Brahe's establishment on the island of Hven.

DAILY LIFE

Normally, honored household members and guests ate at the table of the lord and lady in a large Danish household of Tycho's day, while servants and staff ate in the kitchen, and boon workers received their meals at the grange.[49] The kitchen of Uraniborg was in the north rotunda on the ground floor. A corridor led to the Winter Room, as the lord's chamber was called in a Danish manor house. This chamber was richly paneled under a beamed ceiling, with bookshelves and paintings, benches along the north, south, and east walls, a massive oaken table, a towering tile stove by the west wall, and a curtained four-poster bed in one corner.[50] In this room, Tycho, Kirsten, and their children, learned assistants, senior staff members, and guests took their meals around the oaken table and lingered in conversation until it was time to go to the observatory or to bed.[51]

There were two main meals each day.[52] Everyone normally arose very early. Astronomical observers aside, people were usually up by three or four o'clock in the morning and went to bed around eight at night. On rising, it was common to eat a quick breakfast of warm beer and herring or thick rye porridge (*øllebrød*); then they were off to work until dinner time, which came around nine or ten in the morning. Dinner was normally a full meal of three or four courses, with much more on festive occasions. After dinner, it was back to work until four or five in the afternoon, when the day's labor came to an end and supper was served. Supper was the main meal of the day and also comprised three or four courses on an everyday basis.

Meals were ceremonious at the lord's table, where finer food was served than in the kitchen or grange (scraps were traditionally offered to beggars at the portal). Diners washed their hands and sat on benches along the walls, leaving the other table side open for service. Seating was in order of precedence: The lord and lady, Tycho and Kirsten, sat on the high seat in the middle of the high table. On festive occasions, the table was decked with three layers of colored and white tablecloths, but a single tablecloth was used for everyday. Each place was set with a napkin and a plate or trencher. Diners generally used their own utensils, but for banquets the table was set with knives and spoons; forks were used as serving pieces. In the center of the room was the sideboard, laden with the household's larger vessels of sterling silver, engraved with the family arms. Tycho Brahe and all of his noble kinsmen dined in this elaborate, ceremonious fashion.

A prayer was recited to open the meal. Tycho's chaplain, the pastor of St. Ibb's, could have performed this task. Male pages cut the meat, poured the wine and beer, and served the platters with ceremony. The old custom of eating fish on four fast days a week was not followed, but most Danish households still ate fish on Mondays and Wednesdays. Herring was abundant and was probably eaten almost daily in the grange, the kitchen, and village households of Tuna, but more delicate seafood graced the table of the Winter Room: smoked salmon from Skåne, trout from Herrevad, jellied eels, Norwegian cod and flounder, plaice, hornfish, quenelles of pike, and carp from Tycho's own ponds. Fish was generally eaten with a pungent mustard sauce. Shrimp and crabs were abundant, huge lobsters quite common, likewise mussels, while oysters and crayfish were rare delicacies on the great platters of seafood. The last course on fish days might be bread with thick layers of butter, served with beer or wine, or a great cheese, borne forth and eaten with bread. Several types of wheaten rolls were made, as well as pumpernickel rye and bolted-rye bread.

Meals were served in "sets" of courses, never more than three sets. Several courses would be borne in with ceremony and music, then consumed at leisure over an hour or so, whereupon the next set would be brought on. On one festive occasion in another Brahe household, the first set was soup, carp, pike with turnips, venison with currant sauce, chicken paté, goose liver with cucumbers, sugar cakes, and beef with horseradish; the second was crabs, roast lamb with beets, an almond sweet, and a tart.[53] The carver cut the meat, pages offered the platters, and guests ate with their hands or spoons, cutting with their knives, and wiping on the immense napkins. The cupbearer poured, and toasts were drunk from great tankards that passed around the table, each person drinking in turn and wishing his or her neighbor well, sometimes even singing a snatch of song as the vessel was passed on. Individual *römer* glasses and stemware of Venetian crystal were used on special occasions.

Danes ate immense quantities of food in this era. An Austrian nobleman who visited Denmark in the 1560s said that he had never seen so much food as he saw at a Danish peasant wedding.[54] Years later, he could recall every single item on the great "salt-meat platter," a common Danish course featuring a variety of cold meats, sausages, tongue, chicken, eggs, eel, and much more, from which the diner could choose. All this food was washed down with huge drafts of beer and wine. Beer was the common drink, and normal consumption was four to eight quarts a day. In aristocratic circles, a lot of wine was consumed. Most of it entered the country in Dutch ships through the Tollhouse in Elsinore, where it was preempted in lieu of dues. Distilled aquavit was produced in chemical laboratories and flavored in innumerable ways with berries and herbs, but it was considered a medicament and was not normally served as a beverage.

These immense, leisurely, aristocratic suppers occupied several hours at the end of the working day. Students, guests, learned *famuli,* and musicians were expected to entertain the company with poetry, song, instrumental music, and dramatic presentations. In later years, Tycho's dwarf jester helped to keep the company merry until it was time for bed, or for the observatory.[55] There were numerous occasions to entertain visitors or celebrate events in the life of the island, and all the traditional Lutheran holidays of the late sixteenth century were observed at Uraniborg: the fourteen days of Christmas, Candlemas, the Annunciation of Mary, Easter, Ascension, Pentecost, St. John the Baptist at Midsummer, and the harvest festivals of Michaelmas and Martinmas, as well as the Advent and Lenten fasts and Great Prayer Days.[56]

For Tycho Brahe, these customs provided a social structure that he could adapt to the service of large-scale science, turning the leisurely Danish supper into a daily staff meeting for the scholars, scientists, and leading technicians of Uraniborg. Year in and year out, the whole learned staff gathered once a day and spent several hours together, dining and discussing their common affairs under Tycho's leadership.[57]

The total staff required for such a household is a bit hard to determine. In later years, the staff of Uraniborg included a maid named Live Larsdatter, a chaplain named Jens Wensøsil, a cook named Jonas, a tailor named Nicolaus, a bailiff named Christiern, a scribe named Johan, jesters named Jeppe and Per Gek, and a lutenist named Jacob.[58] These, however, only hint at the full extent of the staff: A generation later, in a more Spartan wartime era, Tycho's niece, Sophie Axelsdatter Brahe, kept careful accounts at Rosenholm Castle. Her husband, Holger Rosenkrantz, had lived for a time with Tycho Brahe, and they too maintained a learned household, so Rosenholm was comparable to Uraniborg.[59] Their *familia* included a chaplain, tutors, clerks, wetnurses, maids, footmen, a cook, cooks' helpers, bakers, brewers, gardeners, a tailor, a buttonmaker, watchmen, coachmen, bailiff, overseers, milkmaids, fishermen, a hops man, smith, and other peasants and workers on the estates.[60] In addition, Sophie Axelsdatter Brahe and her husband often hired the services of goldsmiths, clockmakers, engravers, bookbinders, painters, cabinetmakers, gunsmiths, apothecaries, weavers, cobblers, saddlers, netbinders, ropemakers, wheelwrights, potters, brickmakers, roofers, sawyers, lime burners, charcoal burners, and many others. When the services of some of these people were needed on a regular basis, they were added to the staff. Tycho Brahe's situation was similar, and he must have had the same range of servants, besides the trusted bailiffs and their staffs who administered his lands and fiefs in Skåne, Sjælland, and Norway.

FAMILIA STRUCTURES

Tycho used all the social patterns and networks at his disposal to build a structure that would support his extensive research projects. His position as vassal of the crown allowed him to enlist Steenwinckel and other artists and artisans from Kronborg, negotiating individual contracts for the work they did on Hven. Seigneurial lordship allowed him to utilize the labor of Hven peasants without pay, and in addition, the village of Tuna formed a pool for recruiting household servants of various kinds. Friendship with scholars in Copenhagen and elsewhere, as we have seen, helped Tycho to recruit students whom they recommended to his service.

The idea that people should be paid to live in a household was not popular among the patriarchs of the late sixteenth century. A husband, wife, and children did not receive wages; other members of the household should not expect much either. The normal pattern was room and board, new shoes and clothing at Christmastime, and whatever occasional gratuities the patriarch felt inclined to bestow. The concept of individual privacy was still largely unknown, and nobody got a private room. People ate, slept, and worked together. The students were housed in attic garrets, cooks slept in the kitchen, and servants in the north lodge, while Tycho and Kirsten slept in the great bed in the Winter Room.

Nevertheless, young students considered the rewards of living in Tycho's household to be better than any scholarship at the University of Copenhagen. New arrivals received tours and instruction from Tycho himself, from Peter Jacobsen Flemløse, Gellius Sascerides, and others. They were assigned to research teams working in the observatories, museum, or laboratory. Some of the scientific practices of Uraniborg were transmitted by formal instruction and written manuals, and others by participation in the work. Using Tycho's large instruments, for example, required a seasoned observer, a team leader to read off the positions of sightings, a secretary to record observations, and sometimes a clockwatcher.[61] Room, board, enlightened scientific and philosophical conversation, and the experience of working on the cutting edge of sixteenth-century science: All of these were available at no charge to students who were admitted to Uraniborg by Tycho Brahe. In return, they were expected to do their share of the scientific and secretarial work, and sometimes compose a poem, play a lute, or sing a song at supper time. Their relationship to Tycho Brahe was partly that of *famulus* to patriarch, and partly that of student to teacher.

Tycho Brahe used his access to King Frederick II to act as broker of royal patronage for some of his more important *famuli* and clients. Gemperle and Schardt received royal commissions and the tax-free status of court artists. Gellius received the travel stipend that, at Tycho's request, had been held in abeyance by royal intervention. Labenwolf, Gemperle, and Schardt were commended to the king by Tycho Brahe. On 13 June 1579, on Tycho's recommendation, Flemløse received one of the ripest patronage plums available to a scholar: the promise of a canonry in Roskilde Cathedral – a lifetime sinecure with a residence and substantial income – in return for serving Tycho Brahe as long as he was needed. Royal patronage provided valuable rewards to members of the Uraniborg *familia* without any expense to Tycho; it helped him to recruit and retain some of his key staff members.

Tycho Brahe also made use of piecework contracts and day wages. Master craftsmen like Steffen Brenner, Hans Knieper, Christopher Schissler, Georg Labenwolf, and many others worked on contract in their own shops in Copenhagen, Elsinore, Nuremberg, and Augsburg. Masons, glaziers, cabinetmakers, painters, gilders, and other master craftsmen came with their journeymen and apprentices to work on Tycho's island. They received a day wage or piecework payment, and possibly also subsistence. Others like Steenwinckel, Gemperle, and Tycho's instrument makers joined the *familia* and received their livings from the master.

Administrators of Tycho's extensive lands and fiefs were paid in a combination of benefits, fees, and salary. A bailiff like Christopher Pepler in Tycho's Norwegian fief of Nordfjord, or Tycho's local bailiff on Hven, received free housing, a percentage of goods and payments receipted, and other forms of subsistence. These agreements were negotiable within the range of the normal practices of estate administration.[62] The bailiff in turn hired and paid his own staff.

Daily life at Uraniborg transpired in the context of an aristocratic household taking on additional functions related to Tycho's scientific and technological pursuits. Among the learned activities were instruction, research, construction of apparatus, and the establishment of a large research library. Instruction covered the subjects of geometry, architectural theory, trigonometry, *prosthaphæresis,* observational theory and practice, theoretical astronomy, cosmology, Paracelsian chemistry, and related fields. Research was carried out on the distance to Mars, the comets of 1577, 1580, and 1582, solar and lunar eclipses, the fixed stars, meteorology, and iatro-chemistry. Innovative instruments were constructed on Hven and orient-ed, calibrated, and sighted in new ways for remarkable improvements in accuracy.

Large-scale scientific activity of a new kind was taking shape in the household of Uraniborg on Tycho's island, amid all the affairs of daily life, long and leisurely meals each evening, and a steady stream of visitors. There was no other place quite like it anywhere in the world, and impor-tant things were happening. The time of breakthrough was at hand.

5

BREAKTHROUGH

1584–1587

Uraniborg became a European phenomenon in the middle years of the 1580s, and Tycho Brahe achieved the international renown his name has held to this day: That was the breakthrough. Through correspondence, publications, and travel abroad, Tycho and his associates sent out an attractive force that drew students, scientists, artisans, noblemen, even ruling princes to Tycho's island. Within less than a decade, Tycho Brahe achieved everything King Frederick had hoped for as his patron.

A revolution in scientific organization was taking shape on Tycho's island around 1584. Construction of the underground subsidiary observatory, Stjerneborg, was begun in that year. Tremendous advances were made in the design of precision apparatus as Tycho Brahe and his craftsmen turned out one immense instrument after another [Fig. 16, and see Fig. 25].[1] Early in 1584, Tycho began to establish a scientific press on Hven and sent a member of his *familia* on one of the first scientific expeditions in history. Toward the end of the year, Tycho marked his new position by publishing a series of Latin poems dedicated to friends and reflecting the intellectual program of Uraniborg. In 1586, he established strong ties with astronomers at the court of Landgrave William in Kassel. Finally, he commissioned portraits of himself to use as gifts, publicizing his achievements as he cemented bonds of *amicitia* and patronage. By means of these expanding communication strategies, Tycho linked his magical island into wide European networks of patronage, research, and learning.

Eight years had passed since Tycho and his friends laid the cornerstone of Uraniborg, and four years since the celestial manor had been occupied. Results might have seemed slow in coming, but large scientific enterprises always take time to begin. Wolfgang Panofsky described a double startup

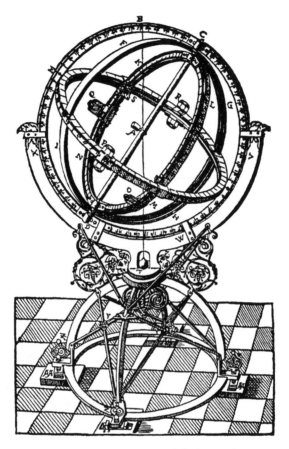

Fig. 16. Equatorial armillary of steel, built 1584. (From *As-tronomiæ instavratæ mechanica*)

process lasting up to nine years for twentieth-century big science: first, the "political process of securing authorization," and then "the physical process of building the machine."[2] For Tycho, the initial funding process took only six months (December 1575 to May 1576), and construction on Hven began in the spring of 1576. Early results during the construction phase were Tycho's astrological analysis of the 1577 comet and his nativities of 1577 and 1579, and they were sensational from the point of view of his royal patron.[3] Tycho frequently appeared at court during these years, and many other "gifts" to his royal patron have undoubtedly gone unrecorded: marvelous clockwork and automata; Paracelsian medicaments; technological and cultural advice on such matters as construction, iconography, maps, and sea charts; and brokering contacts with European scholars and artisans.[4] King Frederick reciprocated by substantially increasing the endowment of Uraniborg in the years 1577–9. Uraniborg was completed by 1580, and Tycho carried out a shakedown of his facilities, then began to reach out in new directions.

THE URANIBORG PRESS

As Tycho prepared detailed studies of the supernova of 1572 and the comet of 1577, he discovered that Copenhagen printers were not equipped to deal with his new kind of technical scientific publication, and he began to consider establishing a press of his own at Uraniborg.[5] Tycho worried about accuracy, the quality of technical illustrations, and plagiarism. The esthetics of power also played a role in his thinking because he believed that the appearance of a book conveyed a message about the author. Tycho intended to use his books as gifts to patrons and friends, and he wanted them to be sumptuous in order to convey the message that Tycho Brahe, lord of Uraniborg, combined great learning with high social standing.

Plagiarism was a new worry in his day. Earlier Renaissance natural philosophers had believed that the aim of philosophy was to recapture ancient wisdom. That meant renovation, not innovation, and they did not think that their work produced anything new. Tycho Brahe grew up in this tradition, but his generation became aware that modern knowledge often surpassed ancient wisdom. They pointed to the printing press, the new anatomy, Paracelsian medicine, the discoveries of unknown lands, and the revolutions in geography, botany, zoology, and ethnography that grew from those discoveries. Tycho himself knew that he could surpass the results of the ancients with his innovative methods of research and experimentation. Because he realized that his work actually created new knowledge, Tycho began to consider its results to be a form of property that belonged to him as the one who produced them.[6] He wanted to protect his intellectual property rights, and this was another strong argument for a printing office on Hven.

Underlying all the reasons was Tycho Brahe's ideal of uniting the fine arts, technology, and science in one single institution dedicated to Urania, Vesta, and the Muses. Tycho believed that the printer's art occupied a natural place in such an institution, and he set out to establish the world's first working scientific press.[7] By February of 1584, he had hired a printer named Joachim and had begun to search far and wide for printing equipment.[8] Joachim was dispatched to Wittenberg for presses and other materials. Tycho wrote to Professor Heinrich Bruɔæus in Rostock, who ob-

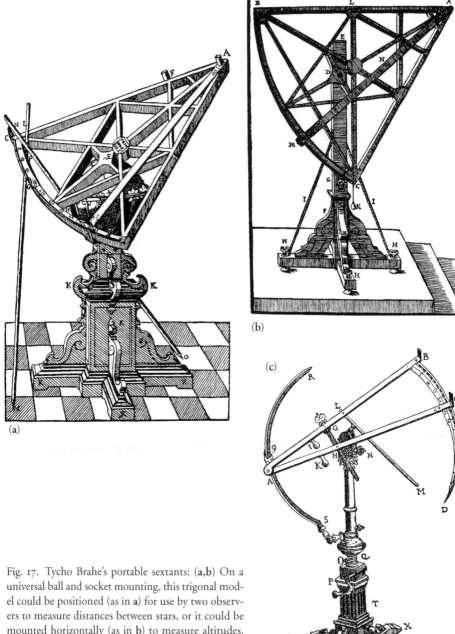

Fig. 17. Tycho Brahe's portable sextants: (a,b) On a universal ball and socket mounting, this trigonal model could be positioned (as in a) for use by two observers to measure distances between stars, or it could be mounted horizontally (as in b) to measure altitudes. Tycho Brahe made three of these sextants, which could be screwed apart and fit into a carrying case. (c) Another sextant, this one in steel. (All from *Astronomiæ instavratæ mechanica*)

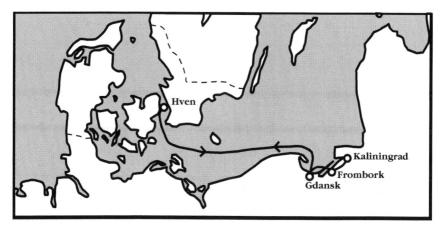

Fig. 18. Route of Morsing's expedition. (Map by the author and Greg Tiburzi)

tained type from a local typefounder and from the printer, Stephan Myli-
ander. Tycho may have contacted other friends and former *famuli* around
the continent. The details are unknown, but the work of preparing to
establish a printing office on Hven continued throughout 1584.

THE EXPEDITION TO FROMBORK

In the spring of 1584, an embassy to the Danish court from the administrator of Prussia, Margrave George Frederick of Brandenburg-Ansbach,
was preparing to return home, and when they sailed from Copenhagen
on 1 May 1584, Tycho Brahe had arranged for a member of his *familia*,
Elias Olsen Morsing, to travel with them as far as Gdańsk (Danzig).[9] In
Morsing's luggage was Tycho's portable sextant [Fig. 17] and meteorological diary, in which Morsing wrote a day-by-day account of his travels.[10]
   Morsing's journey was one of the first scientific expeditions of modern
times.[11] His task was to determine the exact latitude of Copernicus's observatory in Frombork (Frauenburg), on the Zalew Wiślany (Frisches Haff)
estuary, just east of Gdańsk [Fig. 18]. Tycho had found the obliquity of the
ecliptic to be at variance with the figure used by Copernicus, and he had
also noticed that the solar theory of Copernicus deviated considerably
from observed solar positions. Tycho thought these problems stemmed
from an erroneous figure for the latitude of Frombork. He was working
on his own solar theory and needed to get to the bottom of the problem,
so he took the opportunity to send Morsing to Frombork with the Prussian expedition.[12]

Nine days out of Copenhagen, the ships put in to Gdańsk. Morsing set off for Frombork, where he arrived by 13 May 1584. The evening was overcast and remained so until 17 May, when the skies cleared and Morsing set up his sextant near the very tower used by Copernicus. He stayed for three rainy weeks, observing when possible with Professor Matthias Menius of Kaliningrad (Königsberg) and several of the canons. The dean of the Catholic cathedral, Johannes Hanovius, presented an instrument used by Copernicus as a gift to Tycho Brahe. It was a triquetrum (parallactic instrument) of pine, eight feet long, graduated in ink with a quill pen. Morsing also received a portrait of Copernicus, probably a self-portrait.[13] On 6 June, Morsing departed for the seat of Prussian government in Kaliningrad, where the margrave had summoned him to determine the city's latitude. He arrived on 8 June, stayed three weeks, returned to Frombork by 28 June, and departed for home on 4 July. After more than two weeks at sea, Morsing arrived safely at Uraniborg on 23 July 1584. His observations allowed Tycho Brahe to determine the latitude of Copernicus's observatory to be 54° 22′ 15″ as opposed to the value of 54° 19′ 30″ used by Copernicus. Tycho was delighted with the whole expedition. He composed a Latin poem of celebration and gave the triquetrum a place of honor in Uraniborg, while the portrait of Copernicus was hung in the museum.[14] The first planned scientific expedition from Tycho's island had been a success.

REORGANIZING THE RESEARCH TEAMS

By 1584, Tycho Brahe had an excellent staff at Uraniborg. **Morsing** assisted him in the areas of astronomy and meteorology; Gellius and **Flemløse** in iatrochemistry and astronomy; Stephanius, Anders Viborg, and the gifted Hans Buck in a variety of ways; **Gemperle** and **Steenwinckel** in the fine arts (**Schardt** had left around the time Uraniborg was completed); Joachim helped to organize the printing office; and Johannes Bernssøn served as Tycho's courier. There were other *famuli* whose names are not known, such as "N.," mentioned frequently in the margins of Morsing's journal, and the artist with a monogram of the letters J, A, and M, who worked on woodcuts of Uraniborg and its facilities.[15]

Various projects went on more or less simultaneously. Flemløse and his crews had carried out a program of observations of the fixed stars in 1581, using radius, sextant, and quadrant, and Flemløse continued to record many of the astronomical observations. Morsing kept the daily meteorological journal, and Tycho now assigned him to compile an astrological and meteorological calendar for the coming year of 1585.[16] King Frederick, an avid hunter and outdoorsman, had a strong interest in weather signs and their links to astrology: Tycho hoped to publish the calendar as another fine gift for his patron.

Tycho Brahe directed the activities of his learned assistants and attended to numerous other matters. A Norwegian friend from Tycho's student days, Jens Nilssøn, now bishop of Oslo, visited Uraniborg in 1584 and was amazed by the changes since an earlier visit in 1577. Foreign visitors in these years included Professor Johann Seccerwitz from Greifswald; the French diplomat and historian Jacques Bongars; the Scots diplomat Peter Young; the English diplomats Lord Willoughby d'Eresby and Daniel Rogers; and the English Paracelsian Thomas Mouffet.

Erik Lange, now governor of Bygholm Castle in Jutland, visited Tycho Brahe in September 1584 and stayed for two weeks. In his retinue was his secretary, Michael Walter; an Austrian alchemist, Baron Siegfried von Rindscheid; and an ambitious young surveyor named Nicolaus Reymers Bär, whose odd behavior soon attracted the attention of Tycho's *famuli*. Bär (who would later be known as Ursus) was a self-educated peasant's son from the Dithmarsch in Holstein, an able mathematician, former employee of Viceroy Heinrich Rantzau, and a potential candidate to join the *familia* of Uraniborg. He wrote a poem in Tycho's honor to demonstrate his qualifications. The problem was that Bär kept prying about the library and observatories, looking into Tycho's books and manuscripts, fiddling with the instruments, taking notes and making drawings, while Tycho was "having a jolly good time with my close friend Erik and certain other nobles who were here with him."[17] Anders Viborg picked up on Bär's odd behavior, mischievously led him on, and provoked him into preposterous arguments.[18] When crossed, Bär flew into a rage. One day at the dinner table, Tycho cut him off with the remark, "Those German fellows are all half-cracked."[19] Tycho and his phlegmatic fellow Scandinavians found Bär's emotional conduct ridiculous.

Tycho's irritated amusement was coupled with suspicion. One evening at dinner, he later recalled,

My friend Erik wanted me to account for the planets' varying appearances and retrograde motions. I gave him the explanation according to the ancient Ptolemaic as well as the Copernican thinking by drawing it [on the tablecloth]. I also added that the same results could be achieved by a certain other assumption, and I exposed certain absurdities in those earlier views. Erik, with his thoroughly open mind and thirst for knowledge, asked me to show him this arrangement, too. He noticed that I was reluctant to do so while that Dithmarschian servant of his was standing around and watching (for I had previously indicated to him that from the fellow's habits and appearance I had surmised what type he was). Sending him outside to do something else, Erik obtained from me what he wanted. I sketched for him my hypotheses, too, drawn in a general and brief manner, but I erased them right away.[20]

Bär slept in the garret with the *famuli*. Tycho thought he might be pilfering documents and spoke with Anders Viborg. According to Michael Walter,

[Anders] was confident he would find out what was going on, and therefore spent the night in [Bär's] room. Quietly testing whether [Bär] had any such stuff on him, in one side of his pants Andreas found four whole handfuls of tracings and writings. But when he wanted to look on the other side, he was afraid that because [Bär] was stirring, he would become aware, and therefore Andreas could search no further. When [Bär] woke up in the morning, he discovered that not all the surreptitiously written documents were there. Like a raving maniac, he ran around shrieking, weeping, and screaming so that he could hardly be calmed down. It pained him that he had been found out and his secret tricks uncovered, so that the lords and all of us saw his extreme madness, and had a good time over it. Both lords assured him that anything among the papers that belonged to him and did not concern Tycho Brahe would certainly be recovered by him. This also happened, as we all knew.[21]

All of this was both amusing and disgusting, but it also made one wonder whether Tycho's island were really safe from thiefs and plagiarists. Erik Lange dismissed Nicolaus Reymers Bär soon after, and Bär made his way to Germany, where he found a job as a tutor.

Tycho's oldest daughter, Magdalene, was ten years old in 1584, and Sophie was six. Tycho added a chaplain, Jacob Mikkelsen Lemvig, to the *familia* around October of 1584, and one of his tasks may have been to tutor the girls. Lemvig also studied under Tycho and the senior assistants. By this time, Tycho Brahe's learned staff had grown to the point where a more formal structure was needed. Tycho's hospitality was legendary, and he gave numerous visitors the freedom of the island, but it got hard to know who was visitor and who was *famulus*. Moreover, some *famuli* grew restive with no more reward than room and board, new clothes at Christmastime, and occasional gratuities from the patriarch. These standard conditions of service for servants and apprentices were inadequate to master artisans and university graduates, and the normal term of household service (six months or one year) was too short for large research projects.[22] Uraniborg needed to move beyond the pattern of an aristocratic household in order to maintain a stable professional staff. Tycho Brahe hit upon the idea of using written contracts of indenture to specify the term and conditions of service when a new member entered the *familia*. He proposed this to Jacob Mikkelsen Lemvig, who agreed to become the first scholar to enter the household of Uraniborg under a formal written contract, promising to serve as Tycho Brahe's assistant for three years. Written contracts were common, but Tycho's use of them to employ a learned assistant was something new. His staff was evolving toward a more regular organization and greater stability.

URANIBORG METHODS IN KASSEL

Four years had passed since **Paul Wittich** departed from the island of Hven in 1580. Wittich had been teaching in his native city of Wrocław (Breslau), but he visited the court of Hesse-Kassel in 1584, taught the landgrave's clockmaker, Joost Bürgi, how to use the method of *prosthaphæresis,* and showed him some refinements of instrumental design invented by Tycho Brahe, which he passed off as his own.[23] Wittich taught Bürgi and the landgrave's astronomer, **Christopher Rothmann**, everything he had learned about the theory and practice of observation on Tycho's island in 1580. Together, they observed a lunar eclipse in November of 1584, and Bürgi began to construct new instruments incorporating the innovations he learned from Wittich.[24] Unbeknownst to Tycho Brahe, the influence of Uraniborg was being felt far from Denmark. It was an influence, however, which was sadly out of date, for Wittich had departed from Tycho's island too early to know about the mural quadrant, the trigonal sextant, the tripartite arc, and the marvelous new instruments being planned for Stjerneborg.

FIRST URANIBORG PUBLICATIONS

In the fall of 1584, Tycho Brahe's printing office took shape, and he composed a series of Latin friendship poems to be his inaugural publications. The first was dated 23 November 1584 and was addressed to his friend Erik Lange, inspired by their conversations and laboratory visits in September. Lange was a devotee of chemical transmutation, which Tycho called *alchymia* and refused to practice. Tycho's chemistry was what he called *ars pyronomica* or *ars spagyrica,* the pyronomic or spagyric art, which had the aim of healing, both spiritually and by iatrochemical medicines called *medicamenta Paracelsica*.[25] Tycho's poem described Lange's "carnal" obsession

with transmuting base metals into gold and offered love's remedy of spagyric chemistry, which sought not riches but harmony in the human microcosm.[26] In this poem, Tycho Brahe defined the three spheres of activity on Hven as astronomy, chemistry, and the fine arts, and he asserted
that the pursuit of these studies was rooted in the cult of *amicitia*.[27] A
poem dedicated to Tycho's friend and kinsman, the alchemist Falk Gøye,
came next, embellished on the reverse by woodcut emblems of Astronomy
and Chemistry [Fig. 19], based on the Uraniborg statues by Schardt, with
legends from the Emerald Table of Hermes Trismegistus: "By looking up
I see down," and "By looking down I see up."[28] Tycho's third poem was
addressed to the Nestor of his circle of friends, Charles de Dançay, followed by poetic epitaphs to deceased friends at the University of Copenhagen: Johannes Pratensis, Johannes Franciscus Ripensis, who had died
during the summer of 1584, and apparently also Nicolaus Laurentii, who
had died in 1579.[29] At the beginning of 1585, Tycho Brahe addressed a long
New Year's poem to his friend and patron Lord Chancellor Niels Kaas.[30]
These printed broadsides conveyed the intellectual program of Uraniborg
in rich poetic and emblematic allegories. As Peter Zeeberg put it, they reflected "a sophisticated milieu, where the ideal of the literate nobleman has
come to be taken very seriously."[31] The aristocratic splendor of Uraniborg
now equaled the past glories of Castiglione's Urbino and Ficino's Platonic
Academy in Florence, or the Veneto of Palladio and his patrons in Tycho's
own day.

Meanwhile, additional woodcut illustrations were made of two of Tycho's instruments, and possibly also a map of the island.[32] A woodcut of
a brass quadrant was signed with the monogram of "J. A. M." [Fig. 20].[33]
Other new woodcuts of the east elevation and plan of Uraniborg [see Figs.
14, 23a], and a bird's-eye view of the gardens, grounds, and ramparts [see
Fig. 26], may have been drawn by Steenwinckel. Bird's-eye perspective was
a favorite mannerist device: It allowed the viewer to soar magically above
the earth, revealing in a flash what the earthbound eye could never see.

Practical problems still frustrated the press. On 8 January 1585, Tycho
Brahe wrote to Viceroy Heinrich Rantzau of Schleswig-Holstein that he
was running short of paper. Elias Olsen Morsing had prepared the astrological calendar for 1585, but there was not enough paper to print it. The
viceroy's son, Gert Rantzau, had visited Hven and told Tycho that his father's papermill could provide the paper. Tycho's letter said that he intended to print the calendar in two thousand copies.[34] He apparently got the
paper, and in March of 1585, Tycho addressed a poem of friendship to
Heinrich Rantzau, describing Uraniborg and inviting Rantzau to visit.[35]

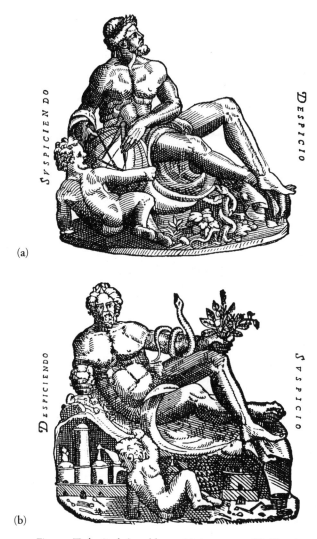

(a)

(b)

Fig. 19. Tycho Brahe's emblems: (a) Astronomy, (b) Chemistry. Their design is ascribed to Johan Gregor van der Schardt. (From *Astronomiæ instavratæ mechanica*)

Fig. 20. Detail of the monogram "J. A. M." in the lower right corner of a woodcut of the brass azimuth quadrant [see Fig. 11b]. (From *De mvndi ætherei recentioribvs phænomenis*)

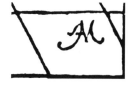

Fig. 21. Map of Hven. Engraving from Braunius & Hogenberg 1572–1617 (1588). A, St. Ibb's Church; B, Uraniborg; C, instrument makers' workshop; D, Stjerneborg in its original shape; E, Tycho's grange; F, Tuna; G, Hundred Thing. This map, based on the sketch in Fig. 6, distorts the shape of the triangular grange. (Courtesy Det Kongelige Bibliotek, Copenhagen)

Around that same time, Tycho Brahe printed the woodcut illustrations of Uraniborg and its facilities.

This completed the initial series of publications from the Uraniborg press. Peter Zeeberg noted that two of the poems honored men who, like Tycho Brahe, had been professional scholars and poets; six were addressed to men who, like Tycho Brahe, were aristocrats in learning and lineage;

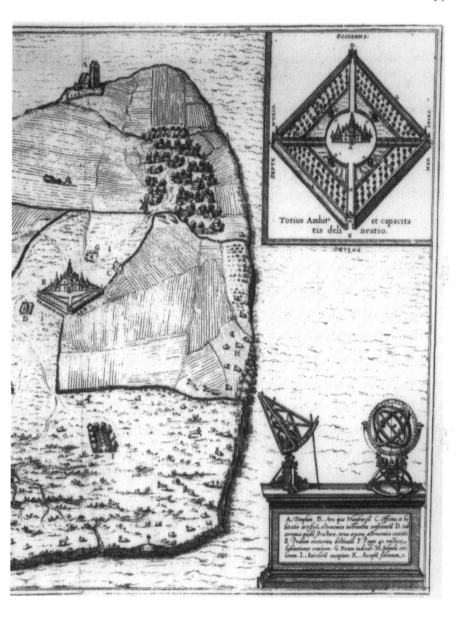

and two were addressed to men who, like Tycho Brahe, held high offices as servants of the crown. The friendship poems thus marked Tycho's lofty view of his own position among Danish men of learning, whom he addressed as friends.[36] At the same time, the published poems, emblems, and icons offered a spellbinding glimpse of Tycho's island to a broad European public. Tycho sent copies to friends and correspondents throughout the

Fig. 22. Apotheosis of the Sound (Øresund).
Engraving from Braunius & Hogenberg 1572–
1617 (1588).

Continent, and Hans Aalborg, now the official bookseller of the University of Copenhagen, distributed them at the Frankfort book fair of 1585,
where he also managed to locate a supply of paper for the Uraniborg
press.[37] In 1588, Tycho Brahe was pleasantly surprised to discover a copperplate engraving showing Hven, Uraniborg, and his two instruments, obviously copied from his works, in the world atlas published by Braunius and
Hogenberg in Cologne [Fig. 21].[38] The printing press had joined traveling
disciples in spreading the fame of Uraniborg.

     Another engraving in the atlas was an apotheosis of the Sound [Fig. 22],
the heart of King Frederick II's far-flung realm of Denmark, Norway, Ice-

land, the Faeroes, Gotland, Ösel, and Schleswig-Holstein. Soaring high over land and water in bird's-eye perspective, the vibrantly colored engraving showed the great castle of Kronborg, its hidden coffers brimming with silver and gold, and Labenwolf's fountain with dancing jets in the courtyard. Tycho Brahe's lighthouse tower on the tip of Kullen, part of the world's first network of lighthouses on seaways (rather than harbor lights), stood to the north. All the cities of the Sound – Elsinore, Helsingborg, Landskrona, Malmö, and Copenhagen – appeared in view, with royal castles and lodges, including Ibstrup in the forests north of Copenhagen. More than fifty ships were shown sailing the Sound: Dano-Norwegian

men-of-war firing salutes in reply to Kronborg's cannons, swarms of mer-
chantmen lying off the Tollhouse of Elsinore, and skiffs bearing captains
and supercargoes ashore with the Sound Tolls. In the midst of all this lay
Tycho's island, with St. Ibb's Church, the village of Tuna, and the celestial manor of Uraniborg in bird's-eye view. This mannerist icon showing
the heart of a great Sea King's realm was probably designed on Tycho's island, possibly by Steenwinckel. It made abundantly clear that Hven, with
all its splendors, was to be reckoned among the glories of a mighty realm.

THE COSMOPOLITAN FAMILIA

In January of 1585, the assistant pastor of St. Petri Church in Malmö died,
and Jacob Mikkelsen Lemvig, who had signed a three-year service contract less than three months before, requested Tycho Brahe's permission to
apply for the call. Tycho allowed it, and Lemvig got the call but must have
worked out a special arrangement to remain on Hven, commuting irregularly to Malmö for the rest of the year. Around March of 1585, a group
of nine students arrived from the University of Copenhagen, including an
Icelander named **Oddur Einarson** and two from western Jutland, Anders
Jacobsen Lemvig and Johannes Varde. The hospitality of Uraniborg was
extended to them, and they stayed for a month, dining at Tycho's table,
participating in discussions, and assisting at times in the museum, laboratory, and observatories.

   Two other new members entered Tycho Brahe's household around 1585.
One was Rudolphus Groningensis, a native of Groningen in the Netherlands, who arrived in June, when Hans Buck departed. Little is known
about him except that he had not attended a university, served for three
years, and apparently signed a contract similar to Lemvig's. Rudolphus
was an artisan, possibly an instrument maker. The other new *famulus* was
Hans Crol, a German goldsmith and instrument maker who was first
mentioned on 9 June 1585 but may have arrived as early as 1584. He was
a native of Westphalia and brought a high level of skill to Tycho's island.
Crol's principal task, possibly in conjunction with Rudolphus, was to build
the large instruments Tycho designed for the Stjerneborg observatory and
to keep all the instruments in top working order. The greatest of these was
the immense equatorial armillary, over eleven feet wide and built of steel
on a stepped stone foundation [see Fig. 25b].[39] This instrument allowed
direct observations of right ascension and declination. It was installed in
1585, aligned on 11 June 1585, and almost immediately taken into use.[40] Besides constructing such marvelous instruments, Crol and Rudolphus at-

tended instruction in various subjects taught at Uraniborg: From the beginning, technicians and scientists had come together in fruitful interaction on Tycho's island.[41] Crol could read and write Latin, which was unusual for an artisan. He had very sharp eyesight and became an excellent observer of the stars. In November of 1585, when Morsing sighted a comet, both Crol and Rudolphus assisted him in observing it.

While they were observing the faint comet, David Sascerides arrived on Hven. He was a brother of Gellius Sascerides but was as simple as Gellius was brilliant. Their sister, Agathe, had just married Jon Jacobsen Venusin, a Hermetic Magus born on Hven when his father was pastor of St. Ibb's. David Sascerides remained in Tycho's service for six months, and Venusinus entered the circle of Tycho Brahe's learned friends.

Tycho Brahe's household was quite cosmopolitan by 1585. Master craftsmen from Wittenberg, Augsburg, Westphalia, Flanders, and Holland were working with scholars from Denmark, Norway, and Iceland in one big *familia* under the patriarchal authority of a noble Dane. Former *famuli* like **Niels Arctander** were moving out of the household and into positions of authority: He was appointed royal chaplain in 1585, succeeding Anders Sørensen Vedel, who moved to Ribe and settled into a new position as Royal Historiographer of Denmark. Less than forty years of age, Tycho had built a remarkable research institute on the island of Hven: a place without parallel in the late sixteenth century, and one that was far different from its ancient model, the Museum of Alexandria.

MUSEUMS

Tycho Brahe's great admiration for Ovid and other Roman authors of the Augustan Age included writers on science and technology like Vitruvius and Pliny the Elder, and he was well aware that the heart of ancient science and mathematics had always remained among the Greeks. From the time of Alexander the Great until the end of the ancient world, their center was in Hellenistic Egypt, the home of Euclid, Apollonius of Perga, Claudius Ptolemy, and many others; where Archimedes came to study; and where a great research institution had existed for over seven centuries in the royal library and Museum of Alexandria.[42]

Alexander the Great had conquered Egypt in the year 332 B.C., and his successors of the Ptolemaic dynasty built the resplendent cosmopolis of Alexandria as their capital. The lighthouse of Pharos rose over the city and guided ships to its quays, as Tycho's lighthouse on Kullen led ships into the Sound. A sprawling palace complex contained all the central institutions of Hellenistic Egypt, including a temple called the Museum, dedicated to Apollo and the nine Muses, goddesses of history, lyric, epic, erotic poetry, hymns, dance and music, tragedy, comedy, and astronomy. The Museum was headed by a high priest appointed by the Ptolemaic kings, later by the emperors of Rome, to carry on activities deemed pleasing to the nine Muses, and this included research and writing. The library of the Museum contained over half a million papyrus scrolls and had a research staff of a hundred scholars, who held property in common and ate in a common dining room. Besides the library, there was an observatory, a zoo, a botanical garden, dissecting rooms, lecture halls, study areas, a covered exedra, and a *peripatos* or grove, where the residents walked, talked, and sang, "beneath the spreading planes and chestnuts, figs and palm trees. The place seemed fragrant with all the riches of Greek thought and song," wrote one classical admirer.[43]

Sixteenth-century scholars revered the memory of the Museum of Alexandria, and the word *museum* was revived by the humanist Paolo Giovio, who called the hall of his palace on Lake Como the "Museum Jovianum," dedicated to Apollo and the Muses and hung with portraits of poets, scholars, artists, and political figures.[44] Tycho Brahe may have visited the Museum Jovianum during his Italian journey of 1575, but his concept of a museum was more strictly classical, though at the same time, it was a late-Renaissance space, neither fully private nor public but always open to a flow of visitors, a space in which occult cultural knowledge and resources could be assembled and given sense and order.[45] In a general way, Tycho conceived of the whole complex on Hven as a museum dedicated to Urania, Apollo, and the Muses, as when he named his palace Urania's Castle and referred to Hven as the Kingdom of Urania.[46] In a more restricted sense, Tycho gave the name "museum" to the south rotunda of Uraniborg, the daily workplace of his scholarly assistants and location of his most treasured objects. This round room [Fig. 23a] was opulently decorated with inscriptions from Plato and other philosophers, as well as Tycho's own elegies to Ptolemy and Copernicus; and it contained portraits of Hipparchus,

Fig. 23 (*facing*). The Museum of Uraniborg and its celestial globe: (a) Plan of Uraniborg, showing the east entrance (A), west entrance (C), Winter Room (D), museum (T) with great celestial globe (W), corridors (θ) as originally built (the south corridor was later made into a chemical laboratory off the Winter Room), Labenwolf's fountain (B), room containing the mural quadrant (E), spare bedrooms (F, G), stairway up (L), stairway to the chemical laboratory and observatory (P), kitchen (H), well supplying running water to all rooms (K), tables (V), chimneys (4), and beds (Y). (b) Tycho Brahe's great celestial globe, constructed 1570–9 by Christopher Schissler in Augsburg and engraved at Uraniborg. (Both from *Astronomiæ instavratæ mechanica*)

OCCIDENS.

MERIDIES.

SEPTENTRIO.

(a)

ORIENS.

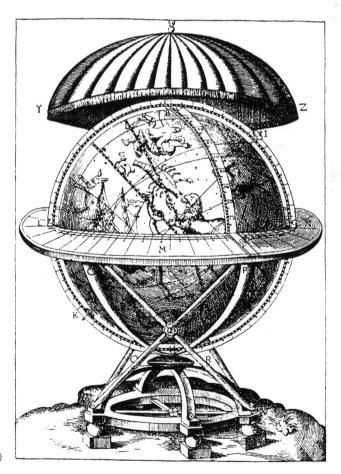

(b)

Ptolemy, al-Battani, Copernicus, William IV of Hesse-Kassel, as well as seven Italian philosophers of the fifteenth century.[47] The room housed Tycho Brahe's library of well over three thousand volumes, seven or eight celestial and terrestrial globes, a number of clockwork celestial models, and Copernicus's triquetrum. The most splendid object in the museum was Tycho's great celestial globe [Fig. 23b], covered in brass, perfectly spherical and stable in all seasons, six feet in diameter on a steel base five feet high, with a steel meridian divided to one minute of arc (sixty minutes in each degree and 21,600 in a full meridian circle). On this globe was inscribed each star of the heavens as its position was determined precisely. The globe stood as a compelling icon of the heavens and of Urania, but it was also used by Tycho and his assistants as an analog computer. On the desks of the museum lay quills, ink, manuscript handbooks of *prosthaphæresis,* and other working materials.

Above the Museum of Uraniborg was an astronomical observatory containing a huge azimuth quadrant (to be replaced with a semicircle in 1588), sextant, zodiacal and equatorial armillaries, brass clock, and other instruments.[48] In the well-lit crypt below the museum was Tycho's chemical laboratory with sixteen furnaces of different types, neatly arranged in niches around the walls.[49] Outside the east and west windows of the museum were aviaries housing a great variety of birds. Orchards and gardens were beyond, resplendent with grassy promenades, pavilions, and an abundance of fruit and nut trees [see Fig. 26].[50] An interlocking geometry of circles and squares, and a series of ratios based on the module of fifteen feet, unified the house, gardens, outbuildings, and ramparts both horizontally and vertically. Thus the museum was centrally located within a complex of working areas organized as a dynamic, spatial microcosm and functioning as the heart of Uraniborg as scientific research institute.

Marsilio Ficino, one of the seven Italian philosophers portrayed in Tycho's museum, had described various ways to construct images of the universe that would function as universal talismans and exert a magical influence over the immanent *spiritus mundi.*[51] Through the rational contemplation of such talismans, the pansophic philosopher could reconstruct the intelligible life-force of the macrocosm within his or her memory and then move through the spheres at will, achieving divine insight and power.[52] In Tycho's day, some philosophers built elaborate curiosity cabinets with many drawers and compartments, suitably decorated with emblematic designs, and within these cabinets, they arranged objects – mineral, animal, vegetable, and artificial – to create magical microcosms with the power of talismans.[53] On Hven, this manipulative scheme had been enlarged into the whole complex centered on the museum. Tycho Brahe believed that the intelligible force (*mens*) that gave structure and life to the universe was both rational and divine, and that the aim of science was to comprehend that force. In the Museum of Uraniborg and its surroundings, he created a magical environment for the study, understanding, and control of the forces of nature, the macrocosm.

THE POWER OF PRINT

Part of this talismanic environment was the printing press: It allowed precise, unchanging data, words, and images to be transmitted in all directions, across any distance. On a more mundane level, however, the Uraniborg press had come to a halt around March of 1585, after the printing of Tycho Brahe's small trial publications. Lack of paper caused the press to stand idle during the remainder of 1585. Joachim may have left around this time, but in June, Morsing wrote two brief references to "the printer" in the margin of the meteorological journal.

Hans Aalborg, as we have seen, attended the Frankfort book fair in 1585, distributed the initial Uraniborg imprints, and located a source of paper. He may also have hired a new printer. The paper arrived around the end of 1585, and the Uraniborg press began to function again. Morsing had recalculated his astrological calendar for the year 1586, and it became the first book published on Hven. A section on the faint comet of 1585 was added to the manuscript. The printed version was dated 1 January 1586 and dedicated to Prince Christian, the nine-year-old heir to the throne.[54] It was a princely gift. The printer is unknown but was probably Christopher Weida.

By this time, Tycho Brahe had formed an elaborate plan for publishing his life's great masterwork: It would be a truly royal favor for his patron, King Frederick, issued from his own Uraniborg press. The work would begin with three volumes under the general title *On the Most Recent Phenomena of the Ætherial World* (*De mundi ætherei recentioribus phænomenis*). The first volume would treat the new star of 1572, the second the comet of 1577, and the third the later comets. The very title asserted that these phenomena occurred in the celestial regions and not in the regions below the moon: These three volumes were to be an overwhelming attack on Aristotelian cosmology, based on massive, irrefutable analysis of empirical evidence and intended to bury the old view once and for all. At the same time, these volumes would begin the herculean task of erecting a wholly new view of heavens and earth, and so Tycho later called them his *Intro-*

duction to the Instauration of Astronomy (Astronomiæ instavratæ progymnasmata). The press was ready to go. Tycho decided to begin with volume two on the comet of 1577, which had already brought forth a voluminous literature in many European languages.[55]

For more than a century, the printing press had been reshaping the literature of astronomy.[56] Various almanacs, commentaries, textbooks, tables, and two editions of Ptolemy's astrological *Tetrabiblios* were among the works that appeared in print before the year 1500. The printing of Ptolemy's *Almagest* in a good edition in 1538 and the appearance of Copernicus's *De revolutionibus* in 1543 raised the published literature of astronomy to a new level. When these two sophisticated, comprehensive works became widely available, comparison was inevitable, which in turn presented Tycho's generation with its great dilemma: Which system was correct?[57]

The sudden wealth of printed books brought new ways of thinking and working because printed books were not the same as manuscripts. Copy a manuscript and its accuracy was bound to decay through scribal error, but every copy of a printed book was exactly the same: This gave printed books a cumulative potential that manuscripts could not have. Moreover, printing quickly led to a quantitative explosion of books, and books became much more widely diffused than manuscripts ever had been. Printing brought standardization, lower cost, and miniaturization as typography put more information into less space.[58] Changes in format like pagination, indexes, punctuation, paragraph indentation, spaces between words, and elimination of scribal abbreviations made printed books easier to read and to cite exactly. Tabulated data, diagrams, maps, and illustrations could be reproduced precisely in print, which meant that they could be tied more intimately to the text.

By the early sixteenth century, scholars had begun to compile personal libraries instead of wandering from place to place in search of unique manuscripts. They began to compare several sources instead of simply writing a commentary on one. Scholars discovered that they could discuss the same printed text by correspondence over any distance. The standardized format of printed books now included a title page, footnotes, list of errata, and bibliography, which facilitated the new methods of scholarship.

Publication of ancient authors like Euclid, Archimedes, and Apollonius of Perga, as well as medieval Arabic sources, stimulated the revival of mathematics. Regiomontanus's "science of triangles" was transformed during the second half of the sixteenth century into plane and spherical trigonometry. In Tycho's day, calculators labored to prepare trigonometrical tables, astronomical tables, ephemerides, almanacs, and practica because the printing press rewarded computation by reproducing its results in clear, exact, durable, diffusable form.

By the 1570s, astronomy had found a wide new audience. Any layperson who could read an almanac could forecast the next new moon, eclipse, or planetary conjunction. The appearance of unusual celestial phenomena called forth an instantaneous literature, like the numerous pamphlets written on the new star of 1572 and the bright comet of 1577. Serious astronomers like Tycho Brahe were among the pamphleteers on both of these subjects; the problem was to make one's own book stand out among the piles and racks of pamphlets and tomes from throughout Europe at the annual book fair in Frankfort-on-the-Main.

In confronting this problem, Tycho Brahe produced a paradigm of the modern scientific monograph: a thorough, systematic treatment of a single subject, written for a scholarly audience. In an elegant format designed to appeal to humanists, Tycho's book on the comet of 1577 was all the more remarkable because it dealt in great detail with a transitory celestial phenomenon, rather than the timeless motions of the stars and planets, sun and moon. The book had a title page and colophon, ten chapters, and a table of contents (in the *procemium*).[59] It contained four technical illustrations, twenty-seven tables, and a grand total of eighty-seven diagrams, each individually cut by Tycho's artist-craftsmen. Key points of each diagram were lettered and explained in the text.

The manuscript for this book had been in the works since 1578, shortly after Tycho and Flemløse had observed the comet. Tycho had worked on it over the course of eight years, and it had grown to the eight chapters that were now turned over to the printers.[60] On a good day, some five hundred sheets could be run through the press, and it took around three days to print eight pages in quarto. The printing went on through 1586 and into 1587, but paper shortages continued to cause delays, and so did Tycho's occasional revision of the manuscript.

Tycho, meanwhile, still worried that Nicolaus Reymers Bär might have seen material describing the new system of the world on which he had been working since 1577, decided to publish the system in order to establish his claim to it. During 1587, he worked on a brief description of the system and commissioned one of his graphic artists, possibly "J. A. M.," to prepare diagrams of it. A new Chapter Eight would be inserted in the book, containing this description of the Tychonic system of the world.[61]

As Tycho and his scribes, engravers, and printers worked on this grand publishing project, the *familia* of Uraniborg continued to change and grow. Jacob Mikkelsen Lemvig finally departed Hven on 27 April 1586 to take up residence as pastor in Malmö. On the same day, his replacement arrived, a nineteen-year-old university student named Christian Hansen Riber, son of a prominent barber-surgeon in Ribe, who may have signed a four-year contract. Another native of Ribe, Jacob Jensen Hegelund, a younger brother of Tycho's old school friend Peter, entered Tycho Brahe's *familia* around the same time. He was thirty years old, not brilliant but lively, convivial, and a good musician. Peter Richter from Haderslev also joined Tycho's *familia* in 1586 and served for around six months. Meanwhile, in Copenhagen, former *famulus* Hans Buck was slain in a student brawl on 5 June 1586.

TIES WITH KASSEL

That same spring of 1586, Nicolaus Reymers Bär traveled to the court of Kassel. Wittich was no longer there, but Bär, or Ursus as he called himself in Latin, met Joost Bürgi, Christopher Rothmann, and the learned landgrave himself, William IV of Hesse-Kassel. On 1 May 1586, Ursus described to the landgrave a new planetary system of his own invention, which he had worked out during the past winter while tutoring in Pomerania. The landgrave was so taken with the Ursine system that he commissioned Bürgi to construct a mechanical model of it.

Wittich's visit to Kassel had whetted the landgrave's appetite to renew his acquaintance with Tycho Brahe, and he had written to Viceroy Heinrich Rantzau after observing a faint comet in November of 1585, requesting him to inquire about Tycho's observations of it. Tycho wrote a long, friendly reply and sent the observations the landgrave had requested. He entrusted the letter to his senior assistant, Peter Jacobsen Flemløse, who was to deliver it to Kassel, remain there long enough to determine the current state of activities there, and then proceed to the Frankfort book fair as Tycho's agent. This would be the second expedition from Uraniborg. Tycho was concerned about the transfer of technology and observational methods from Uraniborg to Kassel, and he did not want his innovations to become public before they were published.

Flemløse left Uraniborg around mid-May of 1586. He and Tycho were not aware that Ursus was in Kassel, but Ursus had left by the time Flemløse arrived. Landgrave William, Bürgi, and Rothmann received Flemløse with open arms, eager to establish cooperation between Kassel and Uraniborg. He found them to be upright individuals who operated in good faith, and he told Tycho Brahe as much when he finally returned to Hven in October of 1586, after an absence of six months. His positive report helped to inaugurate cordial ties between the two leading centers of observational astronomy in late sixteenth-century Europe. Tycho launched a vigorous scientific correspondence with Landgrave William and Christopher Rothmann. He knew about the transfer of technology from Hven to Kassel, and that the sextant, which he had invented, was now being used in Kassel, thanks to Wittich, as well as a version of his parallel slit sights and his transversal arc graduation.[62] What Tycho did not know was that more than technology had been transferred, and that Joost Bürgi was building a model of an early form of his own planetary system, which the Kassel court ascribed to Ursus and not to Tycho. Unbeknownst to him, Tycho's worst fears of the theft of intellectual property had been realized.

QUEEN SOPHIE AT URANIBORG

Tycho Brahe continued to have close relations with the Danish court. He participated in court festivities, prepared an annual astrological calendar for the crown, and drew up a horoscope for each of the sons born to King Frederick II and Queen Sophie. The little queen, with her round face, hazel eyes, and glowing complexion, had come to Denmark as a child bride, and Inger Oxe had been in charge of her court until 1584, when Beate Bille had taken over as Lady Stewardess.[63] By the summer of 1586, Queen Sophie was twenty-eight years of age, regal in bearing, the mother of princes, and fascinated with tales of the wonders of Hven. She decided to visit Tycho's island and see these wonders in person.

Queen Sophie arrived in clear weather on 27 June 1586 with all her courtiers, including Tycho's mother, and Tycho had prepared a splendid reception.[64] All the villagers of Tuna turned out, many undoubtedly clad in the red and gold colors of the Danish royal house. Fanfares and music sounded. A procession came to the portal of Uraniborg. Tycho Brahe was accompanied by his good friend and Royal Historiographer, Anders Sørensen Vedel. The company enjoyed views of the six Danish cities around the Sound, with ships and billowing sails in all directions, and then the queen entered the "emblematic world" of Tycho Brahe.[65]

They visited Stjerneborg [Fig. 24], noted the "living" statue of Hermes/Mercury rotating on the turf-covered roof that symbolized Parnassus (sacred to the Muses), and passed through the ionic portal with its three lions, past the arms of Brahe and Bille, magically going underground to see the skies. One of the lions held a hieroglyphic scepter, indicating that the great instauration of the world was commencing in this place. Tycho read and explained the Latin epigrams, including the inscription dedicating the

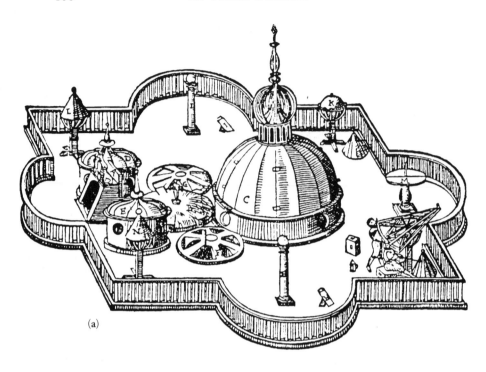

(a)

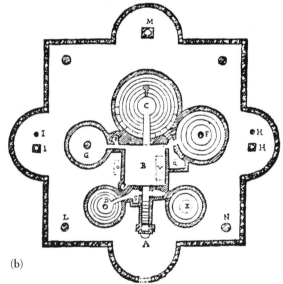

Fig. 24. Stjerneborg: (a) projection,
(b) plan. The Ionic portal (A) led
into the square warming room (B),
which gave admission to crypts containing the great equatorial armillary
(C), the large azimuth quadrant (D),
the zodiacal armillary (E), the largest
azimuth quadrant of steel (F), and
the trigonal sextant on a ball-and-
socket mounting. Alcoves contained
beds for Tycho (O) and the assistants
(Q) to rest during nights in the observatory. The warming room contained a stove (P) and table (V) in addition to paintings and inscriptions.
(From *Astronomiæ instavratæ mechanica*)

(b)

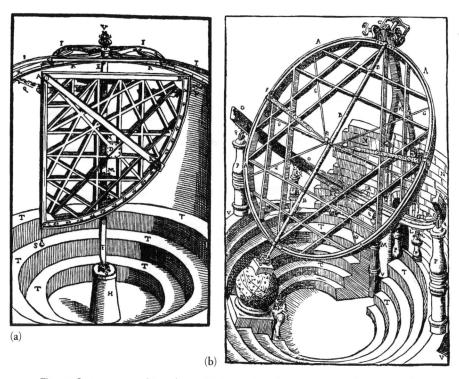

(a)

(b)

Fig. 25. Instruments at Stjerneborg: (a) Large azimuth quadrant, installed 1585–6. (b) Great equatorial armillary, with its Atlas-borne globe. (Both from *Astronomiæ instavratæ mechanica*)

building to Posterity. He noted the relationship of the inscriptions to portraits of famous astronomers marking the history of astronomy and to the portrait of himself, pointing to the Tychonic planetary system painted on the ceiling and asking his predecessors, *Quid si sic?* ("Is this it?"). As the queen looked on, Tycho and his assistants demonstrated the huge instruments, including an azimuth quadrant and the great equatorial armillary [Fig. 25], borne by a sculpted globe on the shoulders of Atlas, immense in scale but reduced to its functional essentials, like an icon of the framework of the universe, swinging on the same axis as the spheres of the heavens.[66]

Ascending from Stjerneborg, they passed the sluices and fishponds in geometrical shapes and went through the eastern gatehouse, the portal of Uraniborg [Fig. 26]. They saw the marvels of the grounds, full of pleasing sights and sounds and aromas, uniting all of nature in a resurrected Eden within the ramparts: the four garden pavilions; the aviary with its singing birds; the orchard promenades with game and pheasants grazing under tall pruned trees, and overhead, clusters of pippin and Rambour and white Gylling apples, white Doyenné and blood pears, yellow and blue and green Reine Claude plums, red-gold and Brussels brown and morello cherries, as well as apricots, figs, quince, and walnuts.[67] This was truly a microcosm, offering beauty, harmony, health, and delight to all the senses, for the abundance of the world was here: If ever it were possible to know the Creator through His Creation, this was the place.

Never before had a building and grounds in northern Europe been so thoroughly integrated and richly embellished. They saw the printing press, housed in one of two ancillary structures that reflected the architectural form of Uraniborg. Paths led through the gardens and into the "philosophical" house from the four corners of the world, north, south, east, and west. On the balconies above, the queen and her courtiers saw the great instruments of the Uraniborg observatories, and higher still, four statues at the cardinal points on the roof, representing the four winds, seasons, qualities (hot, cold, wet, dry), and ages of man [Fig. 14]. Far above these was Pegasus on the highest spire, moving in the breeze and springing skyward from Helicon, mount of the Muses. The strike of his hoof had brought forth Hippocrene, the source of wisdom, the spring in the midst of a garden, which they beheld inside the house as Labenwolf's dancing fountain in the central hall.

The queen and her courtiers descended into the skylit cellars of Uraniborg to see the bright flames of chemical furnaces. They progressed with their host to the museum in the same rotunda as the observatory and laboratory, the working place of humans between the heavens and the depths of earth. They beheld the automata and clocks; the portraits of astronomers, mathematicians, and philosophers; the books and rare manuscripts. They saw Tycho's treasures – the instrument of Copernicus, the great globe, the royal portraits, including one of Queen Sophie herself – and then went to view the mural quadrant, with its arc covering a whole wall. They heard the clocks with sonorous bells; hidden tubes magically brought music and voices from unseen places. They ascended to tour the galleries and observatories of Uraniborg above. Vertically and horizontally, everything was integrated into one microcosmic plan. Finally, they entered the Queen's own chamber, resplendently decorated, which had long awaited her arrival on the heights of Hven.

That evening, there must have been a great banquet in the Summer Room, with music and entertainment. It was the season of evening light in Scandinavia, when twilight lingered throughout the night and the stars were hardly visible in the bright sky.

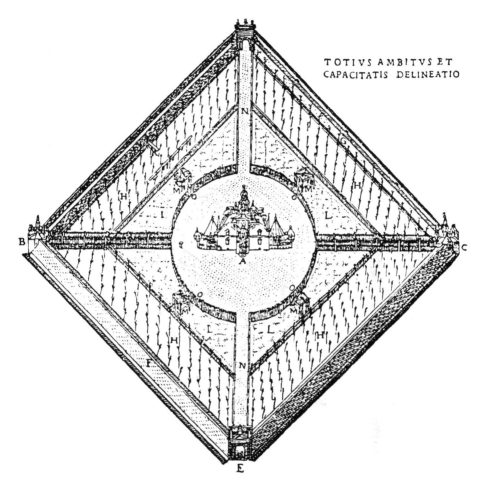

TOTIVS AMBITVS ET
CAPACITATIS DELINEATIO

Fig. 26. The Uraniborg complex: Bird's-eye perspective of the ramparts and gardens as they appeared ca. 1586. Uraniborg (A) was approached through the eastern gatehouse (E), which was rusticated in the Tuscan manner, with English mastiffs in kennels above to announce the approach of visitors. The printing office (B) and servants' quarters (C) were like miniatures of the main house. Paths (N) led from the four cardinal points to the house, and to the fountain of water within. The orchards (H) contained around three hundred trees, and the botanical gardens (L) included both herbs and ornamentals in great variety, as well as four pavilions (M). Wooden palings surrounded the gardens, articulating a square outward and a circle inward, surrounding the house. In the southwestern orchard were facilities for playing. (From Brahe 1913–29, 6: between 348–9)

Clouds piled up, and a storm blew in from the northwest. The next day came rain. The Sound was tempestuous. The queen and her court could not sail, so their stay on Tycho's island was prolonged. Over the banqueting table, as thunder and lightning raged in pounding rain, Tycho Brahe drew the queen's attention to Anders Sørensen Vedel's collections of Danish folk tales and ballads.[68] Queen Sophie was the daughter of a Danish princess with deep roots in the Danish kingdom, and these tales fascinated her. Many of the medieval ballads dealt with the deeds of her own royal ancestors, and there were also ballads of werewolves, demons, bloody deeds, and songs of love and enchantment. *Famuli* sang the old ballads to a lute, and Vedel whiled away the stormy hours by telling the ancient legends of Hven.

Once upon a time, Lady Grimmel had ruled the island from her four castles, now in ruins – Nordborg, Sönderborg, Karlshög, and Hammer.[69] Hven was named for her maiden-in-waiting, Maid Hvenild. Lady Grimmel held a great feast at Nordborg, where she treacherously murdered her own two brothers. Maid Hvenild soon bore a son, Ranke, whose father was one of the slain brothers. Years later, Ranke avenged his father's death by locking Lady Grimmel in her own dungeon, where she starved to death. Lady Grimmel's treasure was buried in the alder fen of Hven, guarded by a dragon, but two golden keys on the bottom of the sea could unlock the trove. Once upon a time, two boys from Tuna had spied the keys through the water. One boy kept the secret, the other revealed it, but when they returned, the keys had vanished.

Vedel told the whole tale that stormy evening at Uraniborg, as Tycho Brahe's musicians played and pages bore in sets of courses to the queen and her court. Queen Sophie was spellbound. Vedel said that he had many more ballads and verses in Ribe. The queen requested to see them, and Vedel eventually responded with a printed collection of one hundred ancient Danish ballads, dedicated to the queen and published in 1591. It was as if the very wind and waves had conspired with Tycho Brahe to broker Queen Sophie's personal patronage for a friend, Anders Vedel.

The next day was mild and clear. Queen Sophie and her courtiers took their gracious leave of Tycho Brahe's island and sailed away. That evening, the whole sky was illuminated with the northern lights, aurora borealis. In those brief days, Tycho had demonstrated to his patroness, and through her to King Frederick II, that the favor shown him had not been for nought. The queen's visit was itself a sign of grace that increased the prestige of Tycho Brahe, and he could rest assured that royal favor would not diminish in the future.

Before the year was out, Queen Sophie returned to Tycho Brahe's island. This time, in addition to her Lady Stewardess, Beate Bille, and her other courtiers, she brought her parents, Duke Ulrich of Mecklenburg and Duchess Elisabeth with their courtiers and the duke's nephew, Duke Sigismund of Mecklenburg-Ivenack. The royal party arrived at half past nine in the morning and was received by Tycho Brahe, possibly accompanied by his sister, **Sophie Brahe,** who was the same age as Queen Sophie. They stayed the whole day and did not depart until four in the afternoon, when they had seen all the marvels of the island.[70] The duke and duchess shared their daughter's interest in astrology, chemistry, allegory, and arcane apparatus. The weather was fair and clear, and this royal visit of 23 August 1586 was also a great success.

PORTRAITS

Toward the end of the year, Tycho Brahe celebrated his fortieth birthday on 14 December 1586 and decided it was time to commission a series of portraits as emblems of his achievements. There was no better gift for a patron, friend, client, or admirer than a portrait, and no better way to spread one's fame or keep one's memory alive.[71] Moreover, a woodcut or engraved portrait could be used as a frontispiece in published works. Tobias Gemperle painted a half-length oil portrait that showed Tycho Brahe with imperious gaze *en face,* blue eyes, dark blond hair and beard, wearing white ruff collar and cuffs, black cape and waistcoat, and a massive double golden chain bearing the royal Danish Order of the Elephant, which had been bestowed upon him by King Frederick. The arms of Brahe and the initials T. B. were in the upper left corner.[72] Tycho later sent this portrait to his kinsman and friend Holger Rosenkrantz. It was destroyed by fire in 1859 and is now known only through a nineteenth-century copy.

A woodcut portrait from 1586 surrounded Tycho with the iconography of a noble natural philosopher [Fig. 27]. The image went back to a woodcut from 1517 of King Christian II under a triumphal arch bearing the arms of his kingdoms and principalities. Other rulers, including Frederick the Wise of Saxony, had also been portrayed under a triumphal arch bearing territorial arms. Parts of this princely image had been transferred to portraits of Martin Luther within an arch with the arms of Saxony and his personal device, while learned physicians and chemists, Leonhard Thurneysser in Berlin (1572) and Georg Tanstetter in Vienna (1582), had been portrayed under arches laden with arcane emblems. Viceroy Heinrich Rantzau had been portrayed in 1574 in full armor between columns bearing the arms of his eight great-grandparents and his wife.[73] Tycho undoubtedly had several of these images in his collections. His own woodcut portrait shows him half-length *en face* in ruff collar, waistcoat, and a high-collared greatcoat with pleated shoulders, wearing the Order of the

Fig. 27. Tycho Brahe. Woodcut, 1586. (From L. Nielsen 1946, 41)

Elephant on a double golden chain. He stands in placid contemplation, eyes turned heavenward, within a simple but powerful triumphal arch bearing the escutcheons of his sixteen great-great-grandparents, all of them of great noble houses. Tycho's motto, *Non haberi sed esse* ("Not to seem but to be"), embellished the arch, and below the portrait was a Latin legend summarizing his accomplishments at the age of forty. A plumed beret was added on the table by his right hand. The portrait sketch was made by Gemperle, no doubt, and he or another skilled artist cut the woodblock. Steenwinckel may have contributed the overall composition. This woodcut transferred the emblems of a Renaissance prince to a learned nobleman and presented Tycho Brahe as majestic "celestial lord."[74]

Fig. 28. Tycho Brahe. Jacques de Gheyn engraving showing the misplaced Bielke ("Belker") arms (see text) on the right column. (Courtesy Det Kongelige Bibliotek, Copenhagen)

Tycho was pleased with the woodcut and wanted it done in copperplate engraving [Fig. 28]. He sent a copy to one of the foremost graphic artists of Europe, Hendrick Goltzius of Haarlem, with instructions to depict additional symbols of royal grace: four more golden chains, including two

Fig. 29. Tycho Brahe. Sketch attributed to Tobias Gemperle. (Courtesy Statens Museum for Kunst, Copenhagen)

bearing a medallion of King Frederick II. Goltzius maintained a large studio and gave Tycho's commission to his most talented assistant, Jacques de Gheyn, who worked with an engraver, Marcus Sadeler, to produce an image that was not in the usual active mannerist style of Goltzius and Gheyn but, rather, in a restrained late-Renaissance mode of profound dignity. The face was handsome, idealized, and youthful, gazing toward the viewer with an expression of hospitable benevolence.[75]

Closely related to these works is a portrait of Tycho done from life: a small pen and chalk sketch, now in Copenhagen, which was undoubtedly made by Gemperle [Fig. 29]. It shows Tycho *en face*, wearing a plumed

Fig. 30. Tycho Brahe. Second engraving by Jacques de Gheyn, corrected to show the royal Vasa ("Stormvase") arms. (From *Epistolarvm astronomicarvm*)

beret but otherwise dressed and posed as in the 1586 woodcut. During the years to come, this sketch would be sent to Haarlem as the basis for a second engraving by Jacques de Gheyn [Fig. 30], because it turned out that there was a mistake both in Gheyn's first engraving and in the original woodcut. A family historian, either Tycho's sister, **Sophie Brahe**, or Claus

Lyschander, the genealogist of the Bille family, probably discovered the error. It stemmed from a mistake in the heraldry of Tycho Brahe's ancestors on the transept arch of Kågeröd church, where the arms of Tycho's maternal grandfather's maternal grandmother, Birgitta Vasa, had inadvertently been transposed with those of an earlier ancestor of the house of Bielke.[76] When Tycho had sixteen of these arms copied for his portrait, he thus got the arms of Bielke ("Belker") where those of Vasa ("Stormvase") should have been. Members of the House of Vasa in Tycho's day were the reigning monarchs of Sweden and Poland, and the Vasa arms needed to find their proper place in the blazoning of Tycho Brahe's ancestry. Gemperle's sketch and the heraldic correction were sent off to Haarlem. The result was a new engraving, more accurate in portraiture and heraldry but lacking the idealized splendor of the earlier version.[77]

In 1587, Tycho Brahe commissioned what he considered to be the artistic masterpiece of Uraniborg: the mural quadrant. The instrument itself had been constructed in 1582 with a solid brass arc, six and a half feet (194 cm) in radius. Within the quarter-circle of this arc, **Steenwinckel** painted emblematic views through triumphal arches into all three regions of the Uraniborg cosmos: laboratory, museum, and observatory. Hans Knieper, the finest landscape artist in Denmark, came from Elsinore to paint a distant vista above Steenwinckel's Uraniborg. In the foreground, **Gemperle** painted a full-length portrait of Tycho Brahe [Fig. 31]. "The likeness could hardly be more striking," commented Tycho with tremendous satisfaction, "and the height and stature of the body is rendered very realistically."[78] The message of the painting was that Tycho Brahe had created a research institute to probe the secrets of heaven and earth, and that the institute itself comprised a microcosm in harmony with the cosmos.

THE BOOK ON THE COMET OF 1577

During 1587, there were no royal visitors to Hven but many others of distinction, including the Scots scholar Duncan Liddel, Baron Siegfrid von Rindscheid from Austria, and Danish noblemen including Gert Rantzau, Hans Spiegel, and Lord Marshal Peter Munk.[79] Falk Gøye visited in September and brought Christian Ebeltoft, whom he commended to Tycho's service. The *familia* suffered a great loss when Gemperle died of the plague in that year. Richter took his leave of Uraniborg, a student named Jens Jensen Wensøsil signed a three-year contract of service, and a lively Englishman, John Hamon of Kent, arrived in July to spend three months as a member of the Uraniborg household.

Fig. 31. Mural quadrant. Engraving of the 1582 instrument and 1587 mural painting by Hans Knieper, Hans van Steenwinckel, and Tobias Gemperle. Tycho Brahe's portrait points toward the front sight of the quadrant. (From *Astronomiæ instavratæ mechanica*)

Gingerich and Voelkel have offered a glimpse into the work going on in Stjerneborg during the early months of 1587. The precise location of Mars and the problem of its diurnal parallax continued to concern Tycho. Observations of the planet in opposition during January of 1585 had shown a much smaller retrogression than he had expected, and Tycho wondered whether refraction was to blame. Once the equatorial armillary and the two large azimuth quadrants were installed in Stjerneborg, he was ready to solve these problems when Mars came into opposition again during March of 1587. Using these instruments and the trigonal sextant, he and his staff set out to establish the effect of atmospheric refraction and separate the planet's retrograde motion from any effects due to parallax. Gingerich and Voelkel reconstructed one night's activity in striking detail:

The great majority of these observations are in Tycho's hand; there are simultaneous observations recorded with the sextant (for example, the distance between Mars and Arcturus), with the quadrant (the altitude and azimuth of Mars), and sometimes with the equatorial armillary (the declination of Mars), so clearly Tycho's staff was hard at work. The evening observations on 10 March 1587 began at 6:27 with a meridian altitude of Jupiter, made with the revolving quadrant, followed by a series of lunar positions made with the armillary, capped off by a measurement of Mars's declination with the same instrument. Just before 7 P.M. they measured distance between Mars and Regulus, then between Mars and Cauda Leonis (β-Leonis), and shortly after between Mars and Arcturus. For the better part of an hour they triangulated distances between these stars, before turning to some measurements of Saturn and Jupiter. At 8:30 everyone took a break until 9:45, when they made a few observations of Regulus and Spica "for learning the refraction of Mars." Shortly thereafter they took a few positions of the moon. For roughly twenty minutes before and following the meridian passage of Mars at 11:42,26, assistants recorded distances with the armillary and trigonal sextant. Then apparently everyone went to bed for several hours until a fresh set of Martian observations was made with the armillary and sextant between 4:14 and 5:16, once more obtaining distances to Arcturus and Cauda Leonis, which by now stood above Mars, as well as to Spica.[80]

After an assiduous program of observation, of which this evening's activity was a part, Tycho sat down to reduce his data. On the basis of excellent observational data, Tycho had already compiled a table of solar refraction, which he now used to correct the position of Mars. The result was a diurnal parallax for Mars that confirmed the theory that Mars came closer to the earth than did the sun. This was good news for Tycho because it supported his own emerging theory of planetary motion as well as that of Copernicus, but in fact, the parallax result was erroneous: Data reduction was flawed due to theoretical errors in his table of solar refraction.[81]

Tycho Brahe's great work on the comet of 1577 was in press, off and on, during much of the year of 1587. It was being printed in fifteen hundred copies, and Christopher Weida was the printer. The first seven chapters had been set in type from a manuscript begun shortly after the appearance of the comet in 1578. Chapter One was a concise summary of Tycho's observations, while Chapter Two established the locations of twelve reference stars, Chapter Three described the daily longitude and latitude of the comet, and Chapter Four calculated its right ascensions and declinations. Notes were added to update the manuscript in Chapters Three and Four. Chapter Five analyzed the path of the comet with respect to the ecliptic, which Tycho presented by seven different methods, appending an ephemeris of the comet. Chapter Six was an exhaustive study of the comet's parallax, verifying that its position was in the heavens far beyond the moon. Chapter Seven analyzed the position of the comet's tail. All in all, the book was profound testimony to the assiduous methods introduced to astronomy by Tycho Brahe, which Victor Thoren characterized as his "unprecedented concern for accuracy and . . . unique ability to detect and eliminate flaws in either observational procedure or mechanical design," to which Thoren added that Tycho "created almost single-handedly the empirical ethic on which modern astronomy was founded."[82]

Now Tycho needed to write a new chapter to describe the position of the comet among the planets, and this meant that he had to wrap up the intricately convoluted problem of describing the whole planetary system. This was where the Tychonic planetary system came in. His solution was not, as some have claimed, a "natural next step" that could have occurred to any astronomer in the generation after Copernicus; rather, as Gingerich and Westman have pointed out, it was a complex technical arrangement to solve interrelated problems growing out of his research, including the geometric possibilities of transferring a heliostatic to a geostatic–heliocentric system, the question of the physical substance of planetary orbs, the intersecting orbs of Mars and the sun, and the diurnal parallax of Mars.[83]

Tycho had wrestled with the general problem of the planetary system since his Copenhagen lectures of 1574, when he had worked out his first model, now lost, for transforming Copernican into geostatic mechanisms. Then the comet of 1577 appeared, and he had tried to fit it into a geostatic system where all the planets revolved around the sun. In 1580, Wittich had shown Tycho his own attempts to solve these problems and had raised possibilities for many new theoretical approaches. A few years later, the instruments of Stjerneborg were taken into use, and the observational activities of Tycho's island raised new theoretical questions about the structure of the universe. In response to all of these challenges, the details of a new planetary system had begun to take shape in Tycho's mind around 1583. During long evening discussions in the Winter Room, nights spent in the observatories, and days of calculation in the museum, this new model was tested and refined. It was still not completely satisfactory in 1584, when Tycho sketched it with chalk on a green tablecloth for Erik Lange and the Austrian baron, but the incomplete system, as we have seen, had been plagiarized by Nicholas Bär and made into a mechanical model by Joost Bürgi the following year.

Now Tycho sat down in 1587 to write a new Chapter Eight for his printers. He wanted his planetary system to solve all the problems raised over the years, and at long last, he thought that he finally had the solution. This chapter was destined, for better or worse, to become the most famous in a remarkable book.[84] In it, Tycho Brahe described what is now called the Tychonic planetary system, with all the planets revolving around the sun and the sun in turn revolving around the earth [Fig. 32]. The orbits of Mars and the sun intersected, and Tycho argued that this was possible because celestial orbits were geometrical and immaterial, not physical, crystalline spheres. Within the context of this new system, Tycho asserted that the comet moved around the sun with an irregular velocity, in the opposite direction from the planets, on an egg-shaped orbit. These assertions defied all the ancient hypotheses about uniform celestial motion, but they were borne out by ten years of research on Tycho's island. By 1587, Tycho Brahe's description of the celestial regions had dynamic, physical elements that moved far beyond the traditional cosmology of Aristotle and the regular, uniform planetary systems of Ptolemy and Copernicus.[85]

Chapter Nine of his volume dealt the with physical size of the comet and its tail, emphasizing the immensity of celestial space and of the comet itself. Chapter Ten established a model for what would eventually become a standard part of scholarly monographs: a critical survey of other literature on the subject. Tycho categorized, summarized, and criticized at length the writings of twenty astronomers, reducing the whole corpus to a series of appendixes that either complemented or were refuted by his great monograph. He concluded with a summary and an illustrated description of the two main instruments used to observe the comet of 1577 – the early steel sextant and an azimuth quadrant of brass – together with an explanation of his method of using parallel lines of sight and his microdivision of an arc by means of transversal points. From now on, there would be only one "standard work" on the comet of 1577. It was a breakthrough publication in the history of science and became a paradigm for the work of later scientists.

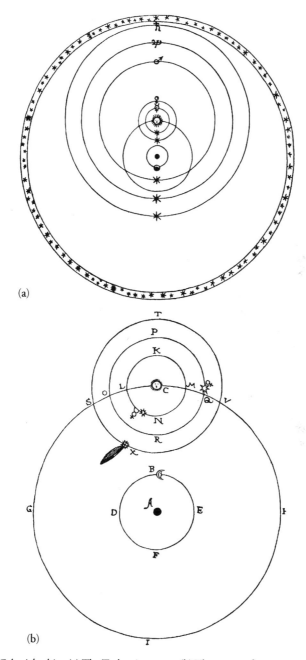

(a)

(b)

Fig. 32. Celestial orbits: (a) The Tychonic system. (b) The comet of 1577 among the planets, shown with its orbit simplified into a circle. (From *De mvndi ætherei recentioribvs phænomenis*)

Christopher Weida finished printing the volume toward the end of 1587, but the colophon was dated 1588.[86] The work ran 465 pages in quarto and was entitled *De mvndi ætherei recentioribvs phænomenis liber secvndvs* (*On the Most Recent Phenomena of the Ætherial World, Book Two*). A few copies were sent to Viceroy Heinrich Rantzau and leading astronomers throughout Europe, and some were offered for sale at the Frankfort book fair in 1588, but most copies were stored unbound until the other volumes of Tycho Brahe's great work had also been printed.

In 1588, Tycho sketched the full extent of his publishing plans in a letter to Caspar Peucer, an old friend and former teacher (and Philipp Melanchthon's son-in-law). The three-volume *Introduction to the Instauration of Astronomy* would be followed by seven tomes establishing the new astronomy: the first on instruments, the second on trigonometry and *prosthaphæresis,* the third a completely new stellar catalog derived (like the stellar positions shown on the great celestial globe) from observations made on Hven, the fourth presenting Tycho's theories of solar and lunar motion, the fifth his theories for the motion of the inner planets, the sixth for the outer planets, and the seventh the latitudes of the planets.[87] Tycho referred to the whole work as *The Theater of Astronomy* (*Theatrum astronomicum*). The plan for this masterwork continued to grow in his mind and take on new shapes for many years to come.

All was well on Tycho's island as the year 1587 drew to a close – except, of course, for a few disturbing signs in heaven and on earth, pointing to the coming year of 1588. Every astrologer knew these dire prophecies, a century old. Moreover, the news from court was that King Frederick II was in ill health.

6

THE PROBLEM OF CONTINUITY

———

1580–1591

Tycho Brahe was a charismatic leader with exceptional political skills, organizational ability, and professional vision. He got things started and kept them going. Huge, tremendously expensive structures had been built up under his direction, and the careers of many coworkers had come to depend upon him. Inevitably, the question arose: What will happen when he is gone?[1] Given the facts of sixteenth-century life expectancy, that question could not be put off too long: Tycho Brahe was forty years of age in 1587, and many people did not live much longer than that.

Tycho Brahe may have broached the problem of continuity to Queen Sophie at Uraniborg on that stormy night in the summer of 1586. The problem was that Tycho's children were not of noble birth, so his sons could not take over the fief of Hven nor carry on after their father; yet the natural assumption of men in the late sixteenth century was that a father's position devolved to his son.

Danish law stated clearly that the privileged status of nobility passed by inheritance only to those whose parents were both of the nobility. Kirsten Jørgensdatter was not of the nobility, and this not only barred their children from inheriting their father's status but also created social problems. Kirsten remained outside noble and court society: When Tycho Brahe was summoned to festive events at court, he was summoned among the bachelors while his wife remained behind.[2] When Tycho went off to attend the great ceremonial weddings and funerals of his noble friends and relatives, Kirsten stayed at home. When noble and princely visitors came to Uraniborg, Kirsten faded into the background and Tycho's sister, **Sophie Brahe**, assumed the role of hostess. Normal family relationships between Tycho Brahe and Kirsten's kin were equally impossible: The reciprocity of equals could not occur between a great noble house and the family of a rural parsonage. At best, her kin might hope to benefit as Tycho's clients.

Common-law marriages like that of Tycho and Kirsten bothered clergymen with tender consciences, and the influence of this puritanical party lay behind a royal ordinance of 11 June 1580 that characterized common-law marriages of men as "an evil, scandalous life with mistresses and loose women, whom they keep in their houses and with whom they openly associate, brazenly and completely without shame, just as if they were their good wives."[3] The ordinance commanded the clergy to separate such couples and put them under the ban if they resisted. The intention was to establish greater public control over marriage, and the ban denied offenders the sacraments and rites of the church, placing them beyond the pale of civil society and under the threat of eternal damnation until they changed their ways.

However, the power of a nobleman in the parish of his seat was considerable, and it was difficult to enforce such ordinances upon them. Tycho Brahe and Kirsten Jørgensdatter may have been challenged, but the evidence is ambiguous. Tycho stopped receiving Holy Eucharist, and this would have been the most immediate effect of a ban, but he may have decided to defy the growing authority of the church by abstaining from the sacrament on his own volition.[4] He and Kirsten continued to live as they had for years, and there was not much the pastor of St. Ibb's could do about that, since he was Tycho Brahe's hand-picked man.

Some Danish noblemen living with commoners, however, had decided it was time to get married, and wed their nonnoble spouses. This had provoked a new royal ordinance dated 19 June 1582, condemning marriage between nobles and commoners and spelling out the consequences in no uncertain terms:

First, when any noble man has taken a commoner woman, or shall take one hereafter, and has children with her, those children shall not be noble children or free folk. Moreover, they shall not bear a coat of arms or noble family name. Likewise, those same children with a commoner woman shall not inherit any land or estate from their father or their father's kin, but when the father dies, his estate and goods shall fall to his nearest noble relatives and heirs according to the law. However, this does not bar the father from giving money and personal property to children he may have sired if he desires to do so. And if these same children have been given money or personal property by their father in his lifetime, it shall remain theirs without hindrance, and they may have and keep it, and no more.[5]

Tycho Brahe and Kirsten Jørgensdatter had five living children in 1582. These children could not bear the name of Brahe but were known by their patronymics in accordance with the common Scandinavian custom of the day. Magdalene Tygesdatter was eight, Sophie four, Elisabeth three, Cecilie two, and Tyge Tygesen less than a year; Jørgen would be born the following year.[6] Their father had inherited half of Knutstorp manor, the advowson of Kågeröd church, and other property, but none of it could ever come to them. Their father was related to the mightiest families in the realm, but his kin were not their kin. Their father held in fee the island of Hven, the vast estates of the Chapel of the Three Holy Kings and the district of Nordfjord, plus Kullen and other minor fiefs, and he drew the largest annual pension in Denmark, but they were commoners who could never hold noble fiefs in fee. Without relatives to plead their case at court or in the State Council, the most they could ever hope for in the line of benefices would be some tithe income, a canonry, pastorate, or professorship. They could emulate their father in one respect: He had married a commoner, and they could also do that when they were grown.

Would that indeed be all? As early as 1584, Tycho began develop plans to link the future of the island of Hven with the fate of his children. His sons were one and three when Anders Sørensen Vedel came to Uraniborg that summer. The two men discussed the problem, and Vedel wrote down the draft of a Latin document that may have been dictated by Tycho. It was a hypothetical royal patent, dated 1 July 1584, granting the island of Hven in perpetual fee to Tycho Brahe and his male issue, providing that they used the island and its facilities solely for the pursuit of mathematical studies and did nothing to injure the crown or kingdom.[7] Lifetime grants in fee were increasingly rare in Denmark, and hereditary fiefs were virtually unknown. An island fief granted in perpetuity to commoners seemed preposterous, but it could be argued that Uraniborg was like an academy or university, and that the office of director was like the superintendent of a secularized monastery, or like a university professor whose income derived from a landed benefice. Commoners held those positions. In any case, it seemed worth a try.

Tycho's children were not nobles, but they had a powerful aristocratic patron in the person of their father. He knew that timing was of the utmost importance in such matters, and he waited for years until the right moment finally came. Sometime between 1582 and 1588, Tycho Brahe personally presented his proposal to King Frederick II. The king [Fig. 33] heard him out and, amazingly, gave his approval, despite the highly unusual nature of the request. The queen was present and heard him sanction the idea. Nothing was formalized, nothing written down: No royal patent was actually issued, but the plan was approved.

Fig. 33. King Frederick II. Polychrome terra-cotta bust
by Johan Gregor van der Schardt, 1577–8. (Courtesy Det
Nationalhistoriske Museum på Frederiksborg, Hillerød)

The king was not well at the time. He was pale, worn out, constantly
drinking, always coughing. After one last round of exhausting celebration,
King Frederick II of Denmark and Norway died on 4 April 1588.[8]

REGENCY SUPPORT

Things moved quickly after that. A Regency Council was appointed to
rule the kingdom during the minority of Prince Christian. It was headed
by Royal Chancellor Niels Kaas and included Lord Admiral Peter Munk,
Lord Treasurer Christoffer Walckendorf, and Councillor Jørgen Rosen-

krantz.[9] Tycho Brahe knew every one of them well. Munk and Walcken-dorf were fierce rivals, unwilling to share power with each other, but they were agreed that Queen Sophie should not be a member of the Regency Council. Prince Christian, heir to the throne, was a child of ten years. He received the title of King Christian IV, and the Regency Council ruled in his name, but there was no question of his sharing power. The Regency Council's authority extended from the Baltic islands of Ösel, Gotland, and Bornholm, throughout the kingdoms of Denmark and Norway, to Iceland, Greenland, and the Faeroes. Heinrich Rantzau, Viceroy of Scheswig and Holstein, held firm control of the royal territories in those duchies on the German border.[10]

Viceroy Heinrich Rantzau visited Uraniborg on 27 May 1588, on his way to the king's funeral in Roskilde. A month later, Councillor Peter Gyldenstierne was on Hven at the same time as a delegation from Mecklenburg. Less than two weeks later, two of the regents, Chancellor Niels Kaas and Councillor Jørgen Rosenkrantz, were on Tycho's island with Anders Sørensen Vedel, now Royal Historiographer. Tycho lobbied hard to ensure that his position did not deteriorate under the new government. He was concerned about the future of Uraniborg but spoke to Kaas and Rosenkrantz also about the tremendous cost of the Hven facilities, claiming he had incurred a personal debt of 6,000 dalers to build and equip them. Vedel had a much more modest request: He needed a grant to travel around Denmark in search of historical and chorographical materials.[11] Kaas and Rosenkrantz moved with remarkable speed, and on 12 July 1588, the Regency Council agreed to pay in cash the immense sum of 6,000 dalers to Tycho Brahe.[12] On 16 July 1588, they granted support to Anders Sørensen Vedel to collect historical materials throughout Denmark, Schleswig, and Holstein.[13]

The aristocracy was back in the high seat, and the new government overflowed with Tycho's friends, relatives, and political allies. In the State Council sat his brother, Steen Brahe; brother-in-law, Christen Skeel; and cousin, Axel Gyldenstierne. Soon after the king's funeral, Gyldenstierne was named to the new office of Viceroy of Norway, with his seat at Akershus Castle.[14] Tycho's brother Jørgen Brahe took over as Lord Lieutenant of Landskrona, while Steen Brahe assumed their father's old fief of Helsingborg. Between Helsingborg Castle and Akershus lay Bohus Castle, under the command of Tycho's brother-in-law Henrik Gyldenstierne. Another brother, Axel Brahe, held the most important fief on the central Danish island of Fyn, and kinsman (as well as alchemist) Falk Gøye was governor of Bornholm. In Jutland, Christen Skeel advanced to the command of Aakær fief in 1588, while Erik Lange held Bygholm fief.

Now was the time for Tycho Brahe to present his plan for Hven. He sailed to Copenhagen on Sunday, 18 August 1588, and stayed for several days, speaking with the regents and calling on the support of his allies. On Thursday, 23 August 1588, the Regency Council issued a patent, proclaiming their desire to perpetuate far into the future the work at Tycho Brahe's observatory on Hven by maintaining the establishment built by Tycho Brahe and endowing it with income from canonries and other ecclesiastical offices, "so good people can be motivated to cultivate this same [astronomical] art, and that it might always be maintained and cultivated here in this realm."[15] Tycho's dream of a permanent research institute was being realized, and when the Regency Council proposed to endow it with ecclesiastical incomes, they were creating a structure that could be led by either a commoner or a nobleman. The patent continued:

And if any of the aforenamed Tycho Brahe's own could be found to be disposed and fit, then this one shall be recognized for this ahead of others, so that they might be able to enjoy the fruits of his labor and effort. And if none of his own in particular, nor of his close relatives or servants are found to be fitting, then others of the native-born Danish nobility shall be appointed, and if there are no nobles who are fit, then other native-born Danes shall be appointed.

The language was awkward, but the intention was clear. The Regency Council established an order of succession to Uraniborg that gave preference to Tycho Brahe's family. This could mean his sons, who were only seven and five years of age at the time, but it could also apply to his future sons-in-law. If none of Tycho's immediate family was suitable as director of Uraniborg, the next choice would be among his numerous nephews, many of whom were already launched upon academic studies. The third choice would be from Tycho's *familia;* after that, from the whole Danish nobility; and finally, all native-born Danes. Given the range of choices, the perpetuation of Uraniborg seemed assured.

The Regency Council did add the standard rejoinder that their authority was more limited than the king's would be when he came of age. Nevertheless, Tycho Brahe had grounds to be immensely pleased as he sailed home to Hven with the patent in hand, and Kirsten Jørgensdatter must have been overjoyed to hear the news. This patent contained the first official recognition of the fact that Tycho even had a family: "Tycho Brahe's own," in the phrase of the patent. Kirsten herself was not mentioned, but the future of her children seemed much brighter now.

STAFF CHANGES

Meanwhile, some important changes were taking place in the staff of Uraniborg. A young French scholar, Rodolphe Méhérenc of Sées, arrived in January of 1588 with a recommendation from Tycho's friend Charles de Dançay.[16] **Peter Jacobsen Flemløse**, who had served Tycho since the very beginning, entered Viceroy Axel Gyldenstierne's service as his personal physician. Before departing for Norway, Gyldenstierne inquired about the canonry promised to Flemløse in 1579, and the Regency Council formally renewed the promise on 25 July 1588, but no canonry was available. Flemløse took his leave of Hven on 14 August 1588. **Steenwinckel** also traveled to Norway that summer to inspect the major border fortresses, including Bohus and Akershus, on command of the Regents.

That summer of 1588, Gellius Sascerides left Uraniborg to study abroad on the scholarship awarded him in 1584. He had served on Hven for five years, and as he traveled into Germany, Tycho assigned him to deliver to Landgrave William an autographed prepublication copy of the opus on the comet of 1577, with its description of the Tychonic planetary system. Tycho was shocked to learn from Rothmann, soon after, that Ursus had passed off a very similar system as his own, and that Joost Bürgi had made a clockwork model of it.[17]

Rudolphus Groningensis left the *familia* of Tycho Brahe that same year, and Tycho used his influence to get him into the University of Copenhagen, for Rudolphus had not previously attended a university. Wensøsil also returned to the university, having served one-and-a-half years of his three-year contract. Weida probably also left in 1588 when another paper shortage brought the press to a temporary standstill. The *familia* of Uraniborg now included only **Morsing**, **Crol**, and Riber, in addition to a large number of miscellaneous servants. To fill the gaps, Tycho brought in two well-qualified scholars, probably on two-year written contracts. Iver Iversen Hemmet, a poet with an M.A. from Wittenberg, was twenty-four when he arrived on Tycho's island. Niels Bertelsen Colding was also a poet and an M.A. who had studied abroad, and both of them stayed out their terms. In March of 1589, Martin Ingelsen from Landskrona, called Coronensis, joined the *familia*. That same year, a farmer's son named Christian Sørensen Lomberg arrived with recommendations from Copenhagen professors. He was still a student at twenty-six, for lack of money had delayed his studies, but he was bright, conscientious, in love with mathematics, and had very sharp eyesight. The Latin form of his name was **Longomontanus**. He learned the theories and methods of Hven under Morsing's and

Tycho's direction and worked for the next three years on Tycho's new star catalog. Tycho Brahe and his family grew very fond of Longomontanus, who may have tutored Tycho's children and also served as his personal secretary. During the autumn of 1590, a learned young nobleman named Jørgen Brockenhuus visited Uraniborg, fell under its spell, and returned to live in the *familia* for a time. Thereafter, he described himself as a disciple of Tycho Brahe.

Tycho kept in touch with his former assistants scattered far and wide. **Oddur Einarson** was elected bishop of Skálholt in Iceland and spent several days on Hven in 1589, when he came to Denmark for his consecration. He presented Tycho with an Icelandic quadrant that was probably made by Bishop Guðbrandur Þorláksson of Hólar. Tycho also kept in contact with Flemløse and Steenwinckel in Norway, Gellius in Germany, **Stephanius** in Sorø, **Arctander** in Copenhagen, and others in scattered locations, though some like Hamon in England seemed to fade out of the picture. When Gellius ran into trouble with the renewal of his *stipendium regium* because he had not taken his M.A. in Copenhagen (a requirement for receiving the stipend), Tycho Brahe wrote to the university on his behalf, and when this failed, he arranged for Gellius to travel to Italy as preceptor for Albert Skeel, Tycho's nephew. The reciprocal obligations between patron and client did not end when a member of the *familia* left Uraniborg.

MAPMAKING

Friendship also created durable bonds. As noted earlier, Tycho Brahe's former preceptor, Anders Sørensen Vedel, had succeeded to the position of Royal Historiographer in 1584; he now lived in Ribe, where Tycho had attended his marriage in 1581 to Mette Hansdatter, the daughter of Bishop Hans Laugesen. Their home, Liljebjerg ("Lily Hill"), became a kind of urban Uraniborg, with collections of books and manuscripts, a printing press, Renaissance gardens, and emblems and epigrams everywhere. Liljebjerg was dedicated to God and the Muses, and a plaque in the library was signed by nine of Vedel's most learned and beloved friends, including Charles de Dançay, Niels Kaas, Tycho Brahe, and Peter Hegelund.[18] This was familiar territory.

As Royal Historiographer, Vedel planned to write an official history of Denmark in twenty-two books, beginning with chorography and ethnography, proceeding through prehistory and history, and concluding with genealogies and chronologies. The chorography required a new map of Denmark because the text would describe places and events associated with specific locations.[19] This approach had great political utility, and King Frederick II had encouraged it. Maps could also be tremendously useful to rulers, as Bruce T. Moran has pointed out: "Maps determined rights to forest lands and fields; offered proof of sovereign power and legal jurisdiction; specified taxable lands; and defined possession of economically productive regions such as mining areas, saltworks, woodlands, and transportation routes."[20] There was something magical about the power of maps. As **David Fabricius** wrote when he sent Count Anton II of Oldenburg-Delmenhorst a map of his principality in 1592, "Your Grace can walk with pleasure through the whole wonderful county in the blink of an eye."[21] Chorographic and historical data reinforced this power of maps to clarify and consolidate royal claims.

Vedel knew something about maps, but he was not a cartographer, so he had taken the matter to Uraniborg. Tycho Brahe – thoroughly familiar with Ptolemaic and Renaissance geographical and cartographic theory in the tradition of Gemma Frisius, Johannes Homilius, and Vedel's old patron, Valentin Thau – had joined the project and begun to collect the best maps available. In September of 1585, King Frederick had even ordered the librarian of Copenhagen Castle to send his map collection to Tycho Brahe.[22] Then, for some reason or other, the project had lapsed, and it was not renewed until Vedel got support from the Regency Council in July of 1588.

At that point, important geodetic and cartographic projects were added to the interdisciplinary activities of Hven as Tycho Brahe threw his extensive theoretical, technical, and financial resources into Vedel's project. **Morsing** was assigned to travel with Vedel and determine the latitudes of places around Denmark, using Tycho's portable brass azimuth quadrant [Fig. 34] and other instruments. **Flemløse** visited Hven during the summer of 1589 and was recruited to assist; during his travels around Norway in 1589–90, he determined the latitudes of many locations. Tycho summoned a craftsman named **Peter Jachenow**, who had made odometers for the courts of Mecklenburg and Denmark, and he commissioned Jachenow to make odometers for him. One was used to measure a baseline from Uraniborg to St. Ibb's Church.

Tycho Brahe was quite familiar with the theory of surveying by triangulation, which had been described by Gemma Frisius but never actually tried by anybody. The giant step from theory to practice was taken when Tycho and his *familia* carried out the first triangulation survey in history to link the island of Hven with the surrounding shores of the Sound

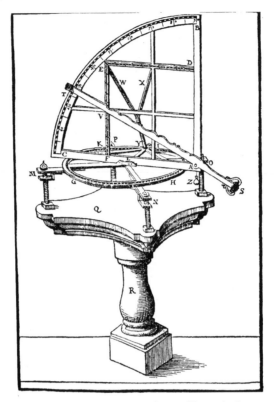

Fig. 34. Portable azimuth quadrant of brass, built 1583. Simpler and sronger than earlier models, it was easily disassembled for transportation. (From *Astronomiæ instavratæ mechanica*)

[Fig. 35a].[23] They gradually learned that the method would take too much time to serve as the basis for Vedel's map, so Tycho sent Vedel some tables of latitudes and longitudes in November of 1590, with instructions for using them as the basis of a simplified but scientifically respectable map. Problems with the data from the survey of the Sound forced Tycho to limit his own mapmaking to Hven. His improved map of the island, drawn by **Willem Janszoon Blaeu** and printed as a woodcut in 1596 [Fig. 35b], was the first map ever based on surveying by triangulation.

PAPERMILL AND COPENHAGEN MANSION

By early 1589, Tycho Brahe had expanded his printing office to include two presses.[24] He must have employed one or two master printers with several journeymen and assistants. One press crew was at work on volume one of his *Introduction to the Instauration of Astronomy*, dealing with the supernova of 1572; the other was printing the first volume of his astronomical correspondence, which Tycho had decided to include as part of his multivolume renovation of astronomy. These letters were really epistolary essays on specific scientific problems, and the volume became a forerunner of the scientific journal.[25] By early 1590, volume three of the *Introduction,* on comets after 1577, was being printed as well, but recurrent paper shortages forced the presses to close down in the fall of 1590. A papermaker named Caspar, provided by Erik Lange in 1589, had been sent on to Lübeck, apparently to collect linen rags, the raw material of sixteenth-century paper; after several such trips, he vanished. By October 1590, Tycho had sent his master printer to Germany to find paper and a good papermaker.

By that time, he had already decided that the only way to solve the persistent paper shortage was to build his own papermill. Viceroy Rantzau had a papermill; so did Lange. Tycho's decision was not made lightly, because such a mill required waterpower. He had previously helped his uncle, Steen Bille, to build a papermill near Herrevad Abbey, but there was no comparable source of power on Hven, where there were a few artificial fishponds but no lakes or streams. Providing waterpower would require a major earth-moving project.

Tycho hired a contractor named Valentin to build an elaborate series of dams and spillways across the southern half of the island. Labor came, as usual, from the villagers of Tuna. Some of the ponds were built in the middle of village fields, so new arable land was plowed, dyked, and fenced in the commons to replace the lost fields. This work went on during 1589–90.

There must have been many suppertime discussions of the writings of Archimedes and Vitruvius, who had dealt with waterworks in antiquity. Many of Tycho's ponds were given geometric shapes, and they were connected by an ingenious system of sluices. The last dam, near the western shore of the island and clearly visible even now, rose to a height of nearly fifty feet. "Valentin promised to have the mill dam finished in four weeks," the diary recorded on 26 June 1589. A huge oaken pipe reinforced with massive bands of iron ran through the dam, leading water under pressure to a twenty-foot overshot waterwheel in the millhouse, just below the dam.

In November of 1589, Tycho wrote to Rothmann in Kassel that

the aforementioned mill is already finished; a high and wide dam provides the necessary amount of water, in the warm as well as the cold time of year. The wheel, which is twelve *alen* in diameter, is driven by the smallest possible amount of water and provides power for two industries besides papermaking. In addition, a large number of fishponds are arranged so they can provide water to the mill if necessary, and just a few years ago, this was all dry land.[26]

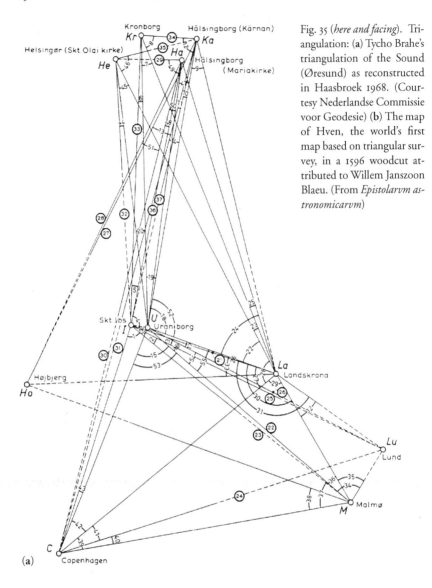

Fig. 35 (*here and facing*). Triangulation: (a) Tycho Brahe's triangulation of the Sound (Øresund) as reconstructed in Haasbroek 1968. (Courtesy Nederlandse Commissie voor Geodesie) (b) The map of Hven, the world's first map based on triangular survey, in a 1596 woodcut attributed to Willem Janszoon Blaeu. (From *Epistolarvm astronomicarvm*)

(a)

Tycho compared it to some of the projects of Landgrave William and considered his massive reshaping of the earth to be a princely enterprise worthy of the classical Augustans themselves.

Some parts of the work dragged on; it was a large project and had to be done right. The millhouse was still not completed. Early in the spring of 1590, Tycho's men felled a mighty oak near Herrevad Abbey to construct beams. It grew into a large half-timbered building, thirty by seventy-five

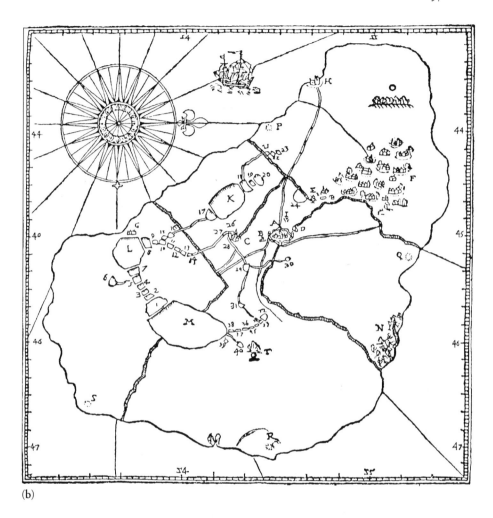

(b)

feet, with a tile roof, containing a papermill (stamp mill), parchment mill,
gristmill, and grindstone, all powered by the large overshot wheel [Fig. 36].
Besides workrooms and storage, the building contained an apartment for
the miller and another chamber with a stove of green glazed tiles, reserved
for the lord, Tycho Brahe. By June of 1590, Valentin and his crews had finally completed the large dam, and Valentin departed the island. Caspar,
the papermaker, was traveling around Skåne and other parts of Denmark
at the time; after he absconded, Tycho inquired of Viceroy Heinrich Rantzau in September whether Caspar was working for him, and if so, whether Rantzau would send him back. By 1591, Tycho had gotten a papermaker, presumably a new one. By 1592, the mill was complete, and once again
a papermaker was scouring the country for linen rags for Tycho Brahe.

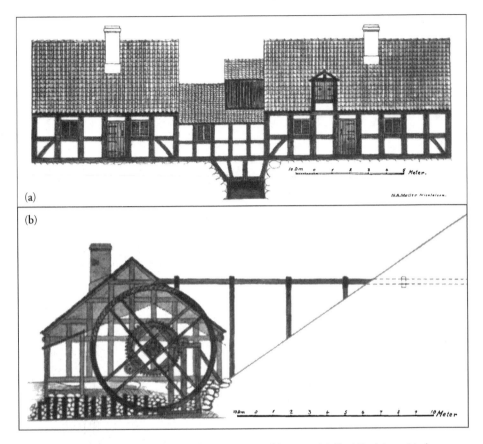

Fig. 36. Tycho Brahe's millhouse, as reconstructed by N. A. Møller Nicolaisen: (a) elevation; (b) cross section showing overshot wheel. (From Møller Nicolaisen 1946)

Tycho erected a free-standing votive stone at the mill and dedicated it with festivities similar to those which had taken place at Uraniborg some sixteen years earlier.[27] Despite problems in retaining skilled papermakers, the mill produced parchment for bookbinding and fine writing paper as well as printing paper for Tycho's publications. Tycho's personal writing paper had a watermark with the arms, helmet, crest, and mantling of Brahe and the initials T. B. Two watermarks were used on the printing paper: One showed the Brahe arms and the initials T. B., and the other, the east elevation of Uraniborg [Fig. 37a].[28] Tycho's parchment bookbindings were of high quality and have remained elegant to this day. He also commissioned a pair of ex libris to stamp on the front and back of books from his bindery. The front showed Tycho Brahe half-length in the style of the first Gheyn engraving, within an oval inscribed with a Latin distich, and

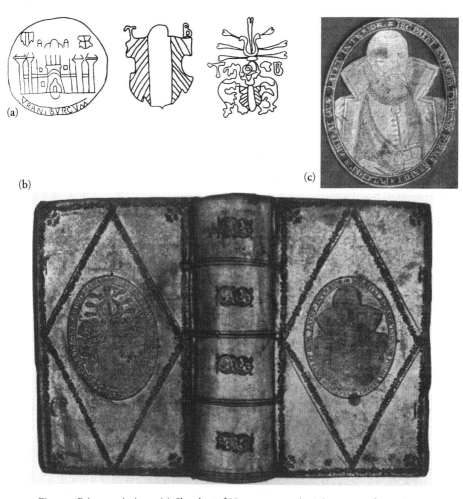

Fig. 37. Private printings: (a) Sketches of Hven watermarks. The image of Uraniborg (*left*) and the Brahe arms (*center*) were used on printing paper; the Brahe arms with helmet, crest, mantling, and initials T. B. (*right*) was used on Tycho Brahe's personal writing paper. (From Nordstrand 1961) (b) Bookbinding from the bindery, in blind-stamped parchment, with Tycho Brahe's personal ex libris on the front and back covers. (From Lauritz Nielsen, *Dansk privatbiblioteker gennem tiderne* 1946) (c) Detail of the ex libris (Courtesy Det Kongelige Bibliotek, Copenhagen)

the back showed the arms, helmet, crest, and mantling of Brahe, elegantly rendered, within an oval with another distich [Fig. 37b].[29]

As his construction projects on Hven reached fulfillment, Tycho Brahe shifted some attention to Copenhagen. He owned a residence in Farvergade ("Dyer's Lane"), where he stayed during his frequent trips to the city. Next to his property was the royal dyeworks and a fuller's shop, and the city wall ran behind, with a bastion called the Windmill Tower in back of the dyeworks. Tycho wanted to acquire these properties, tear down the existing buildings, and erect an aristocratic mansion with an enclosed courtyard and an observatory on the bastion. The city authorities gave him permission to proceed, and the Regency Council turned over the adjacent properties in March of 1589.[30] Tycho probably consulted with Steenwinckel in drawing up plans for a mansion with a bay in front, surmounted by a spire with a clock and bell. Other aristocratic residences were also found in that quarter of the city.

THE POLITICS OF CONTINUITY

The politics of these restless years kept Tycho, his *familia,* and his aristocratic relatives constantly on the move. Queen Sophie's daughters were of an age to enter the marriage market. Princess Elisabeth, fourteen when her father died, was already promised to the learned young Duke Heinrich Julius of Braunschweig-Wolfenbüttel. Princess Anne was thirteen and soon became the subject of negotiations with Scotland, which led to a match the following year. In the autumn of 1589, she sailed from Copenhagen with a splendid entourage headed by Tycho's brother, Steen Brahe, to wed King James VI of Scotland. Fierce autumn storms forced her to seek haven in Norway, where the flotilla put into Oslo and Viceroy Axel Gyldenstierne suddenly found himself with unexpected guests at Akershus Castle. Across the seas in Scotland, King James fretted and fumed, brought Scots witches to trial for stirring up the storms, then impulsively boarded ship and rode the storming westerlies to Oslo to fetch his bride. Their wedding was celebrated by Bishop Jens Nilssøn in Burgomaster Christian Mule's mansion in Oslo on 23 November 1589.[31] After endless festivities, the newlywed couple and their courts departed for Denmark. Along the way, they were entertained at Bohus Castle, where the Royal Chancellor of Scotland, John Maitland, dashed off a laudatory poem in Latin for their host, Tycho Brahe's brother-in-law, Henrik Gyldenstierne.[32]

On 20 March 1590, King James VI and his courtiers visited the renowned island of Tycho Brahe. **Sophie Brahe** and Erik Lange helped to entertain them, and the King of Scotland was given a splendid welcome with a ceremonious tour of all the island's wonders. A great banquet was served, and the company conversed in Latin while musicians and entertainers performed. The wine flowed freely. When talk turned to politics, Chancellor Maitland dashed off epigram after epigram, all in Latin, dealing wittily with grave events of recent years: the defeat of the Spanish

Armada in 1588, the pope's dilemmas, the chaos in France with the deaths of Catherine de Medici, Henry of Guise, and Henry III in 1588–9, and Parma's bootless intrigues with the Catholic lairds of Scotland.[33] The cup went around, and Maitland spontaneously composed a series of Latin epigrams in honor of the host, Tycho Brahe, including this on Uraniborg:

> Muses' royal castle, jewel of the world, rivaling Olympus,
> Nourishing house, your spirit's equal to your name.[34]

King James was in his element and immensely pleased by the visit and the company. Before they departed, Maitland put his poems on the door of the King's Room, where they remained for visitors to copy. King James himself later took a quill pen and composed a Latin verse in Tycho's honor, which ended:

> What Phaethon dared was by Apollo done,
> Who ruled the fiery horses of the sun.
> More Tycho doth, he rules the stars above,
> And is Urania's favorite, and love.[35]

Royal visits took immense preparation by Tycho's staff and all the islanders, but they spread abroad the fame of Uraniborg, attracted students, and helped to ensure that financial support would continue. King James eventually granted Tycho a copyright within the realm of Scotland for his publications.[36]

Uraniborg had another princely visitor when Duke Heinrich Julius of Braunschweig-Wolffenbüttel came in April of 1590.[37] Like King James, the duke was a learned man of considerable wit and was fascinated by the arcane messages of Tycho's island. He married Princess Elisabeth of Denmark that spring. Less elaborate was the visit of the eighteen-year-old Duke William of Courland, brother of the reigning duke, who arrived with a few of his courtiers on 24 August 1592.[38]

Tycho Brahe was still working to guarantee the future of Uraniborg and of his own children, but he found himself compelled to adjust his patronage strategy to a new situation because power was now both divided and disputed. Tycho's aim was to pull the many strands together and avoid any possibility of misunderstanding when authority finally passed to the young King Christian IV. In 1589, he obtained a patent signed by the whole State Council and the Regency Council, reaffirming his patent of 1588 from the Regency Council.[39] Two weeks before the visit of King James, he went to Queen Sophie, who testified in writing that she could clearly remember the late King Frederick II stating, some time before his death, that he intended to install one of Tycho Brahe's own children as head of the observatory on Hven.[40]

Aristocratic factions struggled for control of the government while Queen Sophie fought for her share of authority and Viceroy Heinrich Rantzau beat back challenges to his domination of Schleswig-Holstein. One of the mightiest men in Denmark, Christoffer Walckendorf, was ejected from the Regency Council in 1590 by his enemies; those enemies were the friends and relatives of Tycho Brahe.[41] Times were tense. That summer, witchcraft hysteria broke out in Bergen, Norway, targeting the families of clergymen, including Tycho Brahe's friend from student days, Bishop Anders Foss.[42] Viceroy Axel Gyldenstierne went to Bergen to deal with the problem, and Peter Flemløse accompanied him.

THE WORKING FAMILIA

In these years, the household of Uraniborg continued to change. **Elias Olsen Morsing** died suddenly in March of 1590, less than two weeks before the arrival of King James. Morsing had been Tycho Brahe's steadfast assistant for six years as astronomer, meteorologist, surveyor, teacher, research team leader, author of the first book published at Uraniborg, and the man chosen to go on the expedition to Copernicus's observatory. He expired in the arms of his friend **Christian Longomontanus**. **Hans Crol**, Tycho's ingenious instrument maker, had married in 1587 but seems to have continued to live in the Uraniborg household. His infant son died a few weeks after the visit of Duke Heinrich Julius. Crol and his wife apparently took the remains to Copenhagen for burial and were gone for a month. They may have seen the tomb of Charles de Dançay in Our Lady's Cathedral, for Dançay had expired toward the end of the previous year.

Others arrived on the island of Hven, and Tycho Brahe found time for all of them. Arnold Floris van Langren was nineteen when he showed up on 4 April 1590, two weeks after the royal Scots visit. His father, Jacob Floris van Langren, was a well-known publisher of maps and globes in Amsterdam, and Arnold had worked in his father's shop since the age of nine. The elder van Langren had published a celestial globe in 1586, based on data supplied by Rudolphus Snellius of Leiden. Now he wanted to publish a globe based on the famous Tychonic data. Tycho, pleased with the idea, took Arnold into the museum and showed him the great celestial globe with close to 777 gilt stars, each of them precisely inscribed in turn after repeated observations. He explained, however, that the star catalog was not complete and had therefore not yet been published. He could not give the van Langren press a prior claim to the data, but he promised to send a copy as soon as it was off the Uraniborg press. Arnold Floris van Langren remained for seven weeks in Tycho's *familia,* becoming quite versed in the apparatus, methods, philosophy, and learned treasures of Uraniborg.

Finally, **Christopher Rothmann** of Kassel, with whom Tycho Brahe had corresponded for years, arrived to visit Uraniborg.[43] Tycho went to Copenhagen to meet him, and together they sailed to Hven on the evening of 1 August 1590. Tycho spent the whole month entertaining Rothmann, leaving the assistants to carry on their assigned tasks. Other visitors kept arriving, and Tycho excused himself from ardent astronomical discussions with Rothmann to receive them in his customary hospitable way. Four German and Polish students arrived on 3 August. A party of Scots nobles came on 9 August, and more wayfaring German students arrived that same day. On 13 August, Tycho and Rothmann went to visit the castle of **Sophie Brahe** in Skåne, where her herbal and knot gardens, with their rich allegorical program, made a strong impression on Rothmann. They held long conversations, arguing back and forth about the theories of Copernicus and Tycho. Toward the end of the month, an Austrian envoy visited Tycho's island for three days. The weather was overcast most of the month but cleared on Sunday, 30 August. That evening, Tycho Brahe and Christopher Rothmann joined Longomontanus, Crol, and the other *famuli* in observing the heavens. A splendid display of aurora borealis occurred on 1 September, the day Rothmann took his leave of Tycho's island.

Without a staff of talented assistants and a good deal of continuity in the senior ranks, without well-established procedures and broad research goals that were clearly understood by all, it would not have been possible for Uraniborg to function as a complex, multidimensional research institute in these years. The force of events was transforming Tycho Brahe's role into that of administrator, project initiator, author, and supervisor, and he was also the one who ensured the continuation of financial support for large-scale science on the island of Hven. Much of the day-to-day scientific work fell on the shoulders of his *familia* of scholars and craftsmen, under the leadership of the senior assistants. Short-term guests and visitors helped with the work of research teams, under the supervision of the experienced members.

Niels Colding left the staff in September 1590, and a young Norwegian scholar from Bergen, **Cort Aslakssøn**, arrived in October. Christian Hansen Riber had left the previous spring, but like many former *famuli,* he kept in touch with Tycho Brahe and continued to pursue studies begun on Hven. Riber had a portable astronomical radius and a prepublication copy of Tycho's book on the supernova of 1572. Traveling abroad in 1591 with an old Ribe friend, Ægidius Lauridsen, he taught Ægidius how to use the instrument and instructed him in the book. This was the first case of a former *famulus* transmitting the scientific traditions of Hven to someone who had not visited Tycho's island. In the years to come, there would be many such cases.

For two days in late May of 1591, Tycho and his *familia* enjoyed the sight of a splendid royal fleet, nine men-of-war, maneuvering in the Sound off Elsinore. The young monarch, King Christian IV, and his court were progressing to Oslo, where he was to be hailed as king of Norway. The pastor of Kågeröd, Hans Jørgensen, was visiting Uraniborg at the time.[44] He had recently been appointed by Tycho and his brother Steen, who held the advowson jointly. Hans Jørgensen may have been a brother of Kirsten Jørgensdatter, but there is no evidence that Tycho bestowed more than the usual pastoral patronage upon him.

In June of 1591, the Uraniborg press published a small volume that offered a glimpse into the working methods of Tycho's island. Tycho had managed to accumulate a supply of paper made on the island, and he wanted to use some of it to keep his printers occupied. He remembered a Danish manuscript on weather forecasting that he had assigned Flemløse to write for the late King Frederick, in order to demonstrate the practical applications of meteorological studies. Tycho saw the atmosphere as a transitional zone between heavens and earth but interpreted it far differently than the Aristotelians, who thought that lightning, aurora borealis, comets, and other celestial blazes occurred in a "sphere of fire" in the upper atmosphere, whereas Tycho denied the existence of such a sphere. For nearly a decade, he had compiled daily meteorological observations – the first person in modern times to do so – in an attempt to determine what the atmosphere was really like and how it served to link heavens and earth. Flemløse's manuscript built on the meteorological journal as well as on other sources. Tycho decided to publish it but thought that it needed a preface, so he dictated one to Christian Longomontanus and published it under Flemløse's name.[45] The Copenhagen publisher, Henrik Waldkirch, had visited Hven in April of 1591, undoubtedly to discuss the Uraniborg press.[46] Sometime around 1590, a printer named Hans Gaschitz had joined Tycho Brahe's *familia*, and he was the one who printed Flemløse's book in Danish and German editions. The translator into German was probably a Prussian student known only as Sebastianus Borussus, who came to Hven from the University of Copenhagen.

There was a constant flow of talent to Tycho's island in these years, as Uraniborg became a kind of postgraduate institution for the University of Copenhagen in many areas of the arts and sciences.[47] Several highly competent professors sent Tycho a steady stream of well-trained students. Jørgen Dybvad, the sole professor of mathematics in Copenhagen, gave his students solid training in geometry, surveying, meteorology, and astronomy before sending them to Hven. Dybvad was a quarrelsome and irritating man, but he did not feud with Tycho. He was a productive scholar, the first Dane to publish a commentary on Copernicus, and one who was willing, like Tycho, to accept radical new solutions to old problems.[48] Copenhagen's two professors of medicine also lectured on astronomy in order to explain the effect of astral influences upon the harmony of the human microcosm. Both were academic insiders but not productive scholars, and both had large private practices. Anders Jensen Lemvig, who had succeeded Pratensis, was a protégé of Severinus, sympathetic to iatrochemical approaches, and influenced by Ramism since his student days abroad. Lemvig undoubtedly helped to promote the careers of students from his home area of Jutland by sending them to Hven, and Longomontanus was one of these.[49] The other professor of medicine was Anders Christensen, a native of Ribe and brother-in-law of Anders Vedel. He had studied in Wittenberg, Padua, Basel, Paris, and Strasbourg before succeeding Johannes Franciscus. Anders Christensen was a highly respected teacher and a friend of Tycho Brahe, and he, too, was an innovator. His attempt to introduce anatomical dissection to Copenhagen was frustrated by adverse public opinion, but he eventually established the first botanical garden at the university.[50]

A notebook from Uraniborg around 1589–92, now in the Royal Library in Copenhagen, offers a glimpse into the teaching that took place on Tycho's island [Fig. 38].[51] Latin was the medium of instruction, as it was in universities throughout Europe, and most of the notes were in Latin, with some comments and translations in German. The author was probably an itinerant German student or craftsman, possibly Sebastianus Borussus or a printer, because there is some evidence that the notebook was used in setting type for the edition of Tycho Brahe's astronomical correspondence. The author of the notebook was interested in astronomy and the fine arts but not chemistry. He copied many of Tycho's poems and inscriptions from Uraniborg and Stjerneborg, taking special note of anything that had to do with Copernicus, and he included Chancellor Maitland's political epigrams, dashed off during the visit of the King of Scotland in 1590. He summarized the contents of Tycho's work on the supernova of 1572, heard

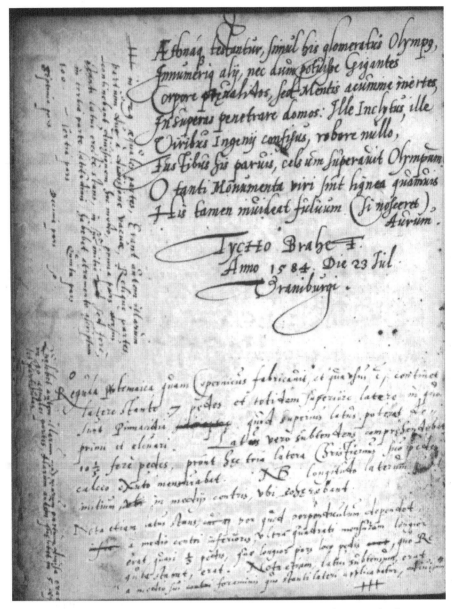

Fig. 38 (*here and facing*). Two adjacent pages from the German student's notebook on Hven. Note the steel quadrant sketched at right. (Courtesy Det Kongelige Bibliotek, Copenhagen)

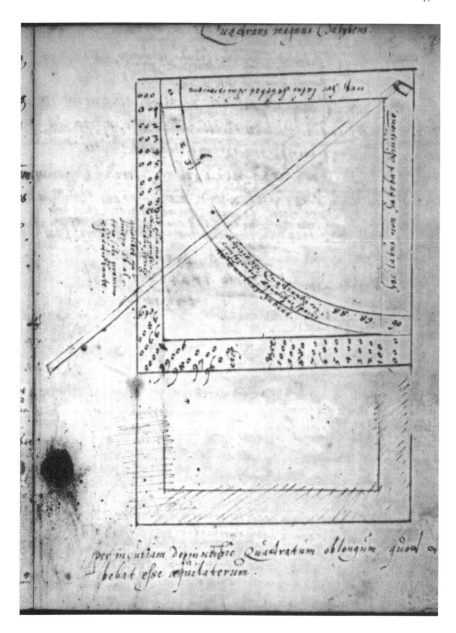

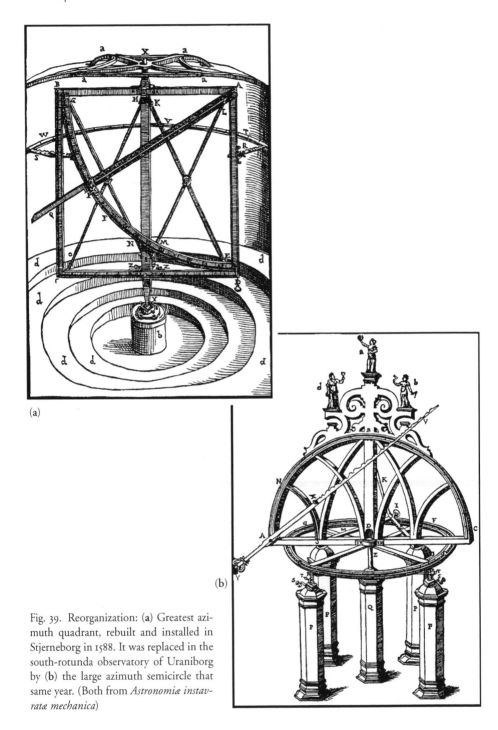

(a)

(b)

Fig. 39. Reorganization: (a) Greatest azimuth quadrant, rebuilt and installed in
Stjerneborg in 1588. It was replaced in the
south-rotunda observatory of Uraniborg
by (b) the large azimuth semicircle that
same year. (Both from *Astronomiæ instavratæ mechanica*)

lectures on refraction by Tycho Brahe and by Elias Olsen Morsing, took notes on the use of Tycho's instruments, and had the opportunity to assist in observational astronomy. This included use of the great steel quadrant, which had been rebuilt in Stjerneborg in 1588 when it was replaced at Uraniborg by a semicircle [Fig. 39].

No wonder students like him came to Tycho's island. The university, with its limited curriculum and lack of research and laboratory facilities, could not compete with Hven, but it could and did cooperate, and the working relationship between the University of Copenhagen and Uraniborg in those years was a fruitful one. No other European university had a similar association with an advanced research institute, though some had better science facilities than Copenhagen. Students and artisans from many parts of the continent were drawn to Tycho's island, and its fame spread far and wide. The dream of King Frederick II had certainly come true, though the king was no longer there to see it.

Tycho Brahe also had a dream: His dream was that Uraniborg would continue, far into the future, as a permanent research institution under the leadership of his own heirs. That dream, too, seemed to have been realized at last.

7

THE SCHOOL OF EUROPE

———

1591–1593

Tycho Brahe's establishment on Hven was embedded in the culture of its
time and place through ties to political, technological, and funding structures, and through the widespread cultural assumptions that Tycho and
his *familia* shared with others. Sometimes the influence of external cultural forces penetrated into the scientific work of Uraniborg to reshape its
structure and objectives for a time. For example, when considerations of
political power created a demand for maps, Tycho Brahe shifted the priorities of his island and adjusted its structures to fit with Anders Vedel's project. When a Danish prince was born, he set aside other matters to compile
a nativity. When the king wanted to know more about weather signs, Tycho assigned Flemløse to write a book on the subject and later devoted the
resources of his press and papermill to publishing it. In these and many
other cases, the political and scientific cultures of Denmark interacted on
Tycho's island as priorities coming from the political culture – which was
the source of funding for science – were able temporarily to reshape the
scientific efforts of Hven.

Cultural differences of another kind penetrated the *familia* when scientists from various countries brought different assumptions to their work
at Uraniborg. The task of integrating them into a single team was as hard
in those days as it was in the late twentieth century, when studies of Jap-
anese and American high-energy physicists revealed striking differences in
leadership styles, decision making, gender relations, recruiting and social-
izing new team members, lifetime career stages, strategies for acquiring
resources and prestige, modes of oral and written communication, procedures for settling disputes, and means of defining the "shared ground on
which the community stands."[1] On Tycho's island, despite a common Lat-

in culture of learning, such differences appeared when scientists, scholars, artisans and others came together from many lands and cultural backgrounds.

THE GERMAN STUDENT

A visitor sailed out to Tycho's island on the evening of 14 May 1591 and stayed for several weeks. His name was Georg Ludwig Frobenius, and soon after leaving Hven on 29 June 1591, he wrote a memorandum to record his experiences. It was an outsider's view by a rather bitter observer who failed to become a part of the *familia*, but it was full of critical insight. This document has lain in manuscript for four hundred years, largely forgotten until I rediscovered it, and it has never before been published.[2]

Frobenius received his M.A. from the University of Wittenberg in 1590, worked for a year as a tutor in Saxony, and decided in the spring of 1591 to visit Scandinavia. "I wanted in particular to visit Tycho Brahe in Denmark," he wrote, "see his instruments and mathematical works, and also work with him myself for a time." He prepared for the journey by obtaining letters of introduction to Viceroy Heinrich Rantzau, Tycho Brahe, and various Danish officials, then set off by foot on the northbound way. He was fortunate to catch a ride in the baggage train of Queen Sophie of Denmark, returning home from the baptism of her first grandchild in Wolffenbüttel. They arrived in Copenhagen on 12 May 1591.

On the following day, I stayed to see the city and to inquire by what means I might best come to Uraniborg on the isle of Hven, where Tycho Brahe maintained his seat and had his celestial observations, since I had not come across any lighter or ship that was traveling there. The next day, I sailed with a lighter from Landskrona in Skåne, first to Landskrona, and when I arrived there towards evening, I immediately hired a small boat with two men who took me and my luggage right over to Hven, which lies two nautical ways from there. I arrived there by night, safe and well, thanks be to God.

I immediately knocked at the castle and asked to be announced to Lord Tycho Brahe, but the porter refused, informing me that his squire had already gone to rest. He did not dare to admit anybody without the squire's knowledge and permission. Consequently, I had to spend the entire night in the field outside the castle, and I was quite hungry because I had not eaten all the previous day. Fortunately, it was a beautiful, clear, and calm night. If it had rained, I would have been drenched, because the island lacked trees to give protection from the rain.

The next day, Saturday, I had myself announced to Tycho Brahe early in the morning, and when he had me brought before him, I presented my letters of recommendation from Dr. Caspar Peucer and Dr. Wilhelm Rudolph Megkbach.

When he had read them and we had conversed at length, he accepted me as a student of astronomy in his Museum.

When I had been there four or five weeks, and had also received my meals at his table, alongside another student from Bergen in Norway by the name of Cort Aslakssøn, he inquired of me through his students whether I would like to remain with him and serve him in the study of astronomy. I declared that I would be pleased to remain with him for one, two, or three years if the terms of employment were tolerable. After a few days, he had me informed of the following conditions: Firstly, that I must sign up for six years with him; secondly, I must promise to reveal nothing whatsoever about his inventions to any person, either during my time in his service or after I might leave him upon the completion of the six years; thirdly, that I would take no notes about his inventions for my own benefit or for my personal use after I had completed my studies with him. My employment would be to serve without hesitation, wherever I could fruitfully be used, in any of his astronomical or pyronomical labors. For this, he would maintain me at his table but would not promise any annual salary or clothing, preferring to grant to me whatever happened to come my way.

I asked for time to consider, and when several days had passed, I declared that the conditions proposed were not acceptable to me. In the first place, I could not commit myself for such a long period of time as six years, because I do not plan to continue studying astronomy but would like to go on in either medicine or law; also, that I wanted to visit foreign places, see the lands and people, and learn foreign languages. I was quite willing to accept the second condition, not to reveal or spread around to any person anything about his inventions or observations. The third condition, however, that I could not at any future time use his inventions to the benefit of my own studies, I found to be entirely unacceptable, since if I learn something and cannot use it in the future, it would simply be thrown away. Moreover, I could not work for board alone, without any fixed salary. I asked him to respond to my reply by changing the conditions, for if he held to the conditions already proposed, I would not accept them, even if it meant I had to leave.

I learned enough about Tycho's wonderful ingenuity, not only through these entirely unacceptable conditions, but also through many other things, in the course of the time I was with him. I was also warned in good faith by his students, especially a German who was born in Westphalia and had been with him for six years without receiving any wages except a few clothes, and who could not get away from him. Since I was not inclined to accept the conditions, I had to watch myself and be careful to seize the first and slightest opportunity to get away.

When I finally came to this conclusion, I decided to use the pretext of the letter of recommendation from Dr. Megkbach to the Royal Viceroy of Holstein, Heinrich Rantzau, which I still had. I would request leave for a few weeks in order to go there and present the letter in person, and at the same time, convey an oral message which I said I was entrusted to bring to him, returning when this had been accomplished. With the help of God, this scheme seemed likely to succeed. Although Tycho raised objections and was not willing to grant me leave, even though I proposed to let my things, which were in a trunk, remain in his castle, he finally did allow me to be away for a time.

After sealing my trunk on all sides as Tycho demanded, I traveled with Tycho to Copenhagen, shortly after Midsummer, when he went to Copenhagen to attend a meeting of the State Council. And as I looked around for a ship, I immediately saw one from Lübeck and rushed over. I had no more luggage than a couple of shirts, a cloak, and handkerchiefs in a black linen satchel.

This remarkable account tells a great deal about the process of negotiation by which young scholars and artisans of many lands were joining Tycho Brahe's *familia* at the height of its fame. By the time Frobenius arrived, Tycho had lost some of his most talented assistants: **Flemløse**, Gellius, and **Morsing**, as well as Groningensis, Riber, Wensøsil, Hemmet, and Niels Colding. According to Frobenius, **Hans Crol** was an embittered man following the death of his son. Gaschitz was running the presses, and papermakers were at work in the mill. **Longomontanus** and **Cort Aslaks-søn** were Tycho's senior assistants, and both had developed a warm personal relationship with Tycho and his family. Longomontanus may have tutored Tycho's sons before Tyge Tygesen went off to Sorø Academy in 1590, nine years of age, to be enrolled among the student commoners, where he was joined in 1593 by his brother, Jørgen Tygesen.[3]

Frobenius found himself in a beehive. Uraniborg had been whitewashed in 1589, and the ramparts had been rebuilt in a new design with semicircular bays on each side and an enlarged pavilion in each bay for meditation, music, or summer dining [Fig. 40]. The orchards were redesigned and the orchard promenades sown with flower-strewn grass, kept short by grazing game and pheasants.[4] Botanical knot gardens were being planted around the house that spring of 1591, presumably in consultation with **Sophie Brahe**, filled with an abundance of life-sustaining herbs: angelica, blessed thistle, bloodwort, elecampane, gentian, juniper, rhubarb, prunella, saffron crocus, sweet flag, wormwood, and valerian for Tycho's medicaments, and many herbs named in the Danish herbals, each known by its signature, all surrounded and contained within boxwood borders in the shape of stars, circles, and squares, the elaborate geometry of a complex microcosmic plan.[5] There was plenty of work to do that spring. Gangs of peasant laborers trudged here and there each day, returning to Tuna at sunset. Innumerable servants and assistants ran in and out, all day long. Gardeners and boon workers labored by day to execute the complex planting plan. At night, observers of the heavens were busy in Stjerneborg and Uraniborg when skies were clear, and assistants sometimes labored through the night by the glowing furnaces of the chemical laboratory ["L" in Fig. 40b].

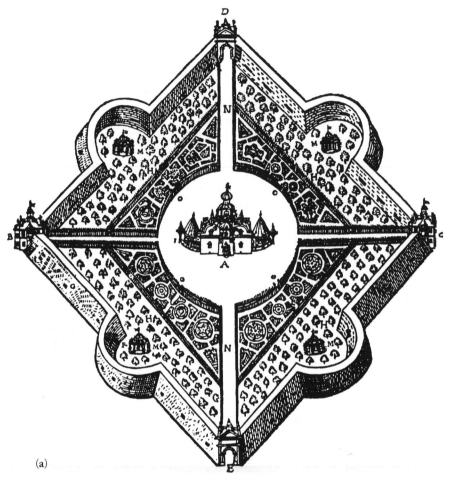

(a)

Fig. 40 (*here and facing*). Renovations: (**a**) Uraniborg ca. 1591. Frobenius must have wait-ed outside the eastern gatehouse (E). The rebuilt ramparts (G), replanted orchards (H), redesigned and replanted knot gardens, garden pavilions (M), and whitewashed house (A) were all quite new. (**b**) The east facade of gleaming white Uraniborg, showing win-dows to the Winter Room (B), a spare bedroom (C), the royal Red Chamber (D) and Blue Chamber (E), with the large green and floral-decorated Summer Room behind, the yellow octagonal chamber (α), eight garret rooms (χ) for *famuli* and the large octagonal

Frobenius arrived at this busy place and presented his letters of recom-mendation, written in Latin. One was from Caspar Peucer, Tycho's friend and former teacher.[6] Tycho read the letters, and they conversed at length, probably in Latin, which was the working language of scholars on Hven. Tycho himself and several others, like Crol and Gaschitz, could speak Ger-man. Dutch, Low German, and Norwegian were also heard in the halls

(b)

Caption to Fig. 40 (*cont.*)

room (ε) under the dome with a clock and bell, surrounded by an open-air gallery. The south observatory contained the azimuth semicircle (O), equatorial armillary (N), trigonal sextant, and other instruments, and it was heavily used. The north observatory contained mainly instruments that were impractical or of historical interest, including Tycho's great azimuth triquetrum (R), another equatorial armillary (δ), his first steel sextant, and, eventually, Copernicus's triquetrum. In the well-lit chemical laboratory (L) below the museum were sixteen furnaces built of fire-resistant Norwegian soapstone, including three bath heaters, a digesting furnace, four large and two small athanors (ash-lined calcination furnaces), two distillation furnaces, two reflecting furnaces of different types, one furnace with a large bellows, and one furnace placed apart from the others. Pantries, larders, and storage rooms filled the remainder of the cellars, with the well (5) in the north rotunda. (Both from *Astronomiæ instavratæ mechanica*)

of Uraniborg, but the most common language was Danish. Frobenius did not know Danish, so he could never establish a cordial relationship with Kirsten Jørgensdatter, Tycho's family, nor most of the others on Hven. From the very beginning, he must have missed a good deal of the casual conversation. Such incomprehension did not bother some people, but apparently it did bother Frobenius.

Tycho Brahe accepted Frobenius as a student at Uraniborg. This meant that Frobenius received free room and board, with an assigned place at Tycho's own table. He accepted Tycho's hospitality for four or five weeks, attending lectures, working no doubt with the others, and becoming familiar with all the marvelous facilities of Tycho's island.

At that point, Tycho Brahe decided it was time to formalize the relationship by offering Frobenius a written contract of service. This had become standard practice for the university-trained *famuli* at Uraniborg, and Frobenius seemed willing to stay. One of the senior assistants, either Longomontanus or Cort Aslakssøn, negotiated on Tycho's behalf. Cort Aslakssøn in particular was a skilled negotiator: In return for his own service for a specific term, Tycho had agreed to help him obtain the coveted *stipendium regium,* which paid for years of study abroad and gave preference to a professorship in the University of Copenhagen.[7]

Most contracts for service at Uraniborg were for three years, but the term was negotiable. Cort Aslakssøn in 1590 apparently signed on for only two-and-a-half years. Hemmet and Niels Colding may have signed two-year contracts in 1588, though Hemmet left with Tycho's blessing after six months. Others like Jacob Lemvig in 1584 and Wensøsil in 1587 signed three-year contracts but negotiated earlier releases. Riber seems to have signed on for four years and served the full term. Tycho was willing to negotiate the length of service, and he always considered the lifetime career opportunities of his *famuli,* rather than the letter of the contract, because he wanted to build enduring networks of loyal former *famuli.* Many scholars served for terms of one, three, or six months and probably did so as guests without a written contract. Some artists and craftsmen like Gemperle and Crol served for longer periods of time, and so did a few scholars like Flemløse (eleven years), Morsing (six), and Gellius (five). All of these men, however, had arrived before written contracts were used for scholarly service on Hven. Longomontanus had been at Uraniborg for around two years when Frobenius arrived. He would eventually remain for eight years, regularly renegotiating the terms of his service.

The offer of a six-year contract to Frobenius was Tycho's opening gambit. If Frobenius accepted, well and good: Uraniborg needed continuity, and the core staff was a bit depleted. If not, Frobenius could make a counteroffer and negotiations would continue. He could propose a shorter term, or rewards beyond room and board, or other modifications, and he did. There was no set form of contract. In fact, however, Frobenius was not willing to negotiate in good faith. He had no intention of remaining on Tycho's island.

Frobenius pretended to take the bitter remarks of Crol at face value, though he must have heard about former *famuli* like Arctander, Colding, Oddur Einarson, Flemløse, Gellius, Hemmet, Jacob Lemvig, Riber, and Stephanius, who had left Uraniborg with Tycho Brahe's continuing favor and patronage. For some reason that he did not explain, Frobenius preferred to skip out and leave his trunk behind, in order to enter the service of Viceroy Heinrich Rantzau.

Diverse personalities and cultural assumptions needed to mesh so that daily life could proceed smoothly on Tycho's island, and this was not always easy. Frobenius was not the only one who had trouble fitting into the work culture of Uraniborg. For some reason or another, a number of Germans who came to Hven – **Gemperle**, Ursus, **Crol**, Caspar, Frobenius, probably some of the printers, and later, **Kepler** in Prague – had more or less serious, sometimes frantic, at times nearly hysterical disagreements with Tycho Brahe, but Dutchmen, Danes, Norwegians, Icelanders, and Englishmen did not. What was wrong with the Germans? Or rather, what was wrong with Tycho Brahe as seen through German eyes?

The Germans came from a tense region torn by conflict. In many parts of central and southern Germany, a larger sense of community had been shattered by religious and dynastic wars. Moreover, German cities were large and highly sophisticated. Many of them were virtually independent city-states. Urban culture flourished in Germany, but it flourished in a kind of vacuum, distant from the countryside and the rural classes of peasants and landed aristocrats. Patronage networks operated differently. Middle-class Germans were not always used to entering the *familia* of great magnates; patrons of science and technology outside the German universities tended to be urban patricians and territorial princes. German scholars and artisans struggled hard to achieve status, prestige, security, and office in this fragmented and particularized world, and professional credentials took on great importance because they served as portable verification of worth: university diplomas, journeymens' and masters' certificates, letters of recommendation. Nevertheless, disputes were common, and the procedures for settling them were complex, in part because there was a maze of overlapping legal jurisdictions. It paid to be extremely careful. In this context, the German lands had developed a *culture of elaboration.*

Tycho Brahe knew Germany. He had lived there for many years as a student, and he spoke the German language fluently; but the Germans did not know Denmark. Scandinavia was developing a *culture of simplification.* One king ruled all of Denmark and Norway. There was one law, and it was comparatively simple and direct, so there were few lawyers. There were no walled cities, and even Copenhagen was a small provincial town by German standards. The key figures in the kingdom were the great landed aristocrats and conciliar families. Men of learning flocked to the castles and manors of magnates like Tycho Brahe to find their patrons.

This was in many ways a foreign world to the Germans, and there was something about Tycho Brahe that made them suspicious. He was imperious, remote, seemed both threatening and friendly at the same time, made jokes they did not understand, surrounded himself with hordes of aristocrats and servants and foreigners chattering incomprehensible languages. Tycho was quick, deft, and skilled in everything he did, and he was a bit impatient. He was rich and powerful. Their German background told them to beware. They misread all of his social signals. Their first thought was, "Who could trust such a man?"

When the Germans negotiated, they tended to be extremely legalistic, sometimes even devious, hiding their true agendas, but at the same time wanting everything spelled out to the last jot and tittle. That was the way it was done at home: It was part of their culture of elaboration. The Danes, however, often operated on the basis of informal mutual understandings and were willing to put personal considerations before the letter of the contract. This was part of their culture of simplification. Danes were a bit amused by what they saw as the petty legalism of the Germans and the pomposity of all the German certificates and diplomas. Moreover, all of that seemed quite inappropriate among a group motivated by Neoplatonic *amicitia.*

To the Germans, on the other hand, the Danish way of doing things smacked of fraud and deception. Frustrated by the Danish old-boy system, they tended to overreact. Ursus, Kepler, and perhaps others threw terrible tantrums. The Danes seemed both amused and repelled by these outbursts. They had their hotheads, too, of course, but they did not condone outward displays of emotion. They seemed to consider the Germans to be too emotional, and therefore, out of control and "half-cracked." Tycho Brahe, however, seemed to understand these strong German reactions, and he could be extremely patient, continuing to negotiate through some incredible scenes. Sometimes there were other complications, as when married men like Crol and Kepler and Johannes Müller found it hard to fit their families into an alien *familia.* Thus German cultural patterns of particularism, legalism, elaboration, and outward emotional display could make it hard for men like Ursus, Frobenius, Crol, and Kepler to work with the subtle Scandinavian aristocrat and man of honor, Tycho Brahe. Obviously, this was not true of every German who came to Hven. The otherwise temperamental **Wittich,** the courtly scholar **Rothmann,** and the old court artisan **Jachenow** were able to get on reasonably well with Tycho Brahe.

So were the Dutch. They were quite different from the southern Germans. Dutchmen were straightforward and commercially minded. They were out to make money, and they came to Tycho's island with a skill or talent to sell, or because they wanted to market science: globes, maps, books, instruments. The northern Netherlands, locked in warfare with Spain, was developing a strong sense of national solidarity, uniting all social classes in the common cause of independence.[8] Differences of class, region, and religion did not divide them to the extent that it did the Germans, and Dutch burghers seemed to move easily among aristocrats. Three of Tycho's brothers had served in the Dutch wars, riding in armor with William the Silent and Jan of Nassau, and Tycho sympathized with the Dutch cause. He also liked the brusque Dutch openness. When Dutchmen came to Hven, they tended to go right to the heart of the matter: I have this globe; I want to mark your stellar positions on it so I can make celestial globes and sell them. If Tycho smiled and said no, they shrugged, sat down to enjoy the fare and conversation, and when they left, asked him to let them know if he changed his mind. Sometimes he did.

The Danes also had a strong national sense of community, based not on resistance to tyranny but on a solid religious consensus, an integrated agricultural economy, cultural homogeneity, and widely shared social assumptions.[9] Networks of patronage were strong, and paths of upward mobility were clearly marked. These vertical social linkages conveyed a sense of justice that brought the lower classes into the national consensus. If a peasant lad was bright, his parish pastor might tutor him and send him off to Latin school or recommend him to the local landlord. Such a lad could end in the university, supported by a noble patron or local bishop, and might even serve on Tycho's island: This was the story of Flemløse, Stephanius, Longomontanus, Cort Aslakssøn, and others. In this and many other ways, Danish society was characterized by simple but effective relationships based on mutual trust and reciprocity that cut across class lines. Formal certification was less important than who you knew and how you acted. The Dutch and Norwegians seemed to understand this kind of casual, straightforward social culture, but many Germans did not.

FAMILY AND FAMILIA PROBLEMS

Of course, no cultural system has ever been perfect, and no society has done away with human flaws. Life was not all roses for Tycho Brahe, despite his soaring international renown. He had plenty of problems with

his staff, and not all of it with Germans. In 1589, his dwarf jester tried to
flee and was beaten. The following spring, two of his trusted servants se-
duced a virgin and had to be dismissed. There was a fire in the kitchen in
August of 1590. During the autumn of that year, Tycho incarcerated his
tailor for three days, and the man later fled the island by night. Around
the same time, the cook was honored for his good work.

A long legal battle with one of his gentlemen tenants, Rasmus Pedersen of the manor of Gundsøgaard, demanded much of Tycho's attention
in 1591. At one stage, he imprisoned Pedersen in the cellars of the Uraniborg complex. That summer, the governor of Bergen brought suit against
Tycho's Norwegian bailiff, Christopher Pepler, for violating the royal trade
monopoly, and the Regency Council reprimanded Tycho Brahe for allow-
ing the Chapel of the Three Holy Kings in Roskilde Cathedral to fall into
disrepair.

Toward the end of 1591, Hans Crol died. Tycho seems to have followed
his remains to Copenhagen for burial, and he praised the memory of this
versatile goldsmith, instrument maker, and observer. A servant named
Theodorus died of the plague during the spring of 1592, striking fear into
the islanders, and two servants fled that summer.[10] Plague, sudden death,
private incarceration, and fleeing servants were facts of life in the late sixteenth century and were not unique to Uraniborg, but they did create situations that demanded the attention of the lord. He could not spend all of
his time publishing books, supervising his staff, or peering at the stars.

Tycho continued to work on the problem of his children's future. He
had done all he could to keep the leadership of Uraniborg in the family,
and he was providing his sons with a solid education. Now he turned to
the vexing question of inheritance and began taking steps to name his children as his heirs. This entailed securing the consent of his mother and siblings, since they were heirs to his landed estate according to the royal ordinance of 19 June 1582. It also entailed liquidating his share in Knutstorp,
which could not pass to his children because only nobles could hold privileged landed estates. Therefore, it entailed shifting all of his assets into
cash, securities, and personal property, which his children were allowed
to inherit. Finally, it entailed appointing legal guardians to ensure that
the provisions of the settlement were carried out in the event of his own
demise.

Around 1590–1, Tycho Brahe drafted a document that named Falk
Gøye, Erik Lange, and his two youngest brothers, Jørgen and Knud Brahe
[Fig. 41a], as legal guardians of his children with Kirsten Jørgensdatter in
the event of his own death.[11] This document revealed a clear understand-

ing of the law. It laid down all the terms of the wardship and provided
that the guardians would serve without remuneration. It stated that Tycho
Brahe had named his children as his heirs, and that he had done so with
the permission of his mother, brothers, and brothers' wives. However, a
clean copy of the document, signed and sealed, has never been found. It
may simply have been lost to history, but its absence may also mean that
the document was never promulgated. There could have been several
grounds: Erik Lange's flight in 1592, the unexpected death of Falk Gøye
in 1594, or failure to obtain the approval of all parties.

The ties within this group had become very close. Jørgen Brahe was a
near neighbor to Hven as Lord Lieutenant of Landskrona 1588–91. Knud
Brahe had married Erik Lange's sister, Margrethe Lange, in 1584. **Sophie
Brahe**, widowed in 1588, had fallen hopelessly in love with Erik Lange,
who was himself desperately enamored with the dream of making gold by
alchemical transmutation and was rapidly burning up his fortune in pursuit of that love.

Sophie Brahe and Erik Lange brought an element of chaos into the family circle at the very time when Tycho Brahe was trying to establish order.
Erik Lange borrowed money wherever he could find it. He got large sums
from Falk Gøye, who was also stricken by the fever of transmutation. In
a desperate attempt to satisfy creditors, Erik Lange signed over his estate
of Engelsholm to Knud Brahe and Margrethe Lange, but he could still not
meet his obligations. He was placed under house arrest for his debts in
1590, then released. That same year, he and Sophie Brahe were betrothed.
In Danish law, betrothal was as binding as marriage. All of her siblings except Tycho opposed it.

Sophie Brahe was living in grand style at her late husband's seat of Eriksholm Castle [Fig. 41b], surrounded by moats, with four high castle wings
around a courtyard and massive towers at two corners. Beyond the moats
were elaborate Renaissance parterres of the ornamental and botanical gardens and orchards that she herself had planned, the finest gardens in all
of Scandinavia, according to Tycho. From her laboratory in the garden,
Sophie dispensed gifts of spagyric medicaments, elixirs, cosmetics, and
"hermetically" preserved fruits. She cast nativities; performed mathematical calculations; compiled extensive aristocratic genealogies on the basis of
solid historical research; and gained a reputation as one of the most learned
women of her day.[12] The estate, however, was held in wardship for her
young son, and the wards held it with a tight hand: Erik Lange could not
pillage her son's wealth to pay his debts, though creditors closed in on her
own substantial fortune.

Fig. 41. Family ties: **(a)** The family of Otte Brahe and Beate Bille. Detail of memorial painting in Kågeröd Church, artist unknown. Behind Otte Brahe (1518–71) are his sons, Tycho (1546–1601), Steen (1547–1620), Knud (1555–1615), Axel (1550–1616), and Jørgen Brahe (1554–1601). Behind Beate Bille (1526–1605) are her daughters, Lisbet Brahe married Gyldenstierne (1545–63), Maren (born 1549), Margrete Brahe married Skeel (1551–1614), Kirsten (1552–66), and Sophie Brahe married Thott and Lange (1559–1643). (Photograph by Viveca Ohlsson, courtesy Landskrona Museum) **(b)** (*facing*) Eriksholm Castle (now Trolleholm). (From Hellstedt 1990)

162

(b)

In 1592, Erik Lange sailed to Hven. He was at his wit's end, his money gone and debts immense. Tycho Brahe helped him to flee the realm and so escape his creditors.[13] He went to Hesse-Kassel, where he continued to run up debts, then fled from place to place across Europe. Sophie Brahe, helpless and distraught, flitted between Uraniborg and Eriksholm. In these circumstances, no wonder Tycho Brahe had second thoughts about naming Erik Lange a guardian of his childrens' inheritance.

There were other cares. Tycho was negotiating with his oldest brother, Steen Brahe, about the sale of Knutstorp Castle and estate. Tycho and Steen were only one year apart and looked almost like twins with their sturdy physiques, immense moustaches, fair hair, and commanding manners [see Fig. 41a]. They had inherited Knutstorp jointly. Steen Brahe had also inherited Bregentved Manor, acquired Næsbyholm Manor by marriage to a Rosenkrantz heiress, and when she died, acquired Barritskov Manor in 1590 by his second marriage to Kirsten Holck. Steen Brahe was Lord Lieutenant of Helsingborg Castle and a member of the State Council. He was the great magnate and the domineering political figure whom King Frederick II had once hoped that Tycho Brahe would become. These two mighty brothers negotiated on good terms. Steen agreed to buy Tycho's half of Knutstorp and allowed his brother to retain the title Tycho Brahe *of Knutstorp*. Steen Brahe paid the full purchase price by the year 1594.[14] This gave Tycho a large amount of cash to build his facilities in Copenhagen, pay off debts, and invest in ways that would benefit his heirs, loaning out several thousand dalers at interest.

Finally, the family problems were being resolved, and life was good on Tycho's island. Its festive side was reflected in a letter to a German friend in March of 1592, which concluded:

Farewell, live well, and drink to my health! I drink to you today, Sunday, on which day nearly everyone here and elsewhere drinks (an activity sacred and pleasing to Bacchus) one or two tankards filled to the brim with the nectar of Lyaeus, a drink which is also bacchanalian, though not philosophical and least of all theological, although this day ought to be holy to Christians. Nonetheless, I drain a tankard to your good fortune, which the great Bartholinus [a pun on Magnus Bartholinus, i.e., Mogens Bertelsen Dallin – see Chapter 8], who is seated here beside me, will gladly verify. I hardly believe he learned this art from you, because it is not entirely unknown in his fatherland of Sjælland. Now it is your turn to repay us, like for like in one voice, also to the tone of music, just as we dedicate these toasts to you to the sound of trumpets, flutes, and strings, and with lively song. Again and again, live well, drink to us, and fare well![15]

Thus the Sunday passed at Uraniborg on Tycho's island.

Fig. 42. Young King Christian IV. Engraving by Crispin de Pas, 1595.
(Courtesy Det Nationalhistoriske Museum på Frederiksborg, Hillerød)

THE YOUNG KING

One of the high points in these busy years came in the summer of 1592,
when King Christian IV [Fig. 42] visited Tycho's island and was shown
all the wonders so fascinating to an intelligent, well-educated fifteen-year-
old. During a long carriage ride from Copenhagen to Frederiksborg that
spring, Chancellor Niels Kaas had described Uraniborg in a manner that
had intrigued the young monarch. Throughout the month of April, Christian had pleaded with his guardians for permission to visit Hven. Just
when the chancellor finally gave his consent, Cort Aslakssøn brought word
that plague was on the island, and the visit had to be postponed. Finally,
on the last day of June, Tycho Brahe came to court and informed the king
that all was in readiness for a royal visit to Hven.[16]

On Monday, 3 July 1592, young King Christian IV and his court sailed to the island of Hven. There they were received by the lord of Uraniborg, Tycho Brahe, accompanied by his *familia,* undoubtedly clad in the Danish royal colors of red and gold. **Sophie Brahe** was at Tycho's side, and Kirsten Jørgensdatter presumably stayed in the background. The king was accompanied by Admiral Peter Munk; Regent Jørgen Rosenkrantz; his guardian, Councillor Hak Ulfstand; and many other courtiers. Inside the threshhold of Uraniborg, as the fountain played in the background, servants with a silver ewer and basin poured water over the hands of the royal guest, and he dried them with a linen towel. Then the lord of Uraniborg drank a toast of welcome and handed the beaker of wine to His Majesty, who returned the honor. These were the ancient Danish ceremonies of greeting, rooted in centuries long past.

Tycho Brahe and his staff showed their royal guest all the sights of Uraniborg: its ramparts, pavilions, orchards, botanical gardens, aviary, cellars and towers, Stjerneborg, the print shop, and all the rest. The young king was fascinated with a gilt planetarium in the museum that showed the daily progression of the heavens, the course of the sun, and the changing phases of the moon. Tycho Brahe presented it to him as a memento of the visit. The weather was splendid: scattered clouds in a blue sky and a breeze from the southwest. There was a banquet in the Summer Room, with vistas of the Sound and surrounding cities. Set after set of courses appeared, with choice wines, fanfares, contrapuntal song, jovial and learned conversation, and the antics of Tycho's jester. His Majesty presented Tycho Brahe with a massive golden chain bearing a medallion with a miniature portrait of the young king. In the economy of honor, chains of solid gold were princely gifts conveying grace and high status, and they also had a substantial market value.[17] Tycho wore two such chains from the king's late father, King Frederick II, as the new chain was ceremoniously placed over his head. From that day forth, on great occasions, he wound all three heavy golden chains, each worth a fortune, around his shoulders to display their royal pendants and the Order of the Elephant on his breast.

Reluctantly, the young king was escorted from the island by his guardians that afternoon, but the memories would remain with him for years to come. For his part, Tycho Brahe could relax that evening, confident that his patronage strategy of the Regency years had borne delicious fruit.

Three days later, Holger Rosenkrantz, son of the Regent, visited Uraniborg with his German preceptor, Master Daniel Cramer. Tycho established a lasting friendship with this brilliant young kinsman, a rising star of the coming generation. In October of 1592, another party of aristocrats spent three festive days on Tycho's island: Tycho's cousins, Birgitte Jensdat-ter Bille and Anne Maltesdatter Sehested with her husband, Ove Lunge, and three daughters around the age of Tycho's children, as well as Chris-tian Friis of Borreby and his wife, the learned Mette Hardenberg, a niece of Anne Hardenberg. Christian Friis was ten years younger than Tycho and rising rapidly in the ranks of government service. Here was another link to the coming generation of leading Danish aristocrats.

The glory of the world, however, is transitory, passing quickly. Merely one year later, Nemesis, the goddess of fate, appeared to punish excessive pride and erase the good will built by the young king's visit and years of loyal service. Tycho's service, though loyal, had not been impeccable. He held an extensive estate attached to the Chapel of the Three Holy Kings in Roskilde Cathedral and derived a large annual income from adminis-trating and maintaining the fief. Like many aristocratic vassals, he took the obligations of administration rather lightly at times, and that was what led to the trouble. Since 1591, the Chapel of the Three Holy Kings had been in bad repair: The roof had begun to leak. Rain threatened to drip down on the alabaster-and-marble sepulchers of King Frederick II and his father, King Christian III. The roof seemed on the verge of collapse. Tycho Brahe, however, had ignored repeated royal requests for repairs.

In August of 1593, young King Christian IV visited the cathedral. When he saw the sad state of the chapel, he was not amused. The young king dictated a command to Tycho Brahe: Renovate the chapel immediately or the king would hire a master builder to do it at Tycho's expense. A post-script demanded Tycho's reply by the same courier. The king meant busi-ness.

Tycho Brahe was a busy man. He had many irons in the fire. Of course, he knew that one should never be too busy to cultivate the favor of the crown. In the days of King Frederick II, he had done so assiduously: He had been young, eager to make a name in the world, and the king had given him the opportunity to do so. Now, Tycho Brahe was renowned throughout Europe, and the king was a child. Power lay in the hands of the Regency Council, and Tycho had their support. He was busy. He ne-glected his obligations as a vassal of the crown.

THE EXPANDING FAMILIA

New faces kept turning up in the bustling *familia* of Uraniborg. In 1592, a bright, sixteen-year-old Jacob Fabricius from the University of Rostock came to Hven and remained for four years. Oluf Brender from Copenha-

gen, who had ties to **Sophie Brahe** and the Sascerides family, may also have
entered the household. Christopher Jensen, the son of Bishop Jens Nilssøn, arrived in March of 1593 from Oslo to spend a year. Johannes Isaksen Pontanus, son of the Dutch consul in Elsinore, came later in 1593 and
spent three years in Tycho's service; he arrived urbane and cosmopolitan
at twenty-two, already a poet, historian, and chemist with an M.A. from
Leiden. Hendrik Floris van Langren spent six weeks as Tycho's guest in
1593 and learned a great deal but was no more successful than his brother
in obtaining an advance copy of the star catalog. Adriaan Metius, son of
the burgomaster of Alkmaar, arrived around 1594 and stayed for a few
months; he knew Pontanus from their student days in Franeker. All these
young men came from urban patrician families in Denmark, Germany,
and the Netherlands. A Dutch nobleman, **Frans Gansneb genaamd Tengnagel van de Camp**, joined the *familia* early in 1595 and remained for a
total of six years. Like the others, he was bright, well-educated, and had
a promising career before him. Toward the end of 1595, a young Dutch
craftsman, **Willem Janszoon Blaeu**, arrived with a blank celestial globe to
copy Tycho's unpublished stellar positions. He remained for six months,
marked the positions from Tycho's great globe or copied them from a manuscript, with the astronomer's permission, and possibly also sketched that
globe's iconography. In addition, he found time to learn Tycho's methods
of instrument making, astronomical and geodetic observation, surveying
by triangulation, cartography, and printing, and to make a map of Hven.

Around that same time, Tycho Brahe's only known female assistant
besides his sister, Sophie Brahe, joined the household. She was **Elisabeth
Paulsdatter**, whose father, Paul Thiisen, was an Elsinore apothecary, originally from Schleswig-Holstein, and whose mother's family came from
Amsterdam. Her father was a regular visitor to Uraniborg and probably
supplied Tycho Brahe with apothecary goods. Elisabeth was literate in
German, possibly also in Danish, Dutch, and other languages, and she was
trained in chemistry. She may have been related to "Doctoress" Barbara
Thiisen, an Elsinore alchemist with ties to Tycho's circle. Tycho took Elisabeth into his *familia,* and she began to assist in the laboratory at Uraniborg. During the summer of 1596, a young Danish scholar named Laurids
Isaksen Sommer, son of a respected Skåne clergyman, also spent some
time at Uraniborg.

Tycho's island was becoming a training ground for scientific and schol-
arly elites from the University of Copenhagen and throughout northern
Europe – Germany, the Netherlands, Denmark, Norway, and Iceland.
Meanwhile, there continued to be problems and tragedies in the house-

hold staff. A maidservant named Tornilla died in the spring of 1593, and another servant, Johannes, died that summer. The butler and cook had to be disciplined by manorial justice in August of 1594. These bare facts were noted in the margins of the meteorological journal without explanation.

Former *famuli* in many parts of Europe remained in touch with Tycho Brahe. **Flemløse** continued to record observations of latitudes as he traveled around Norway, and he probably had a hand in Norwegian maps of unprecedented accuracy that were sent to Tycho around 1594. One was a detailed sketch map of the southwestern coast of Norway, with notes on the reverse side indicating a tie to Bishop Anders Foss.[18] Tycho also received a map of Nordfjord in three colors, dated 1594 and by far the most accurate of any sixteenth-century Norwegian map. It must have been made in consultation with Tycho's bailiff, Christopher Pepler, and the cartographer may have been Flemløse, Pepler, or another from their circle.

In Ribe, Iver Hemmet published almanacs in the Danish vernacular in 1592, 1593, and 1594. Meanwhile, Christian Riber used his portable radius to observe a lunar eclipse of 1592 in Wittenberg and a solar eclipse of 1593 in Zerbst. His friend Ægidius Lauridsen published an oration on the solar eclipse, which lauded Tycho Brahe, while Riber sent his observational notes to Uraniborg.

Gellius Sascerides had arrived in Italy in 1589 with his young charge, Tycho's wealthy nephew, Albert Skeel. They found a whole colony of Danish students in Padua, most of whom had visited Uraniborg before traveling abroad, including Jacob Ulfeldt the Younger and others who had just returned from the Holy Land and Egypt with mummy lids, antiquities, and fascinating tales.[19] Several of them returned home in 1591 and brought letters to Tycho's island. Gellius studied medicine in Padua while Albert Skeel studied law. They visited Venice and Bologna and also traveled to Rome and Naples. Gellius got to know the Bolognese astronomer, Giovanni Antonio Magini, and introduced him to the theories and methods of Uraniborg. Tycho sent autographed copies of his printed works for Gellius to present as gifts to Magini, and soon, the two of them were corresponding directly. Like Tycho Brahe, Magini was a great admirer of Copernicus as mathematical astronomer but rejected his heliocentrism.[20] The correspondence with Magini and reports of travelers to the Levant fired Tycho's imagination. In 1592, he developed a plan for an affiliate observatory in Egypt under the leadership of Gellius.[21] Tycho would have Uraniborg, Stjerneborg, the observatory in Copenhagen, and Alexandria in Egypt, besides his ties with Kassel and scattered observers like Riber, Gellius, and Magini throughout Europe. This would allow coordinated simultaneous observations from many locations, never attempted by astronomers before Tycho Brahe. Magini used his influence to ensure the protection of the Venetian Republic, with its vast trading interests in the Levant. In the long run, however, the political situation did not allow the Egyptian plan to be realized.

In 1593, Pontanus visited Magini before coming to Uraniborg. By then, Gellius was back in Denmark, where he found Tycho's island of Hven to be a veritable school of Europe, swarming with learned men and women and skilled artisans from many lands.

8

MAGDALENE AND CALUMNY

———

1593–1597

Kinship was a key to continuity in the late sixteenth century, but kinship was not simply a matter of common ancestry: It could also be established by fostering and marriage. Both could affect the transmission of property and office from one generation to the next, and both were arranged. Tycho Brahe's four daughters were coming of age, so he needed to look around and think about their marriage arrangements.

One candidate was Gellius Sascerides, who had received his M.D. degree in Basel 1593 and returned home as a young academic lion. Everybody expected great things of him. He spent eleven days on Hven in May and June of 1593. Gellius was the son of a prominent Dutch refugee poet and scholar who had been professor of Hebrew in Copenhagen since Tycho's student days. Gellius had served five years at Uraniborg, and his brother David had also spent time in Tycho Brahe's *familia*. Tycho had helped Gellius obtain the *stipendium regium* and a position as preceptor for Albert Skeel, and Gellius had been Tycho's link to Magini and the Egyptian plan. Here he was, thirty-one years of age, confident, widely traveled, with a medical degree in hand, and ready for great things to happen. The younger *famuli* were eager to meet him, and Kirsten Jørgensdatter was pleased to see him. So were her daughters, who had grown up in his absence: Magdalene was nineteen, Sophie fourteen, Elisabeth thirteen, and Cecilie twelve.

Dr. Gellius had many friends in Copenhagen. His widowed father still taught at the university, and his brother, David, apparently lived at home. His sister, Agathe, had died in 1589, but her husband, Jon Jacobsen Venusin, was pastor of Holy Spirit Church.[1] Venusin, a Ramist and crypto-Calvinist, shared with Gellius an interest in alchemy and Hermetic lore; He was said to have a ring that could summon a genie to his command.

Many of their friends had arcane and medical interests. Mogens Bertelsen Dallin had also studied at Padua and other foreign universities; he was a clever poet and master of amazing Vitruvian devices, a strange, lively personality.[2] Niels Krag was a bit of a contrast, suave and diplomatic, at times even enigmatic, and now the University of Copenhagen's first professor of history.[3] His brother, Dr. Anders Krag, an avid Ramist and Paracelsian, had a medical degree from Montpellier and also taught at the university.[4] These bright, cosmopolitan young men were the rising academic stars of Denmark, and in their circle of *amicitia,* Gellius found his natural place. All of them were frequent guests at Uraniborg, but most of them were already married.

The hospitable manor of Uraniborg was a house full of rich and well-bred daughters. They had no prospect of marrying into their father's noble class, but the academic elite of young physicians and professors enjoyed high social status and shared late-Renaissance culture with learned noblemen. There had been some marriages between the academic elite and the lower rungs of the nobility, though not with great aristocratic houses like the Brahes. Niels Krag's wife, for example, was Else Mule, and her family had noble status, though both Niels and her previous husband were commoners.[5] The best prospects for Tycho's daughters lay within this academic elite. The trick would be to arrange their matches in such a way that the daughters kept their dignity and preserved their aristocratic standard of living.

It was an accepted fact that parents had the responsibility of arranging marriages for their children. Aristocratic Danish parents normally consulted wide networks of relatives and friends in matchmaking.[6] All around them, they could see successful marriages that had been arranged, and they knew that love could grow within marriage; it did not have to come first. A prospective bride or groom might veto a repugnant choice, but children were generally expected to follow the advice of their elders. Tycho Brahe and Kirsten Jørgensdatter had not done so, and they knew from bitter experience the social barriers that faced those who followed the dictates of the heart. They could see the agony love had brought to **Sophie Brahe,** who constantly sought succor at Uraniborg while her beloved Erik Lange fled from place to place in foreign lands. Falling in love was not a fate Tycho and Kirsten wished for their own children. What they wanted was a good match.

Gellius Sascerides began to loom as a prime candidate. He and his brother David visited Uraniborg in September 1593. During the new year, Gellius came again in January and then in May. His father was ill and passed away at this time; otherwise, he might have been on Hven more often.

Chancellor Niels Kaas, Tycho Brahe's stalwart supporter at the head of the Danish government, died on 29 June 1594 and was buried in Viborg Cathedral. Because of his learning and intelligence, his benevolent influence on the patronage of science and learning, his unwavering support for Tycho's island, and not least of all, because of a lifetime of friendship, his death was lamented by Tycho Brahe, who would miss his guiding hand in crown affairs. Manderup Parsberg, with whom Tycho had once fought a duel, assumed the duties of Royal Chancellor; they had long since patched up their differences and become good friends, even kinsmen, since Tycho's brother, Jørgen Brahe, had married Manderup's cousin, Ingeborg Parsberg, in 1580.

COURTSHIP

Courtship and marriage customs were similar in all propertied social classes of late sixteenth-century Denmark, from royalty and aristocracy through the urban middle classes and clergy to prosperous peasants. Differences were mainly in scale. In his pioneering masterpiece of social history, Troels Frederik Troels-Lund described these customs in great detail. First came preliminary soundings through informal conversation within networks of friends and relatives. In the case of Tycho's eldest daughter, Magdalene, this "came of itself without anybody taking the lead," as Tycho put it, sometime before Michaelmas, or around September of 1594.[7]

If a young man's informal inquiries gave reason for hope, he might take the big step of becoming a suitor. He would normally ask trustworthy friends to bring his proposal to the parents of the prospective bride.[8] Gellius had a friend in Mogens Bertelsen Dallin, who had just been appointed superintendent of Holy Spirit Hospital in Copenhagen, and Dallin spoke with Tycho Brahe in Copenhagen, then sailed with him to Uraniborg on Saturday, 14 September 1594. A few days later, Dallin and Gellius came to Hven together with a formal proposal of marriage.[9] The date was Thursday, 19 September 1594.

Normally, the suitor stayed in the background. His spokesman delivered a long oration peppered with biblical allusions, praising the suitor and his distinguished lineage, and concluding with a proposal of marriage. In this case, Mogens Dallin spoke for the hand of Magdalene Tygesdatter on behalf of Dr. Gellius Sascerides. The suitor and his party then took seats and the prospective bride's family stood up while her father or spokesman replied with an equally long speech. The spokesman for Magdalene's party may have been **Sophie Brahe**, who went on to negotiate on their behalf. She would have avoided a direct reply and requested a fortnight to think it over.[10] This way, the proposal was binding on the suitor, but the bride's family remained free to consider options.

The stage was now set for negotiation.[11] Prospective spouses generally came from the same social background, and the aim was to ensure the future position of the bride while striking a reciprocal balance that reflected nuances of wealth and social status. Magdalene's status was a bit ambiguous, but she had always lived like a noblewoman, and her father was far more wealthy than the Sascerides family, so the principle of reciprocity presented difficulties. Moreover, Gellius did not yet have a permanent position. His prospects may have been brilliant, but it was hard to negotiate on the basis of prospects alone and equally hard to support a wellborn wife without a substantial income.

Negotiations determined the amount of the betrothal gifts, dowry, morning gift, and all the wedding arrangements, including who was to pay. Mogens Dallin negotiated with Sophie Brahe. Tycho could provide a considerable dowry and wanted his daughter to continue to live as an aristocrat. In Tycho's circle, much of the dowry customarily came packed in carved oaken trunks and cabinets. It comprised velvet, silk, and damask gowns for a lifetime of festive occasions; pearl-embroidered caps; a massive golden necklace and jeweled brooch; a gold or silver-gilt belt with knife, spoon, and a place to hang the household keys. In addition, there would be hand-woven table linen and bed linen; copper and iron kettles and pans in full array; a four-poster bed with all its hangings, down comforters, and bolsters; bench covers; tapestries; and perhaps even a bridal carriage with two or four horses.[12] Such a dowry ensured that the bride could live as a gentlewoman for the rest of her life, assuming her husband provided the daily necessities. Sophie Brahe described what Magdalene would receive, and the general character of the dowry was agreed upon. The dowry would be ceremoniously delivered to the house of the suitor, a day or so before the wedding, and knowledgeable women would inspect it to be certain that nothing was lacking.

In return for the bride and her dowry, the suitor needed to give assurances that he could maintain her style of life. Here the negotiations ran into problems. Gellius must have been a bit overwhelmed. He did not have a substantial fortune. His prospects were to become a professor of medicine in Copenhagen and pursue a private practice on the side, which would provide a comfortable middle-class living. Right now, however, he needed to come up with a betrothal gift and a morning gift appropriate to his bride and her dowry. Whatever inheritance he had received from his father could vanish in a flash through these negotiations.

In aristocratic circles, the suitor usually gave his bride a massive chain of solid gold as a betrothal gift. It might be two or three yards long and weigh several pounds. Such a chain cost a king's ransom and was a powerful symbol of high social status, as well as a guarantee that the recipient would never be impecunious because it could always be pawned. In addition, the suitor often gave his bride other personal treasures like a silver spoon, silver gilt beaker, belts, gloves, or shoes. She reciprocated by making him a finely woven and embroidered shirt with her own hands, and aristocratic brides gave their noble suitors a bracelet of gold, engraved and set with precious stones.[13] The agreement with Dr. Gellius Sascerides must have been for more modest betrothal gifts, omitting the golden chain, and she in turn omitting the golden bracelet.

The morning gift was presented to the bride on the morning after their wedding night and consisted of jewelry or a purse of money, but aristocrats often included hereditary rights to great landed estates.[14] Once again, the morning gift had to be scaled down to meet the realities of the Sascerides family and Gellius's future prospects.

Moreover, Gellius Sascerides needed some way to support his bride. Therefore, Tycho Brahe proposed that Gellius return to the *familia* of Uraniborg as a son-in-law, wielding an authority second only to Tycho himself. Gellius could take over much of the day-to-day work, freeing Tycho to finish his opus on the instauration of astronomy. This would provide the newlywed couple with an assured standard of living until Gellius was able to move to a professorship in Copenhagen. Tycho may even have hoped that Gellius would stay on as his successor or codirector of Uraniborg with one or both of Magdalene's brothers, but this was not mentioned in the negotiations.

As for the wedding itself, it could be tremendously costly. Weddings typically lasted three days among burghers and rich peasants, and five to nine days among the aristocracy.[15] In this case, not all the Brahes, Billes, Ulfstands, Ruds, Gyldenstiernes, Skeels, Rosensparres, and other relatives of Tycho Brahe would be willing to come, but there were plenty of aristocratic friends, a host of learned friends, and the whole Uraniborg *familia* to consider. Tycho was apparently willing to put on the wedding, but Gellius would have to provide his own new wardrobe for each day: linen collar and cuffs; waistcoats of brocade and other rich fabrics, some trimmed in pearls; knee breeches; silken hose; buckles and shoes; hats and caps; a

new cape, perhaps fur-lined; and a fine sword. He would also need to bor-
row a wedding wreath of pearls or gold to wear when his bride wore her
resplendent wedding crown.[16]

All these matters were negotiated, and the general substance of the nuptial agreement was reached by the middle of October 1594. Tycho was gone
from 24 September until 4 October, finally inspecting his chapel in Roskilde, visiting Gundsøgaard, and spending more than a week in Copenhagen, when the final arrangements were probably made. Gellius came to
Hven on 16 October and remained for several days, helping Tycho's staff
observe a lunar eclipse on the night of 19 October.[17] Everything seemed
to be in place. The proposal party that customarily followed the conclusion of negotiations may have been held at Uraniborg at some point between Wednesday, 16 October, and Saturday, 19 October 1594. In the presence of friends and relatives from both parties, the suitor's spokesman
asked the bride's parents for her hand. Tycho Brahe gave his consent, then
Kirsten Jørgensdatter gave hers, and a small celebration followed.[18]

According to law and custom, the next step was the formal betrothal.
This festive event involved a large gathering of parents, siblings, first and
second cousins, friends and *famuli* of both parties. At the betrothal, the
suitor's spokesman asked for the hand of the bride in the presence of all,
and this time the bride herself, as well as her parents, needed to reply in
the affirmative. Having received the consent of all three, the suitor's party
brought forth his betrothal gift and bestowed it on the bride. She reciprocated with her betrothal gift. Then the suitor extended his right hand, and
the bride's father laid her hand on his with these words: "I give you my
daughter to honor as wife, for half the bed and half the household, for lock
and key and half of all gain, in the name of God the Father, Son, and Holy
Spirit, Amen."[19] All the others came forward and placed their hands on
those of the couple, then everybody gave the couple a festive slap on the
back. A banquet followed the ceremony, hosted by the family of the bride,
and the two families dined together for the first time. From that night
forward, the betrothed couple generally slept together as man and wife,
though this age-old custom was strictly forbidden by the marriage ordinance of 1582.[20]

After the betrothal, the banns were read on three consecutive Sundays
in the bride's parish church, St. Ibb's on Hven, and then the wedding could
take place.[21] Guests generally arrived on a Friday, and the celebration began on Saturday, followed by a festive procession to church on Sunday.
The wedding took place before the church door, and the couple entered
the church as man and wife to receive the Holy Eucharist. The rest of Sun-

day was celebrated with a wild ride home from church, a huge banquet, a wedding dance, and finally, a ribald procession to the bridal bed. On Monday came the festive awakening, the morning gift, and another banquet. Celebrations continued through the coming days, until the couple were finally escorted to their new home, where the groom ceremoniously presented to his bride the keys to the household, giving her authority over all their goods and servants. After one last celebration in the home, the guests finally departed and daily life began for the newlywed couple.[22]

Would the wedding of Magdalene and Gellius be held on Hven, or in Tycho's palatial house in Copenhagen, or in the Town Hall of Copenhagen? In order to have room for all the overnight guests, many aristocratic weddings were held in town. This and all other details had been negotiated. The circle of family and friends began with the bride's parents, three sisters, and two brothers. Magdalene's father's family was not presumed to be her family, but Sophie Brahe and others would certainly be involved. Little is known about Kirsten Jørgensdatter's family: Pastor Hans Jørgensen of Kågeröd may have been her brother and thus a maternal uncle of the bride. Gellius had his brother, David, his brother-in-law, Jon Venusin, and possibly other relatives in Copenhagen. He and the bride's family had a wide circle of mutual friends including Dallin, the brothers Krag, and virtually all other professors of the university. There were old family friends like Vedel and Aalborg, *famuli* and former *famuli,* Bishop Peter Winstrup and other clergymen, printers and master craftsmen, apothecaries, architects, professionals, and officials associated with the Uraniborg circle. Undoubtedly a number of noblemen and ladies would also attend as friends. These plans kept everybody busy on Hven during the autumn of 1594.

THE PANGS OF DISPRIZED LOVE

Tycho Brahe had other matters on his mind as well. In September, he had received another sharp reprimand from young King Christian IV because the Chapel of the Three Holy Kings was now in worse shape than ever. Tycho was ordered to inform the king immediately of what he proposed to do; if the chapel were not repaired by Christmas, the whole fief would be given to another vassal. This threat finally moved Tycho to action: He and Steenwinckel or another master builder went to Roskilde on 26 September, inspected the chapel, and worked out a plan to replace the vaults with a flat beamed ceiling.[23]

Tycho Brahe was away for several days in late October 1594, and when he returned to the island, he received a letter from Sophie Brahe in Copenhagen, informing him that Gellius wanted to scale down the wedding to "no more than a little philosophical wedding" with a banquet for a limited number of guests, rather than an elaborate celebration lasting several days.[24] She also reported that Gellius was willing to stay on Tycho's island until Easter of 1595. Tycho thought the negotiations had just been concluded, yet Gellius, only "two days after I had promised my daughter to him," went to Sophie Brahe and wanted to change things.[25] This irritated him, but he undoubtedly discussed the new proposals with Kirsten Jørgensdatter and Magdalene. Then he replied to his sister on 28 October 1594 that he found the proposals acceptable but hoped that Gellius would not keep changing his mind.[26]

Nonetheless, that is exactly what he did. Tycho Brahe was in Copenhagen for almost two weeks during 6–18 November 1594, when further negotiations took place and the date of the wedding was finally set. Gellius undoubtedly discussed the plans with his friends and asked their advice. He wrote to Tycho and implied that he wanted the wedding to take place as quickly as possible. Some confusion set in at this point. The wedding plans had been scaled down, at Gellius's request, to something like a betrothal banquet. Were we now talking about a combined betrothal and wedding? This was not legally possible according to the marriage ordinance of 1582, but in practice, things could sometimes be more flexible. After continued negotiation with Gellius's spokesman, Mogens Dallin, the date of 15 December 1594 was set for the betrothal – or was it the wedding? The parties themselves seemed confused.[27] In any case, Tycho wrote to Gellius, confirming this date. Gellius came to see Tycho the next day, thanked him for complying with his written request, and promised to act toward him as a son should. When Tycho returned to Hven on 18 November, he could report that the negotiations had reached a successful conclusion and that the date of the betrothal – or the wedding – was less than a month away.

Then more problems arose. On 30 November 1594, Tycho sailed back into Copenhagen to speak with his sister, Sophie Brahe. Gellius had raised new objections. He seems to have demanded a large sum of money in addition to the dowry, and perhaps he even wanted to delay the wedding.[28] On the evening of Tuesday, 3 December 1594, Tycho sent Niels Krag and Mogens Dallin as his spokesmen to meet with Gellius, but Gellius refused to give them a clear answer. Tycho was enraged and decided to establish a paper trail.[29] A full summary of events was drawn up and signed by Mogens Dallin, Niels Krag, Else Mule, Sophie Brahe, and Tycho Brahe himself.

The next morning, Tycho stopped at the home of Niels Krag to pick up two Latin verses that were to be sung contrapuntally on the evening of the wedding. Krag and his wife, Else Mule, invited him to stay for dinner, usually served at nine or ten o'clock in the forenoon. They discussed what had happened, and Tycho said he did not want to bring public accusations against Gellius; he would leave that to the Good Lord, as long as he and his family were not held to be at fault.[30] Tycho admitted that he had been enraged the previous evening, and he was still angry. After dinner, Tycho and Sophie Brahe returned to the island of Hven, where they had to break the sad news to Magdalene Tygesdatter and decide what to do next. Lisbet Bille Jensdatter, a cousin, was there to comfort Magdalene.[31]

Shortly after Tycho had left, Gellius showed up at Niels Krag's house. Krag met him at the door as he was going out.[32] There in the public street, where any passerby could hear, Gellius said that he wished he had simply declined the offer of Magdalene's hand the evening before, and wished he had told Krag and Dallin that her father could do with her as he liked. "You can tell that to Tycho," he said. Niels Krag replied that he was no longer acting as Tycho Brahe's emissary, and he refused to carry the message to him.

A few days later, around 6 December, the brothers Krag had just finished dinner with a nobleman named Sivert Grubbe at the home of Venusin and his second wife, Karine Dankertsdatter. Grubbe had already departed when Gellius came in and began to expostulate with Niels Krag, requesting him once more to tell Tycho that Gellius declined the match. Anders Krag did not want to get involved and withdrew to speak with Karine, but Gellius and the others called for him to join them. Niels Krag again refused to take Gellius's new message to Tycho. He told Gellius that Else Mule thought he ought to contact Sophie Brahe and ask her to speak with Tycho on his behalf. Niels Krag and Venusin both urged Gellius to call on Sophie Brahe and even offered to accompany him to Eriksholm Castle. Gellius became agitated and increasingly irrational until Venusin seized him by the arm and said, "Man! Whither with such words?"[33] Pointing at Gellius's breast, Venusin admonished him to examine his conscience: He had made promises to Magdalene by word and deed in the sight of God; he must beware that he did not act contrary to God and his own conscience. Gellius did not listen but continued to rant. Anders and Niels Krag later testified that he was quite sober at the time, and that he spoke seriously, not in jest.

His friends continued to warn Gellius of the seriousness of his actions. He could be sued for breach of promise, and his whole career could be ruined. Niels Krag advised Gellius to take his case to Sophie Brahe and get her to mediate on his behalf, or else flee abroad until things cooled down, and he warned Gellius not to give the slightest further offense to Tycho Brahe or any of Tycho's relatives.[34]

The next move was up to Tycho Brahe. After lengthy discussions with his wife, sister, Magdalene Tygesdatter herself, and their young kinswoman Lisbet Bille, the decision was made to break off the match. Sophie and Lisbet departed on 5 December. In the next few days, Tycho sat down to draft a formal declaration that the wedding agreement was canceled. He signed and sealed it at Uraniborg on 12 December 1594. On the same day, with a heavy heart, Magdalene Tygesdatter signed a similar declaration.[35] Then the tears could flow. Magdalene knew full well what this would mean: Her reputation was blackened, and no man would have her. Twenty years of age, she was doomed to the life of a spinster. That was what Dr. Gellius Sascerides had done to her.

CALUMNIA

She was like Susanna of old: a victim of calumny, gossip, and slander. Tycho's friend Peter Hegelund had made the biblical story of Susanna into an earthy Danish play, first staged in 1576 and published in 1578.[36] Everybody knew the story: It often appeared in sixteenth-century art, literature, and sermons. The pious and beautiful Susanna was observed in her bath by two corrupt judges, who attempted to seduce her; when she repelled them, they spread false rumors that she has been unfaithful to her husband. Susanna was condemned to death but saved at the last moment by the prophet Daniel. She went free, and her evil accusers were stoned to death. Calumnia, the spirit of gossip and slander, failed to bring about her downfall.

As an interlude to the play, Hegelund had described Calumnia in a long Danish poem based on the personification of Rumor in Virgil's _Æneid_. Calumnia had a thousand tongues, numerous peering eyes, sharp ears, a tail like a fox, rapid wings, and a quiver of stinging arrows. A woodcut showed her in all of her malice [Fig. 43]. Dressed as a woman, mild of manner, she hid her deceptive heart and stalked the land:

> With all my tongues, my words release,
> Nor living nor dead I leave in peace.[37]

Now gossip and slander found new victims in Denmark. Calumnia filled her quiver and pointed her arrows at Magdalene Tygesdatter. Then she paused. She saw a larger target: Tycho Brahe himself.

FAMA MALVM.

Fig. 43. Calumnia, from Peter Hegelund, *Susanna,* 1579. (Courtesy Det Kongelige Bibliotek, Copenhagen)

In his own hand, Tycho Brahe wrote out a full summary of Gellius's strange and insulting conduct since the time when he had first entered into an agreement for the hand of Magdalene Tygesdatter, concluding, "for he apparently believed nobody, and hardly himself."[38] At her father's request, Magdalene also wrote a statement of how she had been treated by Gellius. It tore her heart to do so, but she wrote that she never wanted to think of him again, and she ended the statement, "I am glad that Our Lord has let me be so graciously and well freed and released."[39] The two declarations of cancellation and the two explanations were kept together with a fifth document, the summary of what happened on 3 December 1594, written and signed on the day of the event.

Meanwhile, Mogens Dallin wrote directly to Magdalene Tygesdatter. Tycho intercepted the letter and told her about it, but he did not show it to her because he said the time for her to take advice had run out, and she said that she did not want to see it. When he read the letter, Tycho could see that it would distress her. Dallin put things in such a way that it sounded as if she had acted falsely and was to blame for what had happened. The letter even had a certain threatening tone. Dallin was a good friend, but in his eagerness to repair the match, he had acted unwisely, and the letter backfired. It did not make things better.[40]

The Yuletide could not have been happy at Uraniborg that year. Before the Christmas season was fully past, Tycho sailed to Sjælland on 2 January 1595.[41] He had arranged to meet at the manor of Gundsøgaard with Niels Krag, Jon Venusin, and Mogens Dallin. Gellius knew of this meeting in advance, and he knew that Niels Krag was one of Tycho's best friends. He might have asked Krag, or his own good friend Dallin, or his brother-in-law Venusin, to deliver a message to Tycho Brahe, but he did not ask.[42]

Tycho brought copies of the five documents. He chided Dallin for his letter to Magdalene and Dallin apologized, assuring Tycho that he meant well and had wanted to do everything he could to repair the breach.[43] Tycho gave copies of the five documents to Krag and Dallin, who had agreed to serve once more as his emissaries to Gellius Sascerides. They were to read the documents aloud to Gellius and give him a copy of each one.

As they were about to ride off, Niels Krag, the consummate diplomat, asked what they should do if Gellius indicated that he would abide by his promises after all. Tycho replied: If they had not begun to read the documents, they should not do so; if they had begun to read, they should read no further, but instead, inform Gellius that Tycho would soon be in Copenhagen, and that he would listen to Gellius and do what seemed right.[44]

Niels Krag and Mogens Dallin rode the long way from Gundsøgaard to the residence of Dr. Gellius Sascerides in Copenhagen. They read the documents aloud and presented the copies, but Gellius refused to accept them. He would not even acknowledge in writing that they had been read to him. The documents were returned to Tycho Brahe that same evening.[45]

While these negotiations dragged on and on, a thousand worries burdened Tycho Brahe. He was famous throughout Europe, head of a unique scientific research institute, had discovered new ways of doing science and made many discoveries, but the results of his work were still unpublished. A thousand details needed his attention at once, but how could he attend to them? The long-awaited opposition of Mars had occurred in October of 1595, but Tycho had been too busy to measure its diurnal parallax, even though the observational basis of his planetary system depended upon it.[46] His staff carried on without day-to-day supervision, but no major progress could be made in this way: Large-scale, multifaceted scientific enterprises could not function in the long run without leadership. In 1592, Longomontanus and his crew had finished a catalog of 777 stars, the first independent stellar catalog since ancient times, with little supervision from Tycho except in the very beginning.[47] Now Tycho wanted to solve the excruciating problem of lunar theory. He and his staff had compiled mountains of data over the years, and when eclipses occurred, three independent crews had observed them simultaneously from Stjerneborg and the two Uraniborg observatories, Tycho leading one crew and senior assistants like Flemløse, Gellius, or Longomontanus leading the others.[48] However, the herculean task of deriving a theory from the data was being delayed as Tycho's research on the lunar variation ground to a halt in November of 1594. His attention had to be focused on Magdalene for many agonizing months. Tycho kept a level head despite constant affronts and many pressing concerns, regained his composure after each rebuff by Gellius, consoled his daughter as best he could, and patiently neglected his scientific work as he continued to negotiate.

Tycho spent several weeks in Copenhagen after the meeting at Gundsøgaard. He discovered that rumors were rampant, and all of them put him in a bad light. From his brother-in-law Christen Skeel and his sister Margrete Brahe, he also heard about Gellius's remark in early December: "I wish I had said, 'No, thank you, for your daughter. You can give her to somebody else.'"[49] Tycho could hardly believe his ears. He hastened to visit the witnesses, and they all confirmed that it was true. He asked Niels Krag to write it down for the record, and Krag did so on 8 January 1595, when Tycho was still in Copenhagen.[50]

Three days later, acting quite deliberately in his righteous indignation, Tycho addressed a petition to the Rector Magnificus and professors of the University of Copenhagen. He explained that he had drawn up the five documents and sent Niels Krag and Mogens Dallin to read them to Gellius Sascerides, together with the message that if Gellius wanted to reply, Tycho would be pleased to meet with him in the presence of the rector and some or all of the professors, either publicly or privately as Gellius might prefer. Tycho believed that the university had to become involved because of its close ties to the parties, and also because the rector and professors sitting in Consistory functioned as the high court of the land in cases involving marriage.

I would be most satisfied if the Rector Magnificus and professors, here in the university, would try the case and pronounce on the central issues and main points, which are, whether the fellow has kept his promise, word, and letter, as well as the word and letters of other good people, issued on his behalf and with his knowledge and permission, honorably and uprightly, as ought to be the case, and whether I or anybody on my behalf have given him any valid reason to do what he has done.

Tycho Brahe had good reason to request a speedy hearing: Calumnia was at work.

[Gellius] and his followers have now for a whole month, round about the city and elsewhere, said all kinds of things to make his cause look pretty and put all the blame on me and mine, otherwise than I believe will be found to be the truth, so that many good people have thereby been led to a wrong opinion and misunderstanding.[51]

The university authorities were quick to take action. Instead of trying the case in Consistory, they attempted to mediate a settlement and were surprisingly successful. A new contract was drawn up and dated in Copenhagen, 16 January 1595, signed by Rector Magnificus Nicolaus Theophilus, Bishop Peter Winstrup, who was a member of the faculty of theology, and three additional professors: Anders Krag, Niels Krag, and Dr. Anders Christensen. The terms are not known, except that Gellius promised not to disturb or inconvenience Tycho or his family in any manner whatsoever.[52] Tycho Brahe returned to Hven on 21 January to bring the news to Magdalene and her mother. Before he left town, however, he got Anders Krag to write up his account of what had happened on that day in early December at the home of Jon Venusin.[53]

On 24 March 1595, Gellius finally got a job. He was appointed provincial physician of Skåne with an annual salary of 100 dalers, until such time as a canonry became available to him in Lund Cathedral.[54] The appointment was made on petition from the nobility of the province, most of whom were Tycho Brahe's relatives. This was probably part of the agreement.

One month later, in April of 1595, Tycho Brahe received a disturbing letter from his second cousin, Peter Brahe, whose seat lay in Skåne, at Krageholm Castle. Perhaps the newly appointed provincial physician had been sent there; in any case, Tycho later used this letter as evidence that Gellius was still spreading false rumors.[55] Calumnia had not been silenced. Then on 8 May 1595, Gellius in his cups staggered up to Else Mule at an academic festival in the Brewers' Guild in Copenhagen. He berated her at length because her husband had supported Tycho. She heard him out but finally took offense and walked away.[56]

In the beginning of September 1595, Tycho Brahe spent a week at his mansion in Copenhagen. His friend Bishop Anders Foss of Bergen was in town, and Tycho invited the bishop to dine with him and several others, including Jon Venusin, Anders Krag, and the crown magistrate of Copenhagen, Peter Frandsen, a brother of the late Johannes Franciscus. The conversation turned to Gellius Sascerides, and Tycho complained that Gellius had not lived up to his agreement. He had been spreading all sorts of malicious rumors in order to sow dissension, not only among strangers, but even among Tycho's own flesh and blood, including his own dear mother. When Venusin, Gellius's brother-in-law, expressed his doubts, Tycho produced evidence. After the meal, Tycho read a whole list of false rumors that had been spread by Gellius, and even Venusin admitted that Gellius himself could hardly deny some of them. Tycho said that he was not going to do anything about it for the time being, but he would be forced to take action if the slander continued.[57] The slander did continue. In December of 1595, Tycho received a letter from his kinsman Esge Bille, whose seat was the manor of Ellinge in Skåne, showing that Calumnia had reached those parts.[58]

Tycho Brahe, Magdalene Tygesdatter, and all their household had finally had enough. Malicious rumors were undermining their reputation. People looked askance, whispered behind their backs, and grew silent as they approached. Friends had doubts, strangers became enemies, and old enemies gleefully added rumor to rumor. Tycho had to do something: His honor, his reputation, his very position in society was at stake.

Since Gellius Sascerides was now a resident of the archiepiscopal seat of Lund in Skåne, Tycho began there. His aim was simply to clear his own good name and not to sue Gellius for slander. Throughout this whole affair, Tycho Brahe remained remarkably free of vindictiveness or malice. Accompanied by Peter Frandsen, he traveled to Lund on 10 January 1596 and spoke with officials of the cathedral chapter. Hearings took place before the chapter on 10 and 20 January 1596, and five questions were put to Gellius in writing by Tycho. Gellius did not reply during the hearings but promised to do so within six weeks.[59] Not long after, Tycho received a letter from another kinsman, Lauge Urne, chief justice of Skåne and canon of the cathedral, indicating that Calumnia was abroad in Lund as well.[60]

Immediately after the second hearing in Lund, Tycho Brahe wrote to Jørgen Rosenkrantz and to his son, Holger Rosenkrantz, temporarily home from studies in Wittenberg.[61] Tycho needed to counter the influence of Gellius and his friends at court, and he asked the Rosenkrantzes to ensure that the government did not interfere with his attempts to clear his name. On 5 February 1596, a royal missive commanded the cathedral chapter in Lund to sit as a marriage court and try the case of Tycho Brahe versus Gellius Sascerides. That same day, Manderup Parsberg wrote with evidence that slander had indeed reached the court, and the next two days brought letters from Tycho's kinswoman Thale Ulfstand in Copenhagen and from Lauge Urne in Skåne with more evidence of malicious rumors.[62]

Tycho Brahe did not want a trial. He simply wanted to clear his name and that of his daughter, and to disprove in full public view the vicious rumors spread by Gellius Sascerides and his party. The case was appealed to the Consistory in Copenhagen over Gellius's objections.[63] Meanwhile, evidence of widespread calumny continued to pour in from Tycho's kinsman Oluf Rosensparre in Landskrona; a Dutch visitor to Hven, Nicolaus de Vrint; Hans Mikkelsen in Lund; and from Tycho's cousin Peter Brahe in Skåne. In response, Tycho enlisted Sophie Brahe to write a summary of her view of what happened, and she did so in a statement dated 20 February 1596.[64]

On the morning after Ash Wednesday, 25 February 1596, the appellate hearings began in the medieval Consistory building, which still stands on the campus of the University of Copenhagen. Tycho Brahe was accompanied by Dallin, Venusin, and Carl Bryske, the governor of Copenhagen Castle. In his opening statement, Tycho said that he did not want a judgment in the case but only a clear, written transcript of the proceedings and documents submitted as evidence, as recorded by the notary of the university. He presented in evidence the contract of 16 January 1595, witnessed by many of the professors then present.

Next, Tycho Brahe read and submitted in evidence a list of ten false rumors that had been spread by Gellius Sascerides and his party. These rumors were as follows:

1. Gellius was not allowed to know the wedding date.
2. Gellius could not have his wedding clothes ready by the date that was set.
3. Tycho demanded that Gellius remain for some years on the island of Hven, or as some said, remain there forever, and work for Tycho like a thrall.
4. Tycho demanded that Gellius dress Magdalene in velvet and damask like a lady.
5. Tycho demanded a written statement that Gellius would always provide Magdalene with clothing as good as what she had when she was married.
6. Gellius demanded a large sum of money, and Tycho replied that he would not give him both his daughter and money; if Gellius did not

want her for her own sake, then he would not get her for the sake of the money.

7. Once when Tycho was in Copenhagen, he sent a message to Gellius that the wedding must be held on one of the next two Sundays or not at all.

8. Gellius came once to Tycho and requested that the wedding take place, but Tycho replied that he would not allow it and that Gellius would not get his daughter.

9. Gellius was not allowed to know anything about the wedding plans, even after he had sent two men to learn what they were.

10. Gellius had no job and no place to live, but Tycho would not provide them, and therefore, Gellius would not marry.

Tycho Brahe said these ten false statements had been spread so widely that almost everybody blamed Tycho and held him to be guilty of a great injustice. He demanded to know whether Gellius considered any or all of them to be true. Gellius avoided a direct reply. Tycho again demanded an answer before the court. Gellius asked to see the written list. "It was given to him, and he read through it with his brother, David, who was also present. Then they laid it back on the table and gave no answer."[65]

At this point, Tycho Brahe stood up and declared that whosoever made any of these ten statements, be he learned or unlearned, man or woman, that person was a liar and would remain so until he or she could bring evidence to the contrary. Thereupon, he submitted in evidence a large number of documents to show that Gellius had violated the contract of 16 January 1595 and had spread false and malicious rumors about Tycho Brahe and Magdalene Tygesdatter, including those listed above. Inasmuch as Gellius had flagrantly violated the contract, Tycho requested that it be declared null and void. On the afternoon of that same day, Tycho formally requested that Gellius reply to the five questions he had been asked in Lund, to which he had promised to reply within six weeks. Gellius said that he would do so the next day.[66]

The next morning, 26 February 1596, Gellius did not appear in Consistory, claiming to be ill. Tycho insisted that he either appear in person or send someone with power of attorney. After considerable delay in running back and forth to his residence, Gellius finally sent David Sascerides with his power of attorney, revealing in the process his ignorance of the law by his manner of replying from his "sickbed" and delaying the proceedings.[67] Tycho accepted Gellius's written reply to the five questions, but only conditionally, reserving the right to cross-examine Gellius personally.

Two days later, on Sunday, 28 February 1596, Gellius appeared in Consistory. He began with a very long but unconvincing explanation of his absence at the previous session. He then made a proposal that, in context, must have seemed both shocking and incredible: "then [Gellius] begged the forgiveness of Tycho Brahe for the sake of God and his innate noble piety, and publicly declared himself willing as an honorable man to abide by all the promises he had previously made to Tycho Brahe and his dear daughter, Magdalene."[68] To say the least, Tycho was reluctant to accept this proposal and pointed out that Gellius had already said he no longer wanted Magdalene, but some of the professors urged him to reconsider.

Tycho let the offer pass and moved to Gellius's written reply to the five questions. Had he asked Sophie Brahe to write Tycho in October 1594? Gellius could not remember. Did he state in writing that he wanted the wedding to be held as soon as possible? Gellius gave a vague, evasive reply. Did he know from Tycho's letter that the wedding was scheduled for 15 December 1594? The letter mentioned a betrothal, not a wedding, and he could not remember if he helped to set the date. Did he come to Tycho, thank him for acceding to what Gellius had requested in writing, and say he would be like a son to him? He did thank him but could not remember the exact words. Did Tycho or his daughter ever give him any grounds to go against his own oral and written request? He did not know that he had done so.[69]

Tycho found these replies unsatisfactory. He demanded direct answers.

GELLIUS: "If I have not done what was required of me, let the courts judge between us."

TYCHO (*replying immediately in Latin*): "Ad jus appellasti, ad jus ibis!" which in the vernacular is, "You rush to justice, justice you shall get!"[70]

The Consistory reassembled to continue the hearings on 3 March 1596. On that day, Anders Krag was called to testify. He told what had transpired at the home of Venusin on or about 6 December 1594, and what had occurred at the Copenhagen residence of Tycho Brahe in early September 1595. Gellius became extremely agitated during Krag's testimony and interjected all kinds of excuses: Ale and wine were on the table at Venusin's, implying that he and the others were not sober; people were saying Magdalene's favor had turned away from him; and he claimed that he was only joking. Anders Krag and his brother, Niels Krag, who submitted his written testimony to Consistory on 6 March, denied that any of Gellius's excuses were true.[71] Niels Krag also testified concerning what Gellius said to him outside his house on 4 December 1594 and concerning the

Gundsøgaard meeting and its sequel. Niels Krag told how Gellius had accosted his wife at the banquet in the Brewers' Guild, and Professor Hans Resen, who favored Gellius's side throughout the proceedings, reluctantly confirmed testimony on this point.[72] Not a single word in this lengthy testimony supported the position of Gellius Sascerides: All of it – all the documents entered into the record, and all testimony by witnesses, both friendly and hostile – supported the assertion of Tycho Brahe that he and his daughter had been slandered and maligned. When the hearings were over, Tycho thanked the Consistory in writing and asked for a transcript. For the record, these hearings cleared the good names of Tycho Brahe and Magdalene Tygesdatter.

Calumnia, however, could not but cackle at the sight of the transcripts. What good are records when rumors abound? Can the tongues of Calumnia be silenced, or her words be driven from the memories of all those people? The damage was done, and it proved impossible to undo. Tycho Brahe had failed. He had failed to provide for the future of his eldest daughter. He had failed to tie one of his most talented *famuli* to the future of the research center at Uraniborg. The reasons for his failure were complex, but certainly one factor was the social gulf between Uraniborg and the Sascerides household in Copenhagen. Tycho had failed because the mentalities and life-styles of two social classes, the high nobility and the academic elite, were so different, despite their many points of contact, that they could not be bridged by marriage negotiations. Love alone could bridge the chasm by throwing all conventions to the wind, as Tycho and Kirsten had done in their youth. The elaborate formalities of matchmaking were designed to integrate family with family on the basis of a parity of wealth and status, and they could not do so here. In this sense, Tycho Brahe, Magdalene Tygesdatter, Gellius Sascerides, and all their friends, relatives, and spokesmen, were trapped in the web of the social structures of their day.

In a more concrete sense, Tycho and Magdalene became victims of Gellius's calumny. Terrified as he gradually discovered what was expected of him by his chosen spouse and her family, Gellius tried to escape by evasiveness, then half-truths, lies, and finally, a systematic campaign of slander. The parties were unequally matched, but in the end, the lesser brought down the greater.

Was that all, however, or were there ominous hidden agendas at work? Tycho had friends and allies among the professors, but so did Gellius Sascerides. Hans Resen, Thomas Fincke, and Bishop Peter Winstrup, intellectual leaders of a new generation, consistently sided with Gellius. Were they hostile to Tycho Brahe, or could it be that Tycho and Magdalene had been caught in a crossfire of university politics gone wild? Were the allies of Gellius really trying to get at the brothers Krag, Venusin, or Dr. Anders Christensen, or were they aiming to sever the university's ties to the independent postgraduate center on Tycho's island? Were they making Tycho and his daughter a target for middle-class resentment of aristocratic privilege? These and related questions must have plagued the mind of Tycho Brahe in the aftermath of the Consistory hearings.

Tycho himself had suffered the first great failure of his life, made all the more humiliating because it appeared that he, a great noble aristocrat, had been defeated by a poor middle-class scholar. Worst of all, the affair robbed him of his honor, one of the most precious possessions of any sixteenth-century nobleman. After the Copenhagen hearings, Tycho carefully collected and corrected the transcripts made by the university notary. It took him half the year of 1596.[73] Armed with this evidence, he prepared to bring the case to trial before his own peers at the annual Diet of the Danish nobility in 1597.[74] Gellius, for his part, did not flee the realm but settled into his new position as provincial physician of Skåne.

LIFE GOES ON

Tycho Brahe was in his fiftieth year in 1596, but he seemed older. His hair had grown thin, his moustaches drooped, and his brooding eyes looked weary in a portrait painted the following year [see frontispiece].[75] These last years had been hard, and even good times could seem bitter now. In late November of 1595, a double wedding was celebrated at Uraniborg. Among the guests were Sophie Brahe, the brothers Krag, Venusin, Dallin, and Niels Hammer, but the bride was not Magdalene Tygesdatter. The grooms were a clergyman or doctor called "Johannes" and a chaplain named "Martin"; we do not know the names of the brides.[76] Who were these people? Martin may have been Martin Ingelli Coronensis, who had served Tycho Brahe around 1590–1. Johannes could have been Pastor Hans Jørgensen of Kågeröd or Jens Jensen Wensøsil, who had returned to Hven around 1594 to serve as pastor of St. Ibb's Church. The double wedding was celebrated at Uraniborg, but it was not the wedding that might have been.

The year 1596 brought a series of depressing events. In March, one of the servants committed suicide. In April, a jester, Per Gek, and a student lutenist named Jacob arrived at Uraniborg. They enlivened the long evenings, but the jester tried to flee in May. A month later, both he and the lutenist succeeded in fleeing the island. Around the same time, a dead sailor was found buried on the beach.

Tycho hired a new tailor that spring, a man named Nicolaus, and paid him an annual salary. He kept busy because clothing and shoes were still supplied to all household servants and *famuli*.

Tycho himself was frequently away. He was gone to Helsingborg and Skåne for ten days in January, then nearly a week in Copenhagen. The Copenhagen printer Hans Stockelman visited Uraniborg early in February, in connection with the printing of Tycho's correspondence on astronomy. The Hven press had been inactive since 1592, but now it was in use again, and a native of Tuna called Peter Hvæn worked in the printing office. On 21 February 1596, Tycho Brahe sailed to Copenhagen and was gone for a month, pursuing the case against Gellius before the court of Consistory. During most of April and May, he was on Hven to entertain many visitors, including the noblemen Børge Trolle and Mogens Gøye, the Dutch consul in Elsinore, Isaac Pietersz, Professor Laurentius Scavenius from Copenhagen, and Bishop Anders Foss with his family from Norway. The bishop and his wife were involved in a nasty legal battle with their son-in-law and were also pursuing their case before the Consistory.[77] Early in May of 1596, Tycho commanded that the sturdy little pilot-boat be rigged as a cargo ship; apparently he was planning to haul some heavy freight.

In June, Tycho Brahe was gone for nearly two weeks to attend the annual Diet of the nobility. This great gathering was attended by his brothers and other aristocratic friends and relatives. Steen Brahe was now a member of the Regency Council, and Axel Brahe was admitted to the State Council in that year of 1596. So was Oluf Rosensparre, who had married their niece, Lisbet Gyldenstierne. Axel Brahe now commanded Helsingborg Castle; Steen Brahe had advanced to Kalundborg fief. Jørgen Brahe held Varberg Castle. and Knud Brahe held Aarhus fief. Their cousin, Steen Maltesen Sehested, was Lord Lieutenant of Kronborg Castle. Chancellor Niels Kaas and Falk Gøye were dead, as were Tycho's brothers-in-law Henrik Gyldenstierne and Christen Skeel. Manderup Parsberg held authority as royal chancellor but was about to hand it to Christian Friis of Borreby. Oluf Rosensparre was Lord Lieutenant of Landskrona Castle. Axel Gyldenstierne was still viceroy of Norway, and old Heinrich Rantzau still viceroy of Schleswig-Holstein. Tycho's kinsman Admiral Peter Munk sat in the Regency Council, and another relative, Hak Ulfstand, was the young king's guardian, though Tycho's mother had retired in 1592 as head of the queen mother's court. All in all, the oligarchy of Tycho Brahe's friends and relatives was firmly in control. Their most dangerous rival, Christoffer Walkendorf, was over seventy years of age and had finally been forced out of the inner circles of government. However, the case against Gellius was not ready in time for the Diet of 1596.

The handsome young king was champing at the bit, eager to rule.[78] At nineteen, he was athletic, well-educated, and assertive, and he had toured German courts during 1595, where he met the demure and religious Hohenzollern princess who would soon become his queen. The Diet met in June of 1596 to prepare for a coronation in August, when the Regency Council would be dissolved and the personal reign of King Christian IV would begin.

Tycho Brahe had many visitors that summer. **Peter Jacobsen Flemløse** came in June, accompanied by Johannes Paludan and a Norwegian clergyman, Christopher Hjort. They dined in the garden in endless twilight on Midsummer's Eve and observed the traditional bonfires all around the Sound. Later that summer, Sophie Brahe came with her son, Tage Thott, a bright lad who admired his famous uncle. Bishop Anders Foss and family also returned, and so did Christopher Hjort. An ancient astrologer from the household of Viceroy Heinrich Rantzau came to visit, perhaps with a secret mission. A German student arrived with a recommendation from the landgrave of Hesse-Kassel, and a Danish student named Joannes came from Copenhagen.

On 22 July 1596, three members of the *familia,* including **Elisabeth Paulsdatter,** were dismissed for sedition.[79] The details are unknown, but they may have been using the laboratory without permission. A month later, Elisabeth's father, Paul Thiisen, who was friends with Tycho, came to Uraniborg to make peace.

Tycho Brahe attended the coronation of King Christian IV in Copenhagen Cathedral on 29 August 1596 with several of his *famuli.* Tycho wore his golden chains with the Order of the Elephant and portrait medallions of two kings. His brother, Regent Steen Brahe, bore the royal orb, and all members of the State Council held the crown as it was placed on the royal brow to symbolize the elective nature of Danish kingship, where ultimate power resided with the aristocratic oligarchy. Bishop Peter Winstrup's coronation speech, however, presented the alternative theory of government favored by the young monarch and his nearest advisors, asserting that "kings are gods" who rule by divine right.[80]

Following the coronation, some of the foreign guests came to Uraniborg. Tycho Brahe was visited by the Scots ambassador, Robert Anstruther, and by Johannes Müller, chemist and astronomer to the court of the administrator of Brandenburg. In October, Elisabeth Paulsdatter returned to Hven with her father and others from Elsinore, but whether she reentered Tycho's service is not known. At the end of that month, the wedding of Jonas, the cook of Uraniborg, was celebrated.[81] Life seemed back to normal after the bitter battles with Gellius Sascerides and the transition to a new regime. Then came new and serious problems.

OLIGARCHS CHALLENGED

Tycho Brahe and his relatives were not pleased when Christoffer Walkendorf was called back into the government by the young king. Walkendorf was a dangerous man, and the king had him appointed to the office of Lord Steward, the highest in the realm, which had been vacant since the death of Peter Oxe. Walkendorf was one of those aristocrats who did not support the idea of oligarchy but believed in strong monarchy. So did Christian Friis of Borreby, the new royal chancellor, who was friendly toward Uraniborg but now stood in the service of an assertive young monarch. The oligarchs braced for trouble.

On 21 September 1596, the Norwegian fief of Nordfjord was transferred from the jurisdiction of Tycho Brahe to the Lord Lieutenant of Bergen as part of a general reorganization of fiefs.[82] Tycho had originally received Nordfjord as a temporary substitute for the promised canonry in Roskilde and had always found ways to hang on to it. When the Chapel of the Three Holy Kings had come to him in 1579, he lost Nordfjord but petitioned King Frederick II and got it back. The Norwegian fief had been taken away again in 1580 but restored to him after two months. In 1586, Tycho relinquished Nordfjord in exchange for a pension of 400 dalers. Christoffer Walkendorf had administered all of these arrangements as Lord Treasurer, and he remembered them clearly. In 1589, Nordfjord had been restored to Tycho by the Regency government, even though Tycho continued to receive the 400-daler pension until 1592. When the pension was canceled, Tycho still retained the fief.[83]

As always in the past, Tycho Brahe now petitioned to regain the fief. On 31 December 1596, he wrote to Chancellor Christian Friis and asked to get back Nordfjord or at least be allowed to keep it until the end of the fiscal year on 1 May 1597. He went on to summarize his accomplishments on Hven and enclosed two gifts: a copy of his published astronomical correspondence and a pamphlet of woodcuts of his instruments. He reviewed the strong interest King Frederick II had shown in his work and pointed out that the late king had been prevented by death from endowing Uraniborg as a permanent research institute. Tycho enclosed a copy of the declaration of 1589, signed by the whole State Council and promising to advise King Christian IV, when he came of age, to carry out his father's will by endowing Uraniborg under the leadership of Tycho Brahe's descendants.

Tycho requested the chancellor to bring this matter to the attention of the king.[84]

Christian Friis and his wife had spent three pleasant days on Hven in October of 1592, with a number of Tycho's noble relatives: His wife, a niece of Anne Hardenberg, was also distantly related to the lord of Uraniborg. Tycho clearly expected a favorable response and believed that the long-awaited permanent endowment of the Uraniborg complex would soon be achieved.

Chancellor Christian Friis replied on 20 January 1597. His letter was curt and clear. The surrender of Nordfjord could not be postponed because the Lord Lieutenant of Bergen needed the income. The king could not permanently endow Uraniborg. That was it. His conclusion was a courtly but hollow offer to oblige Tycho's future requests with pleasure.[85]

No chancellor had ever written to Tycho Brahe like that.

A tempest was building over the storm-tossed waters of the Sound. It moved inexorably toward the lofty, unprotected island of Hven.

9

THE TEMPEST

———

1597

ABSOLUTISM RISING

The Sound can be treacherous. It has the shape of a fishtrap, wide at the bottom, fringed with hooks and islands, easy to enter but hard to escape, and it pinches into a tight funnel at the top. A whale entered these narrow waters in January of 1597. The weather was still and cold. Snow covered the ground. Ice jammed the waters. Trapped in the Sound, the leviathan plunged about, confused and weary. On 29 January, it beached itself on the Sjælland coast at Nivå, directly across from the island of Hven. There it died, amid the snow and ice. On 4 February, persistent rumors reached Hven of another beached whale at Malmö.[1] These waters were too narrow for such great creatures.

These same waters had grown dangerous for Tycho Brahe as well. For thrice seven years, the royal sun had shone upon his island. Now it seemed hidden under leaden skies. King Frederick II had counted on Tycho to bring fame to the kingdom with his scientific accomplishments, so that people of other nations would come to Denmark to "see and learn that which they could hardly acquire knowledge of in any other place." Tycho Brahe had done this in full measure, and he had always expressed his gratitude to the royal patron who made it possible. In the days of King Frederick II, support for Tycho Brahe had never diminished. Tycho had been brilliant, ambitious, and young, and the king like a benevolent older brother.

Under the Regency Council, Tycho Brahe had risen to the top. His aristocratic friends and relatives were sovereign in the realm, and Tycho's most fantastic wishes became commands. With charm and sophistication, he had cultivated the ruling elite and strengthened his position within it. Tycho Brahe was no longer young and deferential. He stood at the height of his fame, addressing regents and viceroys as equals and friends. They were flattered by invitations to visit Tycho's island.

Now that phase was past as well. The year was 1597, and there were new rulers in the realm. The king was a headstrong lad of nineteen, confident and eager to show that he was in charge. He expected to receive the deference that any king demands of his subjects, but some proud old aristocrats were slow to respond. The king had not forgotten his dazzling visit to Hven in 1592 – nor the sad state of his father's tomb in 1593–4.

At the young king's elbow stood Lord Steward Christoffer Walkendorf and the forty-year-old Royal Chancellor Christian Friis of Borreby. These men knew how to bring the proud to their knees. Walkendorf was hard as flint. He had been humiliated by his enemies and forced out of the Regency Council, but now he was back. He knew the central administration inside and out from over forty years of experience. Christian Friis was shrewd and erudite, with clients and supporters in the university and ties to Uraniborg, but he was a monarchist, not an oligarch. He had sailed to Tycho's island in 1592. He and Walkendorf knew those waters well, and they knew that the mightiest leviathan could be trapped in such a place.

Opposing theories of government separated Tycho Brahe and his allies from the circle of the young king. King Christian IV believed in the modern sixteenth-century theory that God appointed kings to rule. His views were influenced by those of his royal brother-in-law, King James VI of Scotland, a leading advocate of divine-right monarchy. King Christian and his advisors also believed with the French philosopher Jean Bodin that all sovereignty in a monarchy must be held in one hand. They considered the medieval ideals of universal empire and Catholic church to be outworn, along with the medieval ideal of feudal aristocratic rule by councils and diets. The aim of a progressive ruler like young King Christian IV was to establish a new unity of state and church under the crown. Bishop Peter Winstrup had expressed these very views at the king's coronation.

Tycho Brahe and his kinsmen believed that such theories were contrary to the Danish constitution. Before the state councillors joined hands to place a crown upon King Christian's head in 1596, they had watched him put his signature and seal to a coronation charter that prescribed "mixed monarchy" for Denmark, with sovereignty divided between the crown and State Council. Governance of Denmark-Norway was to be based on a balance of power. If the monarch upset the balance, he became a tyrant, and the State Council believed they would have the right and duty to depose him. Many of them, like Tycho's brothers, were sons and grandsons of state councillors, and they had not forgotten that their ancestors deposed a tyrannical king in 1523. They were confident that they could do it again, if need be.

The king and his ministers had powerful weapons at their disposal, but they could exercise power only in consultation with the State Council, and they needed to heed the annual Diet of the nobility. Within these limits, the king and his ministers could still accomplish a great deal. They could, for example, initiate the transfer of vassals from one fief to another. In this way, they could divide the State Council, rewarding friends and weakening enemies. This was precisely what they began to do. Tycho Brahe lost Nordfjord in the autumn of 1596. Erik Lange was stripped of Bygholm fief at the same time. Tycho's cousin Steen Maltesen Sehested was moved from Kronborg Castle at the heart of the kingdom to Bohus Castle on its fringe. In 1597, Steen Brahe lost Munkeliv Abbey, St. Hans Cloister, and Sæby-gård, though he retained Kalundborg; like Tycho, he kept his major fief but saw his income and influence reduced. Axel Gyldenstierne remained viceroy of Norway, but his authority became hollow when the central government took direct command of Norwegian affairs.[2] Piece by piece, the royal hand reached out and pulled in sovereignty, wresting power from the hands of great aristocrats.

BURDENS

The early months of 1597 were tense ones for Tycho Brahe. The future of his wife and children was in doubt. He had failed to make a match for Magdalene, and his own honor had been impugned. He had failed to endow Uraniborg and ensure that his sons, Tyge and Jørgen, could take over in due course. In January, fire broke out Uraniborg's laboratory, though it was extinguished without major damage. Servant problems recurred, Tuna peasants were restless, and Tycho was short of learned *famuli*. **Blaeu** and Pontanus had left in 1596, though **Longomontanus**, **Tengnagel**, and Wensøsil were still on Hven. To make up for the shortage of skilled manpower, David Pedersen, the bailiff, sometimes helped to observe the stars, and Nicolaus the tailor filled in as Tycho's courier.

Tycho spent much of February 1597 in Copenhagen. Toward the end of the month, he wrote to Viceroy Heinrich Rantzau that he was moving to his Copenhagen mansion and setting up his observatory, laboratory, and printing press there for the time being.[3] His case against Gellius Sascerides had been appealed to the Diet of the nobility. Tycho had to see to this and many other matters. While he was gone, there was a fire in the Winter Room at Uraniborg, but it was extinguished.[4]

In 1595, Tycho had commissioned a series of bronze and silver medallions bearing his portrait on one side and the Brahe arms with his motto on the other. They were intended as gifts to friends and could have played an important part in his strategy to regain honor, but the craftsmanship was not up to Tycho's usual standard. They were much less magnificent than Viceroy Heinrich Rantzau's earlier portrait medallion, to say nothing of the royal medallions and Danish Order of the Elephant that Tycho bore on his golden chains.[5]

Papermaking and printing still caused problems. Tycho had managed to publish his correspondence on astronomy in 1596, and now he wanted to complete his volume on the supernova of 1572, which had been in press for over four years. During February and March of 1597, Longomontanus led a frantic effort to establish the precise positions of over two hundred stars because Tycho wanted to expand the stellar catalog from 777 to 1,000 stars for publication in the forthcoming volume. Tycho also wanted to include his lunar theory but needed to complete it. Two printers arrived on 21 February 1597 to work on parts of the volume.

Everything had grown so large and complex. The range of Tycho's activities was endless, and something always needed attention. His income had been pared considerably by the loss of Nordfjord, but he was reluctant to give up any research or publishing projects. Frustration made him impatient and irritable, and he began to waver between brusque solutions or letting things drag. At heart, Tycho Brahe was a kind and generous person, but he had set great goals in life, and now he was wearing out, beginning to fear that those goals would not be realized.

THEOLOGY AND UNIVERSITY POLITICS

Tycho Brahe was like Joseph with his coat of many colors: Joseph's brothers hated him because he was their father's favorite and flaunted his splendid coat; some (though, happily, not his brothers) hated Tycho Brahe because of the splendors of Uraniborg, or because of his authority over them. Like Joseph's brothers, they veiled their resentment until the time came for revenge. The villagers of Tuna thirsted to be free of the burdens of *Gutswirtschaft*, but they needed an advocate who could stand up to Tycho Brahe. Until 1597, there had been none.

It was now clear that the new government brought new priorities, and Tycho Brahe's science would soon be forced to justify its huge costs. Cultural controversies intruded into the reassessment, and priorities of research and education came to be counterposed against each other. Educational programs seemed much less expensive than large-scale scientific research, with clearer, more immediate social and cultural benefits. Tycho Brahe's Uraniborg was threatened.

Some professors at the University of Copenhagen felt that Tycho's island drained the university of students and financial support. Now Tycho Brahe had built his mansion in the city, its irritating bell tolling every hour, its splendid observatory on the bastion behind the building, and that must have been a sore point in the narrow lanes of the Latin quarter. That this was Tycho's private facility, not part of the university and not a royal fief, must have seemed a direct provocation to a man like Professor Thomas Fincke.

Thomas Fincke was a wealthy patrician by birth, and he had achieved European fame in 1583 at the age of twenty-two by publishing a major work in trigonometry. Fincke was the first to describe the tangent ratio, and he coined the words *secant* and *tangent*.[6] The preface of his book had requested Tycho Brahe's favor, but Fincke never visited Uraniborg. That same spring of 1583, his cousin had married Petrus Severinus, and Fincke attended the wedding before departing for four years of medical studies in Padua. He received the M.D. degree in 1587 but continued to travel until 1590, then returned home, married well, and became physician to Duke Philipp of Schleswig-Holstein-Gottorp. In 1591, he had assumed Jørgen Dybvad's chair of mathematics when Dybvad was promoted to a chair of theology. Thomas Fincke was bright, energetic, and charming, a skilled mathematician, astrologer, and physician who soon developed an extensive private medical practice while also teaching.[7] Not only did he never make it to Tycho's island, he also never became friends with Tycho Brahe. Fincke was a Galenist who opposed Paracelsian medicine and iatrochemistry. Moreover, why should he send his best students to Uraniborg instead of keeping them in Copenhagen?

Christian Friis was both royal chancellor and chancellor of the university. He had ties to academic circles resentful of Tycho Brahe, and his extensive patronage could win the loyalty of scholars, perhaps even win them away from other patrons. Tycho's former *famulus* Johannes Stephanius became a client of the chancellor and was rewarded in 1597 with a university professorship. On the other hand, the loyal **Peter Jacobsen Flemløse** settled in Copenhagen in 1596 but could not obtain a professorship; he lived on incomes from the canonry he had received, at long last, in 1590.

At times, Tycho Brahe even doubted the loyalty of Niels Krag, who became the beneficiary of Anders Sørensen Vedel's fall in 1595. A swift governmental fiat deprived Vedel of everything: the office of Royal Historiographer, his prelature in Ribe Cathedral, all his manuscripts, even his own notes and writings. Niels Krag was named his successor as Royal Historiographer and was given all of Vedel's manuscripts. Clearly, Vedel was at fault for procrastinating too long with his national history, but Krag's role in his fall was unclear and troubling. Niels Krag had visited Hven in the middle of Vedel's humiliating summer. He told Tycho that his commission as Royal Historiographer was to begin with the history of the reigns of King Frederick II and his father, King Christian III. He would not be writing about the years of the Regency after 1588, and he told Tycho he was glad that he could write the "wedding song of Hven, and not its epitaph."8 What did he mean by that? Tycho took it as a veiled threat and was quite disturbed when he heard that Krag was quoting Petrus Severinus. Had his old friend, the royal physician, also turned against him?

Some of the theologians at the university also harbored resentments toward Tycho Brahe that flared up in the spring of 1597. These resentments stemmed from differences between the Philippist and Gnesio-Lutheran schools of Lutheran theology, and those differences had important implications for science. Philippist theology had encouraged the investigation of nature. Philippists argued that every person had a free will, whereas Gnesio-Lutherans emphasized the omnipotent power of God. Philippists put great trust in human reason, Gnesio-Lutherans in faith. Philippists held that matter was energized by a divine spirit immanent throughout the universe, and that mathematical analysis could unveil it: The book of nature was a powerful book of theology to them, and mathematics and astronomy were the means to unlock divine cosmic mysteries. Gnesio-Lutherans thought that science and religion were two separate realms, and that scientists should leave theology and cosmic mysteries to the theological experts. Philippists had a tolerant, irenic view of all religions and philosophies, but Gnesio-Lutherans drew sharp lines between true religion and heresy.

Tycho and Steen Brahe had long been leading patrons of Philippists, and Philippism continued to prevail in Denmark while Gnesio-Lutherans gathered strength in Germany. However, one Copenhagen theologian, Hans Poulsen Resen, had recently returned from abroad with Gnesio-Lutheran views and a determination to infuse them into the Church of Denmark. His strategy was to focus on key issues and fight to enforce conformity to the Gnesio-Lutheran view, one issue at a time.

Marriage was one such issue. Gnesio-Lutherans supported the royal ordinance of 1582 that decreed a marriage had to be performed by a clergyman to be legal. This gave the church a powerful weapon over the moral lives of great lords as well as ordinary people. Philippists, on the other hand, emphasized the binding character of the mutual pledge of betrothal and held that a church wedding was desirable but not necessary.

Another controversial issue was exorcism, the ancient ritual that preceded baptism. Reformed churches had abolished it, and when Philippists proposed to do the same, they were accused of being crypto-Calvinists. In Denmark, the church ordinance of 1539 made exorcism mandatory. Philippist challenges to the ordinance had failed, but Tycho had ordered Jens Jensen Wensøsil to omit exorcism from baptisms performed on the island of Hven.[9] In 1597, the Copenhagen theological faculty was divided on the issue: Bishop Peter Winstrup was a moderate Philippist and did not support exorcism, but Hans Resen considered the ritual to be of vital importance.

Neither Winstrup nor Resen had stood on Tycho Brahe's side when he brought his case against Gellius before Consistory.[10] Within the university faculty in general, Resen frequently opposed the brothers Krag over academic and administrative matters, and one of his most consistent allies in university politics was Thomas Fincke. Dangerous alliances were taking shape in academic as in political circles.

TRAPPED

Tycho Brahe was in Copenhagen from 6–13 March 1597, pursuing his case against Gellius. He returned to Hven with a person known only as "R.," possibly his friend Robert Anstruther, the Scots ambassador. Much was happening in those days, and Tycho's position was crumbling, but we have few sources beyond cryptic notes in the margins of the meteorological journal. The last observation recorded on Hven was made on 15 March 1597. "R." had left the day before but then returned, and a note dated Friday, 18 March 1597, recorded, "R. they go to the king. – Tycho goes to Copenhagen." On that day, Tycho Brahe's annual pension of 500 dalers, which he had received since the founding of Uraniborg, was discontinued; this and the loss of Nordfjord cut his annual income by around one-third.[11] The next Monday, somebody noted in the journal that Tycho's *familia* had completed registration of all his books, over three thousand volumes.[12] Then even the meteorological journal fell silent. Until that day in 1597, it had contained daily weather observations and notes of Hven events from 1582 without a break; then it stopped cold, and for the next three weeks, not a single entry was made.

Meanwhile, a royal patent was issued on Monday, 4 April 1597, ordering Christian Friis and Axel Brahe to visit the island of Hven as royal commissioners to investigate complaints that Tycho Brahe had oppressed the villagers, supported the pastor of St. Ibb's in violating the church ordinance of 1539, and committed other injustices of which Christian Friis was well informed.[13] The commissioners were ordered to submit a report to the crown that they were willing to defend. Tycho Brahe's own brother, Axel Brahe, a member of the State Council and Lord Lieutenant of Helsingborg, seemed compelled by this patent to serve as the instrument of Tycho's foes, for to do otherwise would show disloyalty to the crown; the best he could hope for would be to mitigate the charges. The role of Chancellor Friis was more ambiguous: He must have arranged the commission himself, summoning the Lord Lieutenant of the nearest fief as was customary in such cases, but then naming himself to be the sole other member of the commission. Whose side, then, was the chancellor really on? Was he out to get the lord of Uraniborg, or did he arrange things so he could also work in secret to countervail the malice of Tycho's enemies and mitigate the charges? No historian has ever been able to answer that question convincingly.

The next day, Tuesday, 5 April 1597, the case of Tycho Brahe versus Gellius Sascerides was assigned to a special court headed by Lord Admiral Peter Munk – since 1596 also Lord Marshall – because the Diet did not meet in 1597. Munk was Walkendorf's archrival and Tycho's stalwart friend.[14] Once again, the crown's advisors put an ally of Tycho Brahe in the awkward position of sitting as his judge – or was this more evidence that the chancellor was manipulating events to favor Tycho?

The royal commissioners came to Hven on Saturday and Sunday, 9–10 April 1597, and found the island in a state of turmoil. Tycho Brahe had packed his movable goods, including his library, laboratory equipment, printing presses, and all instruments capable of being disassembled. Many things had already been transported to Copenhagen. The commissioners, Christian Friis and Axel Brahe, met with the hundred bailiff, village alderman, grands, and others, collected their depositions according to Christian Friis's carefully prearranged plan, stayed long enough for the chancellor to assess the state of Tycho Brahe's affairs in detail, probably attended services at St. Ibb's Church, and then sailed off to prepare their report to the king.

The next day, Monday, 11 April 1597, was clear and fair. On that day, Tycho Brahe and his *familia* moved into the new "museum" in Copenhagen. He would never return to the island of Hven. His bailiff, David Pedersen, remained in charge at Uraniborg. Longomontanus, Tengnagel, Wensøsil, Johannes Eriksen, the housekeeper, cook, butler, maids and other servants, perhaps even Nicolaus the tailor, departed with Tycho Brahe,

Kirsten Jørgensdatter, and their children. They established themselves in Tycho's private complex in Farvergade.[15] The complex was large, occupying three lots. On the bastion behind the mansion, an observatory had been erected for the instruments brought from Hven.

The report of the royal commissioners resulted in a summons to court to answer charges brought by the peasants of Hven. Chancellor Christian Friis put questions to the peasants in the presence of King Christian IV and Tycho Brahe. The peasants stated that Tycho Brahe had not appointed churchwardens to maintain St. Ibb's Church but had pocketed the incomes and tithes and let the church deteriorate. Instead of maintaining the parsonage farm as a pastoral living, they charged that Tycho had expropriated the glebe lands, torn down some of the parsonage buildings, paid the pastor a few shillings and given him meals with laborers at the manor, appointed and dismissed pastors at will, without a call from the congregation and without due process of dispossession.[16] They confirmed that the words of exorcism had been omitted from the ritual of baptism in St. Ibb's Church for many years, and that Tycho Brahe had done nothing to correct the omission.[17] Tycho apparently brought countercharges that the peasants had damaged the Stjerneborg instruments with malice aforethought. This would explain why the proceedings came to a halt until more evidence could be presented.

Meanwhile, on Thursday, 14 April 1597, charges were filed against Jens Jensen Wensøsil, the pastor of St. Ibb's, for violating the ordinances of the Church of Denmark. Wensøsil was to be tried before Bishop Peter Winstrup on 22 April in Roskilde. To ensure that the trial went as it should, the king's true man, Ditlev Holck, now Steen Maltesen Sehested's successor as commander of Kronborg, was ordered to be in Roskilde to "assist" the bishop.[18]

Tycho Brahe traveled to Roskilde on Sunday, 17 April 1597. The night was stormy, and a blizzard blew in the next day. On Wednesday, 20 April 1597, he met with his brothers, Steen Brahe and Axel Brahe, both members of the State Council.[19] The next day, Chancellor Christian Friis gave his instructions to Hans Olufsen, chancery secretary, who was to prosecute the case against Wensøsil.[20] The trial took place as scheduled on Friday, 22 April 1597. Hans Olufsen brought two charges against Jens Jensen Wensøsil: first, that he had omitted exorcism from the ritual of baptism in violation of the church ordinance, and second, "that he has not punished and admonished Tycho Brahe on Hven, who for eighteen years has not taken Holy Eucharist but has lived an evil life with a mistress."[21] Wensøsil was found guilty on both charges.

The charges made it clear that the real target was Tycho Brahe, not his poor *famulus* pastor. They also revealed precisely how the two broader issues, Philippism versus Gnesio-Lutheranism and royal sovereignty versus mixed monarchy, were involved in the fate of Tycho Brahe. The fact that Tycho had ordered Wensøsil to omit exorcism revealed his strong commitment to Philippist theology. Bishop Winstrup was also a Philippist, but he was a timid and vaccilating man. Wensøsil was on trial not for his theology but for breaking the law by violating the church ordinance of 1539, which prescribed exorcism as an essential part of the baptism ritual in the Church of Denmark. Furthermore, he was on trial for failing to enforce the ordinances of 1580 and 1582 against common-law morganatic marriages. According to the 1580 ordinance, Tycho Brahe should have been denied the sacrament of Holy Eucharist for living with a woman with whom he had not been betrothed and wed before a clergyman, and that part of the ordinance had indeed been enforced, but it was not enough. The sentence of guilt established that a pastor must also demand that morganatic marriages be broken up.

They caught the little fish but could not yet trap the leviathan. Nothing in the ordinances could be used against Tycho Brahe. He had not taken Holy Eucharist in violation of the ban. The prosecutor might accuse Tycho Brahe of living "an evil life" and describe Kirsten Jørgensdatter as a "mistress" not a wife, but crown lawyers knew that the marriage of Tycho and Kirsten was perfectly legal under Jutish Law, and that any nobleman could appeal to be tried by Jutish Law before the high court of King in Council. Mixed monarchy in Denmark guaranteed the individual rights of the privileged aristocracy. Unable to trap Tycho directly, the advocates of royal sovereignty got at the big fish by scooping up the little ones. They also continued their campaign to impugn the honor of Tycho Brahe.

Jens Jensen Wensøsil was cast into a dungeon for a month. Many died under such conditions; according to Tycho, Wensøsil would have been beheaded had powerful friends not intervened on his behalf.[22] Could this have been a veiled reference to the subtle chancellor? By the end of May, Wensøsil staggered into the sunlight and returned to Tycho Brahe's mansion in Copenhagen. By then, the town constable had knocked on Tycho's door and, in the name of the king, forbidden Tycho Brahe and his assistants to use the bastion or city walls as an observatory; he came on orders from Lord Steward Christoffer Walkendorf.[23] This was a direct attack on the honor of Tycho Brahe and clearly intended to subject him to ridicule.

There must have been mirth that evening in Copenhagen Castle, just across the water from Tycho's mansion in Farvergade. They could see from the castle window that the instruments were gone and the bastion stood empty. The lord steward and his minions had done their work well. If the young king had been there, he might have danced with glee. It was like seeing a great whale thrashing in shallow waters.

Tycho Brahe was finished. On 1 June 1597, he gave **Christian Sørensen Longomontanus** a letter of recommendation for eight years' faithful service and sent him on his way.[24] Tengnagel also departed for the Netherlands around this time, if not earlier. The next day, Saturday, 2 June 1597, Tycho and the remainder of his household, more than a score of people, took their leave of Copenhagen and headed south.[25] A few days later, they saw their last of Denmark. Tycho Brahe sailed away from his native land. Sorely plagued, he had been forced at last to make a choice, and he made it in frustration and anger: Tycho Brahe chose exile over dishonor.[26]

On the very day he left Copenhagen, a commission of two professors from the university was dispatched by the crown to investigate Tycho's charge that peasants had damaged the Stjerneborg instruments. One was Iver Stub, and the other, Thomas Fincke. Together, they bobbed across the waters of the Sound toward Tycho's island. David Pedersen met them, showed them around, answered their questions, and explained everything. What they saw and reported is known only as Tycho Brahe told it from exile, presumably on the basis of information sent by David Pedersen.[27] Into an Ovidian elegy, written later that same year, Tycho Brahe wove this picture of Thomas Fincke viewing the wonders of Stjerneborg:

> He they sent with one to Hven to spy,
> Urania, has probed your sacred site,
> Was staggered by the splendor he did see
> Of instruments, though I left only few.
> So ignorant a man, how can he help,
> Admitting he's not seen nor heard of such?
> Puzzled, stares and asks where they were made,
> And how they work, and shows himself the fool;
> So that his journey will not seem in vain
> Criticizes, full of hate, what he does not
> Understand. Of course: my foe had sent him,
> Long he planned evil, scheming in secret always.

The inspection completed, Thomas Fincke and Ivar Stub took their leave of Tycho's island. What tales they could tell in the Latin quarter, after they had reported to the mighty ones at Copenhagen Castle!

On 10 June 1597, the crown took action on another matter. The canonry of the Chapel of the Three Holy Kings in Roskilde Cathedral, with its

lucrative estate of manors, farms, cottages, mills, and advowsons, was removed from the jurisdiction of the former canon, Tycho Brahe, and bestowed upon a new lord: Royal Chancellor Christian Friis of Borreby.[28]
Was this a fine reward for all he had done to break an overmighty subject,
or was it a bitter irony for one who had conspired to protect his kinsman,
Tycho Brahe, from the wrath of Lord Steward Walkendorf and the young
monarch?

10

EPILOGUE: IN SEARCH OF MÆCENAS

———

1597–1599

ROSTOCK

Uraniborg stood empty under the watchful eye of David Pedersen in the late spring of 1597. The four great Stjerneborg instruments left behind were silent, and David lived in the bailiff's quarters of the grange. Laborers were still summoned from Tuna each day to work on the manor. When they failed to appear or caused trouble, David could still appeal to the crown, and Christopher Walkendorf even came to his aid to maintain order.[1] Thomas Fincke and Iver Stub had undoubtedly discovered everything under control on Hven. Tycho's complex in Copenhagen also stood empty. Esge Bille, a relative and friend, took general charge of Tycho's affairs in Denmark.[2] Whether a caretaker staff lived in the mansion is unclear.

Tycho Brahe and his *familia* traveled the southern road to Germany. In addition to teamsters and escorts, the party consisted of more than a score of people: Tycho, Kirsten, their six children, Wensøsil, and various other servants.[3] Tycho and Kirsten were around fifty years of age, their daughters twenty-three, nineteen, eighteen, and seventeen, and their sons sixteen and fourteen. Most of them would never see their native land again.

International transfer of a complete research establishment like Tycho Brahe's involved a large-scale relocation of personnel, archives, and equipment.[4] The move from Hven to Copenhagen had brought disruptions, but the distance involved was short. The strategy of moving overseas was much more complex because of the need to deal simultaneously with a large variety of cultural and scientific options while convincing individual staff members to remain with the project and, at the same time, making provisions for long-distance transfers of material. Mounted and armed men now accompanied the train of heavy wagons carrying Tycho's books, manuscripts, printed sheets of unpublished treatises, laboratory equipment, astronomical instruments, and the immense globe, in addition to the family's clothing, household goods, and ironbound chests holding a fortune in gold, silver, and jewels. How to continue working during the transfer was a major problem: Much activity would have to wait until everything could be reassembled in a new location, and even then, it would hardly be possible to continue as before. New political, economic, social, and technical structures would inexorably reshape priorities.

The caravan lumbered on, along the road to the continent, southward to Køge and on to Vordingborg, where Tycho Brahe had spent part of his youth in the sprawling castle commanded by his foster father. They were ferried across narrow straits to the island of Falster and rode on to Nykøbing, where Tycho's foster father had commanded another royal castle. Finally, they came to the port of Gedser on the southern tip of Denmark.[5]

As he traveled this road, Tycho Brahe must have thought of many others who had left their native land in order to recapture lost power. His foster uncle, Peter Oxe, had been driven into exile, forty years earlier, by enemies at court, but had then been recalled to become the most powerful man in the kingdom.[6] His own brother, Knud Brahe, had fled Denmark in disgrace some eighteen years past – Tycho had helped to arrange his flight – then returned after two years to regain positions of high honor.[7] Tycho's Swedish kinsman Lord Chancellor Erik Sparre had passed through Denmark in March of 1597, three months since, fleeing the tyranny of the Swedish regent, Duke Carl, and heading for the court of his sovereign, King Sigismund III of Sweden and Poland.[8] Finally, there was Erik Lange, who had fled in 1592 with Tycho's assistance and still wandered in search of the philosopher's stone while his betrothed, **Sophie Brahe**, pined for him at her castle in Skåne.

Tycho and his *familia* sailed to the free city of Rostock, where they settled into temporary quarters around the middle of June 1597. Tycho had spent several student years in Rostock. Lucas Bacmeister still lived in the house where Tycho's duel with Manderup Parsberg had taken place, and he still took in Danish student boarders. There were many Danes at the university, and Tycho had numerous friends, including the most famous Rostock professor, David Chytreus. His friends greeted him warmly.

Tycho Brahe in the summer of 1597 was far different from the hot-blooded Danish student he had been in 1566. Through a long process of self-fashioning, he had become by stages a solitary scholar, an aristocratic courtier, a royal favorite, and finally, a unique persona: lord of Urania's Castle and Prince of the Muses, whose splendor of life-style, learning, and intellectual and technical achievement drew monarchs and scholars to his presence. He had achieved a fame that would last for centuries, and his persona was not a mask or an assumed identity but something that went to the core of his being.[9] Tycho was no Machiavellian, full of prudent cunning, but one who strove sincerely to bring his inner self in harmony with outward appearances, taking as his motto *Non habere sed esse* ("Not to seem, but to be").[10] In 1597, Tycho was truly a famous, Apollonian figure whose cultural accomplishments were legion, and he was accepted as such. In Rostock, his honor was fully intact.

At the same time, Tycho Brahe had much work to do. His broad aims in exile were the same as they had been at home: To regain his honor in Denmark, finish and publish his work in astronomy, and provide for the future of his wife and children. Voluntary exile was simply a new strategy to accomplish these aims. He devised a fourfold plan. First, he would challenge King Christian IV to live up to his obligations. Second, he would invest his liquid assets abroad where they would be safe. Third, he would bring foreign pressure to bear upon the Danish government until his enemies relented and he was restored to honor and glory. Fourth, his fallback strategy, was to seek a new patron among the greatest princes of Europe.

On 10 July 1597, Tycho Brahe turned to the first and most difficult of those tasks: to remind the king of Denmark of his obligations and challenge him to live up to them. Tycho wrote a long letter to King Christian IV in the Danish language, stating his reasons for going abroad without first taking leave of the king, as a courtier should have done. He reminded the king that his father had promised to endow Hven in perpetuity as an observatory and research institute, and that this had been confirmed in formal documents signed and sealed by Queen Sophie, the Regency Council, and the whole State Council. Tycho stated that he had invested much labor and money to improve the facilities on Hven in the hope that King Christian IV, when he came of age, would let "me and mine" enjoy the fruits of his effort. Things had turned out differently than expected: Tycho's income from royal fiefs had been cut off, and he did not have the resources to continue at his own expense, "for I had already come so far astern, even when I had the previous fiefs, that I found myself compelled to dispose of my paternal estate." He asserted that he would rather serve the king of Denmark than any other master, "insofar as this may be done on reasonable terms, and without damage to me." If that were not possible, he said that he would seek a patron abroad.[11]

This letter did not have a deferential courtly tone. It was the letter of a proud aristocrat to a stripling youth, or of a tired old man who had been baited and humiliated beyond endurance and would suffer no more. It expressed genuine feelings but was also somewhat less than ingenuous in parts. Its overriding flaw was hubris: Tycho Brahe wrote as a man who stood above kings, and he was not. Had he forgotten this, or was he deliberately trying to erase it from his mind?[12] Tycho worked carefully on the letter, sent a draft to Holger Rosenkrantz, considered his comments, then signed and sealed the clean copy, gave it to the courier, wished him Godspeed, and sent him on his way.

The same day, he wrote again to Holger Rosenkrantz. Tycho was eager for news from Denmark and wondered what people were saying about his departure. He awaited the arrival of a Danish embassy in which his brother, Steen Brahe, was a leading member.[13] Steen was a state councillor and former regent, and he was well qualified to play a key role in negotiating Tycho's return to Uraniborg. Tycho wrote Holger that he had been advised to wait upon the intentions of the king of Denmark, but he was confident that he could find another royal patron if that proved necessary.[14]

Tycho Brahe spent the whole summer of 1597 in Rostock. In this strongly Lutheran city, he could enter St. Mary's Church with his family to receive Holy Eucharist for the first time in eighteen years. This civilized and learned city was a fit abode for all the Muses.[15] It could have been such a pleasant place, if only one could forget those nagging questions, always lingering in the back of the mind. What will happen next? When will the king reply?

At the end of July, Tycho Brahe wrote to Duke Ulrich of Mecklenburg, whose court was in Güstrow, and to the duke's chancellor. Tycho explained that he was in Rostock and asked for the duke's favor in completing his work in astronomy.[16] Within days, he received an extremely cordial reply in which the duke recalled his visit to Hven in 1586 and the gifts of publications he had received from Tycho Brahe. Duke Ulrich enclosed a favor of his own: the draft of an intercession to his grandson, King Christian IV, graciously requesting him to carry out the will of his late father and the State Council so Tycho Brahe could return to Hven and complete his work in astronomy, to the glory and honor of the kingdom of Denmark, for Tycho was a man whose equal could not easily be found and his work was famous in all lands.[17]

Soon Tycho Brahe and the duke were involved in other matters. The future rulers of Mecklenburg were two children, Duke Adolf Friedrich and Duke Johann Albrecht, aged nine and seven, descendants of Duke Ulrich's elder brother and heirs, respectively, to Mecklenburg-Schwerin and Mecklenburg-Güstrow. Their guardians were the current regents of their lands, Duke Ulrich and his nephew, Duke Sigismund August of Mecklenburg-Ivenack, who had also visited Tycho's island in 1586. The guardians were in urgent need of cash for public matters, and they negotiated with Tycho Brahe for a loan of ten thousand thalers (the German thaler being worth roughly the same as the Danish daler). Several thousand dalers of Tycho's own money were tied up in loans to Axel Gyldenstierne and other Danish friends, but Tycho had borrowed ten thousand dalers from Heinrich Rantzau at 6 percent interest, so he had plenty of cash on hand. He agreed to loan the same amount to the young dukes at 12 percent interest, but only with very strict guarantees of repayment: the bond of ten landed noblemen and the whole county of Doberan as surety until a binding contract might take force. The guardians agreed to the terms, and Tycho made the loan in early September.[18] By that time, plague was threatening Rostock, and Tycho Brahe was planning to leave. Viceroy Heinrich Rantzau had invited him to Schleswig-Holstein. King Christian IV still had not replied to his letter, written nearly two months earlier.

On 7 September 1597, Tycho wrote to Lord Chancellor Erik Sparre in Poland about obtaining a Baltic island from King Sigismund. He suggested Dalan near Riga or Werden near Gdańsk, among others.[19] Apparently Tycho had offered medical advice and Paracelsian medicaments to members of Sparre's party during their brief sojourn in Denmark, so they had experienced his active wisdom at first hand, but the timing of his request was not good. Warfare was brewing between King Sigismund, who held Poland-Lithuania, and Duke Carl, who held Sweden-Finland and was demanding the return of Erik Sparre so he could hold him as a hostage. Nothing came of Tycho's request.

WANDSBURG

Tycho Brahe and his *familia* took their leave of Rostock on 8 September 1597 and proceeded westward through Doberan and Wismar toward the duchy of Holstein. By 17 September, they had had arrived at Segeberg Fortress, seat of the viceroy of Schleswig-Holstein, only to learn that he was in Bramstedt. Tycho sent him a letter. What I need, Tycho wrote, is a residence close to a city like Hamburg or Lübeck, where I can work with printers to complete my astronomical publications. Tycho and the *familia* proceeded to Bramstedt, and Viceroy Heinrich Rantzau received them with open arms.[20]

Viceroy Heinrich Rantzau [Fig. 44] was as famous as Tycho Brahe, and much more rich and powerful.[21] In his youth, he had supped at Martin Luther's table and served at the court of Holy Roman Emperor Charles V.

Fig. 44. Viceroy Heinrich Rantzau. Painting by unknown artist, 1598.
(Courtesy Det Nationalhistoriske Museum på Frederiksborg, Hillerød)

He corresponded in Latin with learned men across Europe, kept a staff of
scholars and learned assistants as Tycho Brahe had done, commissioned
works of art and scholarship, and published a number of books in his own
name. His library at Breitenburg Castle included over six thousand volumes. His learned interests ranged from history, chorography, and genealogy to poetry, astrology, religion, and the occult. Heinrich Rantzau had
technical interests and owned nearly forty mills, including oilmills, paper-
mills, and coppermills. He was a great builder in the Renaissance style,
richly embellishing his works with Latin epigrams, pavilions, pyramids,
and emblematic art including portraits of princes and warriors from all
lands. On numerous occasions, he struck medallions to commemorate his
own accomplishments. Despite all these learned and material splendors,
Heinrich Rantzau was first and foremost a politician. Now over seventy

Fig. 45. Wandsburg. Woodcut by unknown artist, 1590. (From Lorenzen 1912)

years of age, he had been Viceroy of Schleswig-Holstein for more than forty years. He had been careful to do proper homage to the young king, and Christian IV did not challenge his viceregal authority.

Tycho Brahe and Heinrich Rantzau discussed strategies of patronage and the need to find a suitable residence until Tycho and his household could return to Denmark. The Rantzau clan owed some seventy castles and manors, of which Heinrich Rantzau and his sons owned no fewer than eighteen.[22] Among these was Wandsburg Castle in Wandsbek on the northern outskirts of Hamburg [Fig. 45], built in 1568 and currently leased to one of Heinrich Rantzau's former *famuli*. Tycho recognized the name immediately: It was none other than his onetime guest on Hven, Georg Ludwig Frobenius. Frobenius had married the niece of Rantzau's principal secretary, Detlev Wolders, and his career had blossomed.[23] Despite the lease, Heinrich Rantzau graciously put the castle at the disposal of Tycho Brahe, and Tycho was pleased to accept, knowing that a creature like Frobenius would always adjust to his patron's will.

Despite his quasi-princely status, Viceroy Heinrich Rantzau was not prepared to become Tycho Brahe's new patron but merely to loan him a castle; and Tycho was not prepared to become the client of a viceroy, or for that matter, of a landgrave or duke, but only of a king or emperor. The delicate status protocol of patronage required Rantzau to receive Tycho Brahe as a friend and guest who was welcome to stay as long as he chose.

The Brahe caravan rolled into Wandsburg soon after. It was late September of 1597, and the road to the castle ran through an autumnal wood. The portal led into a quadrangular courtyard, surrounded on three sides by half-timbered outbuildings.[24] Directly ahead was the moat and drawbridge; beyond that, the massive brick castle itself, banners flapping from many spires above stepped gables. Swans and ducks swam in the moat. The horses clattered across the cobblestone courtyard, over the drawbridge, through a Renaissance portal decorated with the arms of Rantzau and Halle, into the court of honor within the castle. High castle wings enclosed three sides of the court, with a brick wall on the fourth side. Here in Wandsburg, Tycho Brahe could live in a manner appropriate to his station until the patronage of King Christian IV could be ensured.

Wandsburg Castle no longer stands, but its spaces reflected the typical style of Holstein castles and manors. The entrance undoubtedly led into a large hall, decorated with weapons and trophies of the hunt.[25] From the hall, a spiral stairway led to the Knights Hall above, with carved paneling, Gobelin tapestries, stucco ceiling, and immense fireplaces at each end. The chambers of the lord and his household filled the main wing. Guest rooms, servants' quarters, kitchen, bakery, brewery, dairy, and storage rooms occupied the side wings of the castle. Windows in the main wing offered views of a Renaissance knot garden behind the outer courtyard. The top of the high stairway tower afforded a clear view of the heavens with room to set up instruments.

As Tycho Brahe and his household were getting established at Wandsburg and the servants were unpacking personal gear, printing presses, and instruments, a courier arrived with a letter from King Christian IV. Tycho set everything aside. His secretary broke the royal signet, unfolded the letter, and read it aloud. This was the moment Tycho had long awaited, but it was not the message for which he had hoped. In sharp, precise language, the letter stated that King Christian IV took great offense at the tone of Tycho's letter, written "audaciously and not without great lack of understanding, as if We were to render account to you concerning why and with what cause We made changes in Our and the crown's estates."[26] The king let Tycho Brahe know in no uncertain terms that "from this day on, We shall be otherwise respected by you if you expect to find in Us a gracious lord and king." King Christian objected to Tycho's going abroad to beg for patronage, accompanied by his "woman and children" – he would still not acknowledge Kirsten Jørgensdatter as a true wife – as if King Christian IV and the realms of Denmark and Norway were too poor to support him. The king answered Tycho's assertions, point by point, well aware that Tycho's actions were motivated by family concerns as well as science, and asserted, "it is extremely doubtful that you have used the money from the estate you sold to pay for astronomical instruments, for it is said that you have many thousands to loan to lords and princes, to the benefit of your children and not to the honor of the realm or the advancement of [astronomy]." The letter was dated 8 October 1597, signed and sealed at Copenhagen Castle.[27]

It was concise, brutal, and effective. Tycho Brahe was crushed, and yet, he should have known better. During the years since the death of King Frederick II, Tycho had gradually refashioned his "courtier" persona into that of an "overmighty subject." A courtier sought the favor of the ruler in order to obtain patronage. An overmighty subject tried to compel the ruler to act. A ruler had but two choices in facing an overmighty subject: confront and crush him, or else become his puppet. Young King Christian IV, surrounded by sagacious and loyal advisors, was determined to be no man's puppet.

The move against Tycho Brahe was part of the young king's strategy to consolidate power at the beginning of his reign. For the time being, he would abide by the coronation charter and share sovereignty with the State Council, but at the same time, he worked to strengthen his supporters and weaken his opponents among the state councillors. In religious affairs, he did not interfere with the ambitions of the Gnesio-Lutherans to take control of the Church of Denmark, aware that this would also strengthen royal control. In the patronage of learning, he moved to build up the university at the expense of Tycho's island, and he concentrated iatrochemistry at court with other arcane studies that had political applications. Of all the oligarchs who stood in his way, Tycho Brahe was the most vulnerable because of his unorthodox marriage, his malfeasance with the chapel in Roskilde, his stubborn pride, and the enmity of some professors and noblemen toward him. The king and his advisors therefore decided to make an example of Tycho Brahe.

Tycho read and reread the letter in his chambers at Wandsburg Castle. What now? As so often in times of great intensity, Tycho's thoughts found expression in words inspired by the Augustan poet, Ovid, who had also known the bitter fruit of exile. One autumn day, not long after receiving the king's letter, Tycho Brahe sat down in Wandsburg, took paper, stirred the ink, sharpened a quill, and let his emotions flow in Latin verse, modeled on the elegies of Ovid. The first word he wrote was *Dania:* Denmark, his native land, the land he longed for. The words came freely after that. When he was finished, his "Elegy to Dania" ran to 102 lines, beginning as follows.[28]

> Denmark, what is my offense? How
> Have I offended you, my fatherland?
> You may think that what I did was wrong,
> But was it wrong to spread your fame abroad?
> Tell me, who has done such things before
> And sung your honor to the very stars?
> And now, who can carry on, who can use
> The precious treasure I have given you?
> He they sent with one to Hven to spy,
> Urania, has probed your sacred site,
> Was staggered by the splendor he did see
> Of instruments, though I left only few.

The poem went on to assess Tycho's achievement in terms that were far from modest, but also far from untrue. Tycho emphasized his unparalleled achievements as an observer of the heavens, but also his work as a physician who dispensed medicaments from his laboratory without charge, as an educator who trained young scientists, and as a nobleman who extended the hospitality of Uraniborg to all. The elegy moved from dejection to an optimistic view of the future and ended with thanks to Heinrich Rantzau for providing a place of refuge at Wandsburg.

> Only a few of the Danes honored my work.
> Herculean it was, for as they say
> That Hercules held up the fainting Atlas
> And hindered heaven's fall, from pole to pole,
> You, Ptolemy, Alfonso, Copernicus,
> I gave a hand: you slipped, but I stood fast.
> The motion of the stars you could not grasp
> As I have done. In truth, my work was great:
> New pillars raised for heaven's sparkling dome.
> . . .
> And Machaon's art and wisdom I have searched,
> Who helps the sick, restores the weak to strength.
> If Danes are silent, Norway, Sweden speak,

And tell of many ill who sought my help!
And never did I ask for gain, although
I served a hard apprenticeship to learn it.
And that, perhaps, aroused an envy, hate,
Which grew until it drove me into flight.
Hatred grew secretly, never I knew it was there
Until it appeared, strong, and did its deed.
. . .

Many by me have been led into life's deepest secrets
And many a man I gave goods and food in my home.
More I have done, which I omit to name,
For I have said enough and will not boast.
Ye gods above! For this, they drive me out!
Six children share my fate, their mother too.
But rightly seen, though driven out, my wings
Are free at last. And Denmark was my exile.
My wings are spread, my will is free, and I
Have lost my home to win a wider world.
. . .

So fare thee well! My fatherland now lies
Wherever, humbly, men behold the stars.
. . .

Hail thee, Heinrich, sprung from Rantzau stock,
Urania found refuge in your home,
Here, beside the mighty Hamburg walls,
Your Wandsburg towers up, but few years old,
And built, when he had reached his forty years,
By that heroic man whose fame will live.
May God, who steers the starry heavens high,
Allow that I may rightly use it now,
So all the world may see the heavenly sight
That I do see, long hid in ether's night.
O Rantzau, truly, long will live your name
And when you die, eternal burn your fame.

Tycho Brahe was never one to mope. He sent a copy of the elegy to
Viceroy Heinrich Rantzau, then threw himself into his work with three
aims in mind. He would transform Wandsburg into a new Uraniborg
where he could carry on his research and publication projects. He would
keep pressure on the Danish government to restore him to the island of
Hven on a permanent basis. At the same time, he would put his fallback
plan into action and search for a new patron among the greatest rulers of
Europe, one who was willing and able to support him in his accustomed

style in return for the glory that Tycho could bring to a patron and his realm. The Prince of the Muses set out to find an imperial patron.

PATRONAGE SEARCH

Twenty-two years earlier, Tycho Brahe had attended the coronation of Archduke Rudolf of Austria as king of the Romans, and this same prince was now Holy Roman Emperor, with his court in Prague. At the coronation in 1575, Tycho Brahe had become acquainted with Archduke Rudolf's personal physician, Thaddeus Hagecius, and they had remained friends and faithful correspondents over the years. Hagecius was still the Emperor's physician and one of his most trusted personal advisors. A powerful patronage broker – it was he who had commended Paul Wittich to Tycho's service in 1580 – Hagecius was now recruited to lay the groundwork for Tycho Brahe himself to enter the service of the Emperor, advising Tycho by frequent letters whom to approach and how.

Tycho had other contacts among imperial courtiers. Hugo Blotius, the imperial librarian, had been an acquaintance since student days in Basel. Jakob Kurz von Senftenau, the late imperial vice-chancellor, had shown favor to Tycho and obtained an imperial copyright for him in 1590.[29] Kurz had died in 1594, but Hagecius would soon put Tycho in touch with the new vice-chancellor, Rudolf von Coraduz. Moreover, two Austrian noblemen had visited Uraniborg, spent time at the Danish court, and later worked with Hagecius to acquire **Paul Wittich's** library for Tycho.[30] They were Baron Felicianus von Herberstein and Baron Johann Septimus von Liechtenstein, whose nephew, Baron Karl von Liechtenstein, was a rapidly rising star in the imperial administration. Another great Bohemian magnate with arcane interests, Baron Ehrnfried von Minckwitz, had also been in Denmark as imperial legate to the Danish court and eventually became Tycho's firm friend.[31] There was a fourth Austrian baron, an alchemist who had visited Uraniborg with Erik Lange in 1584 – what was his name? He had married a Danish noblewoman and settled on her estate in Jutland. Perhaps he could help as well, but Tycho had lost track of him.

In the meantime, Viceroy Heinrich Rantzau provided Tycho Brahe with a place to live, but the daily maintenance of the household was at Tycho's personal expense. Tycho needed to rebuild his staff. He thought of those he knew, scattered far and wide across the continent of Europe. Some of his former *famuli* held good positions and would not be available: Arctander was bishop of Viborg; Oddur Einarson bishop of Skálholt; Peter Hegelund bishop of Ribe; Stephanius professor in Copenhagen; Flemløse a canon in Roskilde; Hemmet lector of theology in Ribe; Christoffer Jensen a pastor in Norway; Anders Lemvig, Jacob Lemvig, and Jacob Hegelund pastors in Denmark; Steenwinckel still active as an architect; Elisabeth Paulsdatter married and living in Elsinore; Peter Hvæn and Hans Stockelman presumably working as printers in Copenhagen; Blaeu recently married and established as a publisher in Alkmaar; the brothers Langren back in their father's shop in Amsterdam; also in Amsterdam, Christoffer Pepler, his bailiff of many years in Nordfjord, now a goldsmith; and David Pedersen keeping track of things on Hven. Where was Martin Coronensis? Tycho probably knew. Gemperle, Crol, and Morsing were dead; so were Wittich, Niels Colding, and possibly also Anders Viborg. Gellius and his brother David Sascerides were out, of course. Tycho had lost touch with John Hamon in England and had not heard from Christopher Rothmann since 1594.

Frans Gansneb Tengnagel van de Camp had been visiting his kinsmen in the Low Countries when Tycho left Denmark, but he rejoined the *familia* in Rostock. He, Wensøsil, and possibly Johannes Eriksen were Tycho's only learned assistants at the time, and Wensøsil wanted to go home. Tycho did not hold him but gave him travel money and sent him on his way. That left Tengnagel and Johannes Eriksen. There may have been others, but if so, their names are unknown.

Several of Tycho's former *famuli* were studying at foreign schools and universities. **Cort Aslakssøn** was in Siegen, serving as preceptor to Tycho's nephew, Otte Steensen Brahe, and another young nobleman, and he kept in touch with Tycho by correspondence. Jacob Fabricius, Longomontanus, and Metius were at German universities. Riber was in Italy. Pontanus had spent two years in England but returned to Leiden in 1597. Tycho kept track of them all. He began to write to some of them, inviting them to join him at Wandsburg.

Meanwhile, Tycho Brahe made contact with a young Hamburg printer, Philipp von Ohr, who had done some work for Viceroy Heinrich Rantzau. Tycho needed favors to captivate a new patron, and he thought of a book describing his past accomplishments, his prodigious scientific apparatus and facilities on the island of Hven, and his future plans under a new patron.[32] There could be no better demonstration of the superiority of Tycho's instruments and methods than such a book. Tycho also decided to prepare hand-lettered copies of his great achievement, the catalog of a thousand stars: the first new stellar catalog since the birth of astronomy in ancient times, and far more accurate than any other current data. These two works, the book and the manuscript, luxuriously bound and autographed by Tycho Brahe, could be carried by couriers of noble blood as gifts to important scholars, princes, and potential brokers of imperial patronage. The most sumptuous copies would be presented by Tycho himself to his new royal or imperial patron.

Tycho Brahe set his assistants to work, completing the observations necessary to catalog one thousand stars.[33] He drafted his publication plans with Philipp von Ohr. Tycho had a good collection of woodcut printing blocks at Wandsburg, besides his presses, fonts, and other equipment, and he had already published a pamphlet in 1596 with eighteen woodcuts of his instruments. He also had Blaeu's woodcut map of Hven and the woodcuts of Uraniborg and Stjerneborg used to illustrate the volume of his correspondence, but he had no copperplate engravings except the portraits of himself by Gheyn and Sadeler. Rantzau or Ohr may have helped him locate artists to make copperplate engravings of Uraniborg, the mural quadrant, and the large celestial globe. Copyists were put to work on the star catalog, hand lettering the finest exemplars on vellum parchment.

Meanwhile, Viceroy Heinrich Rantzau was planning to attend the wedding of King Christian IV to Princess Anne Catherine of Brandenburg. Tycho wrote to the old nobleman, addressing him a friend and kinsman, sending suggestions for guarding his health on the journey, "and if anyone asks whether I will return to Denmark, the Lord Viceroy should simply reply that one does not know."[34] In case anyone persisted, however, Tycho sent suggestions for brokering his return to Hven.

The parents of the royal bride were Margrave Joachim Frederick and Margravine Catherine of Brandenburg-Küstrin. They arrived for the nuptials with a splendid throng of two hundred courtiers, including their court astrologer, Johannes Müller, who had visited Hven in 1596. Müller had looked forward to visiting Tycho again and was surprised to learn that he had left Denmark. On their way home, the Brandenburg delegation passed through Schleswig-Holstein, and Viceroy Rantzau arranged for Tycho Brahe to meet the margrave and margravine in Bramstedt on 22 December 1597. They were extremely accommodating and agreed to plead Tycho's case with their new son-in-law, in the hope that Tycho could be restored to Uraniborg.[35] The meeting also gave Tycho a chance to renew his acquaintaince with Müller.

Returning to Wandsburg, Tycho Brahe worked on the book describing his instruments. He took the long description of Hven from his scientific correspondence, rewrote it, added a brief autobiography, and composed a foreword dated 31 December 1597, which he addressed to the Holy Roman Emperor Rudolf II, but the work did not come off the press until the spring of 1598. Ohr set the type from his own fonts, but he used Tycho's illustrations and printed it on Tycho's press.

In January of 1598, the margrave became Elector Joachim Frederick of Brandenburg upon the death of his father. With his elevated title and dignity, he and Electress Catherine wrote to Denmark on behalf of Tycho Brahe.[36] In order to penetrate the phalanx of enemies in the inner guard of the Danish court, they applied a strategy of transmission using four letters. The first was a cover letter addressed to Royal Chancellor Friis and Lord Steward Walkendorf, requesting them to present the enclosed letters to King Christian IV. No official could deny a direct request from the king's parents-in-law – although any official knew how to delay matters. The two enclosed letters were from the elector and electress, making short, direct appeals to restore Tycho Brahe to his position for the good of the Danish kingdom. The fourth letter was from Electress Catherine to her daughter, the queen. This letter took the form of private family correspondence in order to circumvent the channels of court officials, but the content was exactly the same as her letter to the king. However, Tycho's enemies found ways to frustrate the strategy of transmission, and all the letters were delayed until they could no longer be effective.

Wandsburg was on the direct route from Jutland to the continent, and a stream of young Danish scholars stopped there during the winter of 1597–8, just as student preceptors and their aristocratic pupils had formerly called at Uraniborg before going abroad. Students liked to have famous men sign their autograph books, and they wanted Tycho's advice, perhaps even his letters of recommendation to famous professors who were his friends and correspondents.[37] Another visitor that winter was Holger Rosenkrantz, twenty-three years of age and one of the brightest young aristocrats of his generation. After several years of study in Rostock and Wittenberg, he had entered the Danish Chancery and soon became one of Tycho Brahe's main sources of inside information from court. Rosenkrantz was about to marry Tycho's niece, Sophie Axelsdatter Brahe, but a personal religious crisis drove him to seek out Tycho at Wandsburg, where his quick mind, erudition, and earnest integrity soon endeared him to the astronomer. Rosenkrantz opened his heart and laid out his orthodox, German-inspired religious views, while Tycho listened with care and gently replied with his own philosophical theology of Hermetic and Platonic Philippism, touched with Ramism and unified by an emblematic world view.[38] Tycho taught Holger the empirical methods of Uraniborg, and on 10 February 1598, they stayed up most of the night, observing a lunar eclipse.[39] The whole experience was both inspiring and terribly disturbing to the young man because it shook his religious orthodoxy to the roots, and he went on to ponder their conversations for many months. Meanwhile, the two of them worked out a secret code for future correspondence, and Holger Rosenkrantz returned to promote Tycho's cause in Denmark.

There was a solar eclipse on 24 February 1598. The sky was overcast at Wandsburg, and Tycho Brahe was unable to observe all of it, but now there were trained observers, using his methods, in many parts of Europe. Christian Riber observed the eclipse from the little Danish town of Hobro, David Pedersen from Hven, and Christian Longomontanus from Rostock. Tycho also heard from Melchior Jöstelius in Wittenberg and Joachim Radenicius of Rostock. All of them sent the results of their observations to Wandsburg, and Tycho entered them in his journal.[40] Collective empirical endeavors like this spanned the continent, centered on the old master, Tycho Brahe.

Around the time of the eclipse, Tycho Brahe received a letter from Electress Catherine of Brandenburg: She was sending Johannes Müller to Wandsburg to study astronomy and chemistry. Müller arrived by March. He was a man after Tycho's own heart, and they got on very well. Tycho put Müller in charge of the astronomical work and made him supervisor of the students in the *familia,* but he did not stay nearly as long as Tycho would have liked, because the Electress summoned him home by early summer.

Tycho was receiving frequent reports from the courts in Prague, Copenhagen, and other places, sent by informers who had access to courtly circles but did not possess the power to influence decisions. They provided the flow of information he needed to refine his strategy of seeking patronage.[41] Tycho wrote to Holger Rosenkrantz, inquiring about the Austrian baron who had married a Danish noblewoman, and Rosenkrantz replied that Siegfried von Rindscheid, Baron zu Friedberg, had returned in 1596 to his family estates in Styria, an imperial Austrian province dominated by his kinsmen and their allies. Soon, Tycho and the baron were corresponding in the Danish language, and Tycho had recruited one more advocate with influence at the imperial court.[42]

Tycho had hoped to see **Peter Jacobsen Flemløse** that spring. Flemløse was planning to travel to Basel and receive the M.D. degree when he died quite unexpectedly in Copenhagen. What a disappointment that must have been to Tycho. Another Danish physician visited in the early spring of 1598. He was Jens Mule, a brother of Else Mule and a former Uraniborg visitor who had studied in Padua and Siena and had just received the M.D. degree in Basel.[43] Traveling scholars like him kept Tycho informed of mutual acquaintances and others throughout the continent, and they served as letter couriers. Perhaps Jens Mule was the traveler who arrived from Helmstedt early in March of 1598, bringing two books and a letter for Tycho Brahe. The letter was from someone named **Johannes Kepler**. It solicited Tycho's opinion of one of the enclosed books, Kepler's very first publication. The other book was by none other than Nicolaus Reymers Ursus. It was a vicious attack on Tycho Brahe, and surprisingly, it contained a letter from this same Johannes Kepler, praising Ursus to the high heavens. Who was this Kepler? Tycho wanted to know. He was very upset by the scurrilous attacks of Ursus upon his professional competence, his personal dignity, and the honor of his wife and daughters, and he was also deeply disturbed by the fact that Ursus had somehow become imperial mathematician to Rudolf II in Prague. Ursus stood in the way of Tycho's own attempts to establish a patronage relationship with the emperor, and Tycho immediately concluded that Ursus must fall before he could rise at the imperial court.[44]

Christian Longomontanus was in Wrocław that winter, where Tycho's friend Jacob Monaw arranged for him to meet the sister of **Paul Wittich**, and she showed him her late brother's library. Tycho had never forgotten his spirited conversations with Wittich, and he remembered Wittich's copies of Copernicus, brimming with annotations and drawings of Wittich's planetary, lunar, and solar theories. Wittich also wrote manuscripts on *prosthaphæresis*, which he had developed further in the years after visiting Uraniborg. Tycho wanted to acquire these works, but the sister was a hard bargainer: Hagecius and the two Austrian barons had already tried to get them for him in 1589, and Hagecius and Monaw had tried again in 1595. Now Longomontanus and Monaw made a third try and eventually succeeded.[45] Tycho also had Ursus on his mind and asked Longomontanus to recall whatever he could about the visit of Ursus to Uraniborg because Tycho himself had more or less forgotten it. He wanted Longomontanus to come to Wandsburg so they could discuss Ursus and other matters.

When the book on Tycho Brahe's instruments appeared that spring, it was one of the most elegant scientific works ever published and set a new standard for description of technical apparatus. Its twenty-two woodcuts and engravings of instruments were masterpieces of technical illustration, based on the mannerist theory that art should imitate nature in such a way that the art "came alive." These illustrations certainly did so, drawing the viewer into the picture to imagine what it would be like to handle and use the instruments. Used in conjunction with the text, the illustrations allowed readers to reproduce the instruments in precise detail. In addition, there were bird's-eye views, plans and elevations of Uraniborg and Stjerne-borg, a precise map of Hven, illustrations of technical details, and the Gheyn portrait of Tycho with his golden chains and the blazoning of his sixteen noble quarterings. This instrument book, accompanied by a hand-lettered copy of the stellar catalog, showed that Tycho Brahe's large-scale activities had achieved a revolution in empirical science surpassing all ac-complishments of the past. Learned readers might wonder whether this achievement heralded the beginning of the Great Instauration of Wis-dom, when humans would live on earth like gods, controlling the natural world. Tycho entitled the instrument book *Astronomiæ instavratæ mecha-nica* – the *Mechanics of the Instauration of Astronomy.*[46]

As copies of the book's pages were pulled off the press, they were turned over to Tycho's bookbinders. The early copies, like the one sent to Holger Rosenkrantz in May, were bound in leather. The finer copies took more time. Some were bound in vellum, and the most sumptuous were in pale silk with metal clasps. Tycho sent an autographed copy to his twenty-year-old nephew, Otte Steensen Brahe, studying in Geneva with **Cort Aslaks-søn.** Christian Riber visited Wandsburg that spring on his way abroad, and Tycho enlisted him to bring a copy to Joseph Scaliger in Leiden. Others went to friends like George Rollenhagen and **David Fabricius.**[47]

In June 1598, Claus Mule, another brother of Else and Jens Mule, ar-rived with letters from several of Tycho's friends in Denmark. Tycho need-ed couriers, and he persuaded Claus to join the *familia.* Claus Mule and **Frans Gansneb Tengnagel** van de Camp were both noblemen, and their social rank assured them a smooth entry to the courts of rulers. That sum-mer, Tengnagel delivered luxurious copies of the instrument book and star catalog to Prince Maurice of Orange. On Viceroy Heinrich Rantzau's rec-ommendation, Tycho Brahe also sent copies to Archbishop Elector Ernest of Cologne, a great patron of the Paracelsian movement, cousin of Em-peror Rudolf II, and an influential broker of imperial patronage. The elec-tor was deeply moved by Tengnagel's tale of Tycho's tribulations and told him to tell "that good, honorable man" that he would "treat him as if he were his own dear brother."[48] He presented Tengnagel with a gold medal-lion and a riding horse as tokens of his grace. German relatives of the Dan-ish royal family also received copies of the instrument book and star cat-alog, including Duke Ulrich of Mecklenburg, Duke Heinrich Julius of Braunschweig-Wolfenbüttel, and the Lutheran prince-bishop of Bremen, John Adolf of Gottorp.[49] The first two had visited Uraniborg, and all three were princes of the empire who could broker contacts with the imperial court as well as with Denmark.[50]

The results of this patronage strategy were immediate. The heads of the Dutch government, Prince Maurice of Orange and Grand Pensionary Oldenbarneveld, promised to seek public support for Tycho in the Netherlands, though Scaliger warned that it might be slow in coming.[51] Elector Ernest of Cologne wrote to Emperor Rudolf II that the "whole German fatherland" would be grateful if he granted liberal patronage to Tycho Brahe, the "unique and most laudable restorer of the sciences." The elector also wrote to Johannes Barvitius, the emperor's closest adviser, urging him to press the case of Tycho Brahe to a quick, favorable conclusion.[52] Finally, Duke Ulrich of Mecklenburg wrote to Emperor Rudolf and urged him to support the illustrious studies of Tycho Brahe.[53] These letters were sent to Tycho so that he could present them to Emperor Rudolf in person.

There was one court, however, that Tycho Brahe could not penetrate, and that was the court of Denmark. Tycho learned that the letters sent on his behalf by German rulers had been intercepted and never reached the king. Tycho wanted to present the king with elegant copies of the star catalog and instrument book. Knowing that his couriers would never get past the Lord Steward and his minions, he sent the gifts to Holger Rosenkrantz, whose lineage and wealth ensured him ready access to the crown, but even young Rosenkrantz wavered in the face of this delicate task of brokering in a hostile court environment. He never quite found himself in a situation to make the presentation and eventually offered to return the books to Tycho. This was more than Tycho could take: He scolded Rosenkrantz like a prodigal son, then turned to others. Two of his kinsmen had been named state councillors in 1596: Steen Rosensparre, the stepson of Peter Oxe, married to Tycho's niece; and Henrik Ramel, the king's former preceptor, now head of the foreign ministry (called the German Chancery), and married first to a Rantzau, then to a Brahe. These two were fearless, experienced, and adroit in the ways of the court, but even they had problems in finding the right moment. Ramel eventually made the presentation, but not until the late summer of 1599, when he gave the book and manuscript to the king in the presence of the Royal Chancellor. The gifts were accepted, but not with grace and favor.[54] By then, Tycho was no longer at Wandsburg.

As long as he was there, he lived in great style. Splendor was an essential part of his patronage strategy and helped to maintain his image as a man of great worth. At times, it caused cash-flow problems, but Tycho's excellent credit overcame them. A noble Danish student, Hans Stygge, wrote that Tycho traveled to Hamburg with the equipage of a prince in a coach drawn by six horses, surrounded by retainers. His wife and children lived in the style of counts and countesses. When Tycho attended church, he sat in a special loge and was shown all the honors of a sovereign lord. Hans Stygge also reported that many said Denmark had lost her eyes when Tycho left, and some described the cause of his departure as a barbaric attack, while others said it was "Machiavellianism."[55]

Meanwhile, Tycho's contacts with the imperial court were growing by the day. An imperial astrologer and gem artist, Caspar Lehmann, acted as Tycho's informer. He sent the welcome news that Ursus had displayed his own dishonesty and lack of skill to the emperor and was no longer in favor.[56] Meanwhile, Tycho's main broker of patronage, Hagecius, had enlisted the support of Vice-Chancellor Rudolf von Coraduz, and Johannes Barvitius was also helping; these powerful men comprised the inner circle closest to the emperor. Wherever he turned, Emperor Rudolf heard the name of Tycho Brahe: from his physician, his librarian, his gem cutter, his vice-chancellor, the great magnates of Styria and Moravia. Tycho also made contact with the influential Baron Vilém Slavata of Chlum and the Archbishop of Prague, Zbyněk Berka of Dubá.[57] Then there were the books: Princes and scholars throughout Europe were turning the pages of those marvelous volumes, learning the secrets of Tycho Brahe's great instruments and his celestial island. Rudolf knew this, yet he had not seen the books. Was he not emperor, the greatest of all princes? Why should he alone be denied these arcane pleasures? Rudolf grew impatient. Lehmann reported that the emperor had a strong desire for Tycho to come to Prague, and that he would grant him the splendid castle of Brandeis near the city.

Tycho's patronage strategy was proceeding as planned. He worked to prepare even more marvels and gifts for his future patron. He corresponded busily with Hagecius, with princes and scholars and learned friends in universities throughout Europe, and with astronomers like **Kepler**, Mästlin, and **David Fabricius**. The volume of his extant letters from this period is large, and we know that many other letters have been lost.

David Fabricius was a Protestant clergyman in East Frisia, where he pursued his interests in astronomy and astrology on a highly sophisticated level. He had corresponded with Joost Bürgi, who advised him how to make an iron quadrant and sextant in the Tychonic style. Using these instruments, he had discovered the variable star Mira Ceti in 1596 and had sent his observations to Tycho Brahe, touching off a regular correspondence. He visited Wandsburg during May of 1598 and got on very well with Tycho, who gave him the new instrument book, the volume of scientific correspondence, and the preliminary printing of the book on the supernova of 1572. Fabricius went home, studied them carefully, and became a thorough disciple, inundating Tycho Brahe with observations and notes from his well-equipped observatory in the village of Resterhave, which began to function as a hyperactive affiliate of Tycho's own more ambulatory observatory.

David Fabricius had left Wandsburg when Claus Mule returned with letters from Denmark in August 1598, but Christian Longomontanus had arrived. Early in September, Mule and Longomontanus traveled to Helmstedt on an errand for Holger Rosenkrantz. They were back by the middle of the month, in time to help Tycho Brahe and his *familia* prepare to leave Wandsburg. The *familia* had been living at Wandsburg Castle for a year, but Tycho's patronage brokers in Prague now advised him to come because the emperor was inclined to accept him into the imperial service. Just after Michaelmas, the traditional day of moving, 29 September 1598, the wagons and horses of a great noble household rumbled out of Wandsburg. Three splendid carriages led the train, including Tycho Brahe's personal carriage drawn by six horses. They moved along the Elbe River on the road to Prague.[58]

At the ducal seat of Harburg near Hamburg, Tycho called on Duke Otto II of Braunschweig-Lüneburg, whose daughter, Duchess Elisabeth, was married to Tycho's Swedish kinsman, Count Erik Brahe. Tycho presented the ancient duke with an illuminated copy of the instrument book, bound in green silk, which he inscribed on 5 October 1598. The duke responded with a letter of recommendation to the imperial High Steward, Wolfgang Rumpf von Wullross. The duke, duchess, and their sons were fascinated with Tycho's work, and the duke was eager to have the great man cast his nativity.[59]

MAGDEBURG

The caravan progressed as far as Magdeburg, where they stayed for a week. Tycho met there with Erik Lange and was appalled to learn that Erik wanted to reveal to the emperor his exorbitantly expensive but "still incomplete" method of transmuting base metals into gold. Tycho had no faith in goldmaking, and he could no longer afford to be patient. He did not dwell on the fact that Erik had brought "that foul beast," Ursus, to Uraniborg, had caused Tycho grief without end, and had heaped heartbreak and humiliation upon his beloved sister, **Sophie Brahe**. Tycho's primary concern at this point was for his imperial patronage strategy, and he did not want it ruined by Erik's fantasies. He made it clear that Erik would not be allowed to accompany him to Prague. Then he summoned a notary and required Erik to testify about Ursus's conduct on Hven. The testimony was recorded in a public document.[60]

A more pleasant meeting was that with with his correspondent of many years Rector George Rollenhagen, a prominent citizen of Magdeburg. It was Rollenhagen who had first tipped off Tycho about the plagiarism of Ursus and about Ursus's terrible recent book. Through his brother-in-law, Caspar Lehmann, Rollenhagen had supplied Tycho with inside information from the imperial court. He had advised Tycho not to lower himself to brawling in the dung with Ursus, and indeed, Tycho did not plan to do so. The great gulf in social status between Ursus and Tycho meant that Tycho would leave it to his lawyers, clients, and friends to refute and crush Ursus, as aristocrats normally did in the patronage of disputation with commoners.[61] Furthermore, as Lehmann had reported, the star of Ursus was on the wane, while Tycho's shone brightly on the imperial horizon.

Nevertheless, new strategy sessions in Magdeburg led to a change of plans. It was hard to travel with the whole household. Tycho had earlier considered going ahead with a small entourage of assistants and a few instruments, and now he returned to this plan. He and his sons and a few others would pass through the mountains to Prague. When and if the patronage of the emperor was assured, he would summon Kirsten Jørgensdatter and the others. That would leave a path of withdrawal in case things did not go right in Prague: Tycho could always claim that he needed to return to his *familia*.[62] Kirsten Jørgensdatter with her daughters and servants returned to Wandsburg with Christian Longomontanus. When they arrived, Philipp von Ohr was printing Tycho's lunar theory for the volume on the supernova of 1572, which was to be a gift for Emperor Rudolf II, but Longomontanus found it so full of errors that the whole press run had to be scrapped. Eventually, Longomontanus headed for Denmark with letters to Tycho's friends and relatives, and a *famulus* named Andreas arrived to handle affairs at Wandsburg.

Leaving the instruments and much baggage in Magdeburg, Tycho proceeded with his sons, Tengnagel, and a few others to Dresden, where they remained for a month. He reported his location to the imperial court and awaited instructions. After some delay, Hagecius conveyed the reply of Barvitius and Coraduz: The emperor was pleased to hear that Tycho Brahe was on the way, but he should stay away because a dangerous epidemic was raging in Bohemia. The court had withdrawn from Prague, winter was setting in, and Tycho would not be able to present himself to Emperor Rudolf II because of the epidemic.[63] He therefore decided to spend the winter in Martin Luther's old city of Wittenberg, not far from Dresden.

Tycho meanwhile continued to explore new patronage channels. He sent Frans Tengnagel to present copies of the instrument book and star catalog to the rulers of Venice, Tuscany, and Parma, as well as to Tycho's friend and fellow astronomer Giovanni Antonio Magini. Tycho's accompanying letter to the doge of Venice, which made reference to the plan to establish an observatory in Alexandria, Egypt, was dated 28 November 1598 in Dresden.[64] Tengnagel set off across the Alps, presented the gifts, and made his usual good impression. In Venice, he personally received the favor of a knighthood of the Order of San Marco. The doge of Venice and Grand Duke Ferdinand de' Medici in Florence were in close contact with the imperial court, and the Farnese of Parma were Habsburg clients. Even Magini was a learned client of Emperor Rudolf II. From him, Tycho requested a Latin tribute from Italy to print in his forthcoming book among letters and poems from humanists of many lands.[65]

While in Dresden, Tycho Brahe had an audience with Duke Frederick William of Saxe-Altenburg, administrator of Electoral Saxony and guardian of the young Saxon princes, Elector Christian II, Duke John George, and Duke Augustus. Tycho presented illuminated copies of his instrument book and the star catalog to the duke, the young princes, and the duke's chancellor, Samuel Mosbach. As a token of ducal grace, he received a golden chain with three portrait medallions, the whole worth over 130 guldens.[66] On formal occasions, Tycho Brahe now bore a tremendous weight of golden chains, princely medallions, and royal insignia over his shoulders and breast.

WITTENBERG

From Dresden, Tycho and the others proceeded back along the Elbe to Wittenberg, where they arrived on 4 December 1598. Tycho was well acquainted with this small university town, which was the cradle of the Reformation and shrine of Lutheranism. They lodged with Professor Johannes Jessenius (Jan Jesenský de Nagh Jessen), a Slavic nobleman who had many contacts in Prague. Jessenius lived in Philipp Melanchthon's former house, a handsome Gothic Renaissance building with high, stepped gables and a shaded garden where Luther and Melanchthon had loved to converse in days gone by. Melanchthon's son-in-law, Caspar Peucer, had lived in the house when Tycho was a student in Wittenberg. The plan of the house was not unlike that which Andreas Libavius would describe in 1606 as the ideal laboratory-dwelling.[67] Service rooms and storerooms faced a vaulted hallway running the length of the ground floor, and Tycho soon set up a chemical laboratory, probably in one of these rooms. A spiral stairway led to the main hall above, and off the hall was a study with beamed ceiling, deep window niches, and heavy oaken furniture, much like the Winter Room at Uraniborg.[68]

Tycho Brahe settled in for an enjoyable winter of work and study, entertainment, academic and religious ceremonies. Tycho loved the atmosphere of any university town, but this place, Luther's and Melanchthon's old home, was something special. He soon had a circle of friends including his physician host, the theologian Ægidius Hunnius, the learned poet Friedrich Taubmann, and the mathematician Melchior Jöstelius. In his laboratory, Tycho undoubtedly prepared gifts of Paracelsian arcana for his friends, and he also handed out copies of his publications. He and Jöstelius succeeded in solving some of the last problems of *prosthaphæresis*.[69] Those were happy days in good company. Once again, Tycho Brahe was Prince of the Muses.

At the end of December 1598, Tycho wrote to Longomontanus and invited him to come to Wittenberg. At that very time, Viceroy Heinrich Rantzau lay dying, and he expired on the eve of the new year, but the news took a while to reach Wittenberg. When Tycho wrote to Longomontanus again in January, he was confident and robust, dispensing promises of patronage, even offering to help Longomontanus get a professorship in Wittenberg if he wanted one. Tycho's son, Tyge (or Tycho) Tygesen, now seventeen years of age, was sent with Andreas to fetch Kirsten Jørgensdatter, her daughters, and their servants. They all left Wandsburg on 6 January 1599 and arrived in Wittenberg some days later.[70]

Tycho Brahe devoted much of that winter of 1598–9 to printing his great work on the supernova of 1572, volume one of his *Introduction to the Instauration of Astronomy* (*Astronomiæ instavratæ progymnasmata*). It had been in press for over a decade, delayed by innumerable problems. He had progressed to the section on his new lunar theory, which he discussed in detail with Jöstelius. Tycho Brahe's science was on a large scale, and collaborative efforts frequently resulted in publications by Tycho's clientele, based on data and theories controlled by Tycho. That winter in Amsterdam, for example, the Langrens, **Blaeu**, and Jodocus Hondius the Elder, working with Petrus Plancius, were racing to construct the first published celestial globes based on Tycho's data.[71] Tycho (like later leaders of big-science projects) frequently permitted his results to be used by others or turned over controversial or questionable matters to friends and students. Tycho was reluctant to risk his own reputation, but younger collaborators were willing to stick out their necks because, for them, it was a chance to make a reputation, not lose one. Jöstelius was such a person, and Tycho encouraged him to publish a pamphlet using Tycho's lunar theory and data to predict the circumstances of a lunar eclipse due to occur on 31 January 1599. Tycho wanted to test the accuracy of his material, publicly and in print, but not in his own name. Jöstelius was willing. He did all the calculations and wrote a pamphlet that compared Tycho's results with three alternatives. The eclipse came, but something was seriously wrong: It arrived almost half an hour too early! This was embarassing to both Tycho and Jöstelius, and it was a serious blow to Tycho's patronage strategy – he had hoped to present the completed volume with the lunar theory to the emperor during his first audience. The lunar theory clearly needed revision, and that could not be done in time to execute the patronage strategy. Instead, Tycho decided to prepare a manuscript ephemeris of daily solar and lunar positions for the year 1599 as an alternative gift to the emperor.[72]

There were many festive occasions that winter. Tycho's patronage strategy required him to live as Prince of the Muses and a serious student of astronomy, chemistry, natural philosophy, and the arts. However, he could not spend all his time conferring with Jöstelius, dictating correspondence to Hagecius and others, or supervising the calculators working on the ephemeris. It was essential that he maintain a splendid public life as well. On one occasion, the rector of the university invited Tycho Brahe to Wittenberg Castle with the faculty. In his student days, Tycho had never formally matriculated in Wittenberg, so in February 1599, he did, together with his sons, Tycho Tygesen and Jørgen Tygesen, and his *famulus*, Johannes Eriksen.

By March 1599, Christian Longomontanus was in Rostock, and Tycho wrote urging him to come and discuss the problem of the eclipse because he knew the lunar data so well. Tycho said that he had been summoned to Prague by Emperor Rudolf II and urged Longomontanus to accompany him.[73] Throughout the winter, Tycho remained in contact with the imperial court, where Hagecius was working with others in the inner circle: Coraduz, Barvitius, and Rumpf.[74] The summons came at last from Vice-Chancellor Coraduz, who graciously informed Tycho Brahe that His Imperial Majesty desired to receive him in audience in Prague.[75]

Tycho rushed to get ready. His odometer was attached to his finest carriage, and he set out to buy some outstanding carriage horses. That took time. A grave illness of his daughter, Magdalene, brought more delay. Then came the spring thaw, followed by heavy rains, making the roads impassable. Tycho still did not have all the favors he intended to present to the emperor. Sumptuously bound and illuminated exemplars of his instrument book and star catalog were ready, dedicated to the emperor in his own hand, but apparently the ephemeris for 1599 was still not complete. Tycho's scribes and bookbinders prepared a presentation copy of his solar theory and revised lunar theory. From his laboratory, Tycho planned to bring potent medicaments against plague and other dread diseases, and in his head, he would bring the secrets that only he, Tycho Brahe, with his unparalleled command of cosmic forces, could whisper into the imperial ear. Impatiently, with meticulous attention to detail as always, he prepared for the day of departure, when everything would be ready.

PRAGUE

Tycho Brahe and all his *familia* took leave of Wittenberg on 14 June 1599, in the long days and lingering twilight of summer. The train of wagons, carriages, and outriders proceeded to Castle Pretzsch, where they were entertained by the noble lord, Johann Löser. Then they traveled to Dresden, where they arrived on 16 June. Tycho left his wife, daughters, and their servants in the care of Jöstelius's brother-in-law, Johann Tauchmann, a high Saxon official.[76] He proceeded on to Prague in his finest carriage with his son, Tycho Tygesen, and a few servants. As they drew near, Ursus fled the city.

Upon their arrival in Prague, the emperor's close personal adviser, Johannes Barvitius, greeted Tycho Brahe cordially on behalf of the emperor and the Privy Council. He assured Tycho that the emperor was favorably disposed toward him and would soon receive him. The weather was fair, and they conversed in a garden. Tycho showed Barvitius the three volumes he had prepared as favors for the emperor. He produced his letters of recommendation from Elector Ernest of Cologne and Duke Ulrich of Mecklenburg. Tycho also had a letter to Barvitius from the elector. Barvitius replied that he would confer with the emperor and learn who should make the presentation of Tycho's favors.

Tycho Brahe and his party were lodged in suitable quarters. The following day, Barvitius reported that the emperor wanted to receive the favors from the hand of Tycho Brahe himself. Tycho would be summoned to the palace very soon. Tycho then showed Barvitius his letter of recommendation to High Steward Rumpf from Duke Otto II of Braunschweig-Lüneburg. Barvitius advised him to show this letter to the emperor as well.

Soon after, Tycho Brahe rode to Hradčany Castle [Fig. 46] in his carriage. He was admitted to the presence of High Steward Rumpf, who received him warmly, and they conversed at length in Latin. Rumpf ex-

Fig. 46. Prague, showing Hradčany Castle and the huge St. Vitus Cathedral on the heights. Detail of an engraving by Egidius Sadeler, 1606. (From Norlind 1970)

pressed the astonishment of the imperial court that King Christian IV would allow such a man to leave his kingdom. Perhaps it was because the king was young and preferred the deeds of war to those of peace. With courtly grace, Tycho Brahe defended the Danish king, praising his intelligence and ability, and said he did not doubt that the king would have supported his work as liberally as the late King Frederick II had it been rightly presented to him by his advisors. Perhaps a special act of divine providence brought him to Prague, Tycho said, where his labor of many years could redound to the glory of the emperor. At length, Tycho presented Rumpf with the letter from Duke Otto II. In parting, the High Steward drew him aside and promised that he would personally press Tycho's case with the emperor.

When Tycho Brahe came down from the castle, Johannes Barvitius met him and escorted him to the magnificent Italian Renaissance palace of Tycho's old correspondent, the late Vice-Chancellor Kurz von Senftenau. It was not far from the castle, and its grounds were beautiful. Tycho could have it as his residence, Barvitius said, or if he preferred a castle in a more secluded rural setting, the emperor would gladly put one at his disposal; in addition, the emperor was determined to bestow a substantial annual grant upon him. Tycho would hear more when he appeared in audience before the emperor in a few days.

Meanwhile, the imperial Privy Council met to discuss Tycho's case and decided that High Steward Rumpf himself, the president of the Council, would formally introduce Tycho Brahe to the emperor. At the appointed time, Tycho Brahe rode once again to Hradčany Castle in his carriage, accompanied by his son, Tycho Tygesen. Emperor Rudolf II observed their approach from a high window and noticed the odometer, ticking off the miles. When they entered, Barvitius met them and escorted them to the imperial antechamber. Tycho's son bore the favors for the emperor, but the emperor had decided that he wanted to receive Tycho Brahe in complete privacy, so High Steward Rumpf was not present, and the others stayed in the antechamber.

"I went in to the emperor alone," Tycho wrote to Holger Rosenkrantz soon after, "and saw him sitting in the room on a bench with his back against a table, completely alone in the whole room, without even an attending page." This was the Holy Roman Emperor Rudolf II [Fig. 47], forty-six years of age, radiating dignity, his broad face smiling, with short, dark hair and a full beard streaked with gray over the ponderous Habsburg jaw. He was dressed in rich imperial garments. Tycho Brahe of Knutstorp was fifty-two, quite bald, his hazel eyes watery, his odd nose drawn flat, his beard and long moustaches a pallid reddish blond. He wore a silk brocade waistcoat, lace collar, and several golden chains, and he bore himself with the air of a polished courtier.

"After the customary courtly gestures," Tycho continued, "the emperor immediately called me over to him with a nod, and when I approached him, he graciously reached out his hand to me." They shook hands. In this context, the handshake, like the private audience, was a gesture of profound imperial grace.

Fig. 47. Emperor Rudolf II. Oil portrait by Hans von Aachen, ca. 1603–4. (Courtesy Kunsthistorisches Museum, Vienna)

I then drew back a bit and made a short speech in Latin, in which I said that I had been called at his gracious command by a letter from Vice-Chancellor Coraduz and was now here – and would have been here quite a bit earlier if it had been possible – in order that, if he might graciously condescend to excuse the delay, he might, with his imperial favor, support and patronize me and the research I had conducted long and well. Therefore, I wanted humbly to leave with him personally some documents from the Most Serene Archbishop of Cologne and the Most Illustrious Duke of Mecklenburg, and he benevolently received and opened them in my presence. He then laid them on the table without reading them and immediately responded graciously to me with a more detailed speech than mine, saying, among other things, how agreeable my arrival was to him, and he promised to support me and my research, all the while smiling in the most kindly way, so that his whole face beamed with benevolence. I could not take in everything he said because he naturally speaks very softly. I humbly thanked him for this proof of his grace and mentioned the three books I had brought with me to present to him with the utmost deference. When he graciously responded that he would accept them, I immediately fetched them from my son, Tycho, who had them where he was waiting in the antechamber. I then entered again and left them with the emperor, who was still sitting at the table. When he took them and laid them out on the table, I reviewed the contents of each briefly. Then the emperor again responded with a splendid speech, saying most graciously that they would please him greatly. I then withdrew in the proper courtly manner.[77]

Johannes Barvitius later reported that the emperor stayed up far into the night, studying these books from Tycho Brahe. He was also fascinated with Tycho's odometer and borrowed it in order to have it replicated.[78]

Tycho Brahe had found his Mæcenas. In the days to come, rich tokens of imperial grace were showered upon him, and Tycho became a member of the narrow circle of advisors immediately surrounding the emperor. Big science had come to the big court. Tycho's island of Hven was behind him forever, and a glorious new phase of his life was beginning.

LEGACY

THE FRAGMENTED LEGACY

Tycho Brahe's big science flourished briefly in Bohemia, but only briefly. Weary after many strenuous translocations and exhausted by the demands of imperial court life, his body heavy with quicksilver from numerous Paracelsian elixirs, Tycho Brahe died on 24 October 1601.[1] Almost immediately, his integrated, team-based enterprise began to come apart. His legacy survived, but Tycho's strong hand was no longer at the helm, and it lost the large-scale unity he had always imposed upon it. Aristocratic big science gave way to a brilliant scattering of individualized science by middle-class scholars.

Tycho Brahe's legacy to his family was ensured by the time of his death. Kirsten Jørgensdatter and her children achieved the status of nobility in Bohemia, and the children married into great aristocratic houses. Tycho's sons lived as Bohemian squires; his youngest daughter, Cecilie, became a baroness; and Elisabeth married **Frans Gansneb Tengnagel**, Ritter von und zu Camp, who soon became a key figure in Habsburg international affairs.[2] Magdalene never married but remained in contact with friends and relatives in Denmark. The heirs sold their father's instruments to Emperor Rudolf for a princely sum that was never actually paid, but the interest alone provided a good income for several decades.[3] The family remained clients of the Habsburgs, and Tengnagel assumed Tycho's former role as imperial councillor, though his sphere of activity gradually shifted away from astronomy into high politics. The eldest son of Frans Tengnagel and Elisabeth Brahe continued on the political path. By 1641, Rudolf Tycho Gansneb genannt Tengnagel, Ritter von und zu Camp, was Commandant of Hradčany Castle and all of Prague west of the Vltava (Moldau) River. With his stocky build, thick sandy hair, and sweeping moustaches, he was the very image of his Danish grandfather. He bore the aristocratic legacy of his ancestors, and his descendants remained leading magnates of the Habsburg empire for generations to come.[4]

When he returned to Prague a year after Tycho's death, Frans Tengnagel quickly took charge of his father-in-law's scientific legacy. Tengnagel sold the instruments to the emperor, turned over Tycho's observational protocols to Kepler, sold the printed sheets of Tycho's astronomical correspondence and remnants of the instrument book to a Nuremberg bookseller, and personally directed the editing and publication (with Kepler's assistance) of Tycho Brahe's posthumous masterworks on the supernova of 1572 and the comet of 1577. These works issued from the Schumann press in Prague as *Introduction to the Instauration of Astronomy* (*Astronomiæ instavratæ progymnasmata*, 1602) and *On the Most Recent Phenomena of the Ætherial World* (*De mvndi ætherei recentioribvs phænomenis*, 1603).[5] Together with the scientific correspondence and a reissue of the instrument book in 1601–2 by Levin Hülsius of Nuremberg, these posthumous publications made the astronomy of Tycho's island available to natural philosophers throughout Europe.[6] Still remaining was the task of publishing tables based on Tycho's data, which Kepler finally accomplished in 1627, and the publication of that data itself, which had to wait until the twentieth century.[7]

Tengnagel and Kepler were two of more than fifty former *famuli* scattered across Europe when Tycho Brahe died in 1601. Half a dozen were at the imperial court, a like number in the Netherlands, nine or ten in Germany, one in England, one in Iceland, three in Norway, and twenty-eight in Denmark. The number would grow and the nationalities expand to include Scots and Frenchmen if one were to count individuals like Duncan Liddel and Rodolphe Méhérenc.

Collectively, the former *famuli* carried Tycho Brahe's diverse cultural legacy into the seventeenth century, but after his death in 1601, their contributions were not unified by the guiding hand of a single patron. The most spectacular achievement was clearly that of **Johannes Kepler,** who was appointed to the office of imperial mathematician and built his contribution to astronomy on the data and epistemology of Tycho's island.[8] Kepler, however, never assumed Tycho's aristocratic roles of imperial councillor and large-scale patron of science; nor did he ever employ more than one or two assistants at a time (including Matthias Seiffart and Johannes Eriksen). In the Netherlands, Willem Blaeu, Willebrord Snel, Adriaan Metius, Johannes Pontanus, and the Langren brothers integrated Tycho Brahe's approaches into the mainstream of seventeenth-century cartogra-

phy, geodesy, and chorography by working through the established structures of university and printing office.[9] **David Gans** introduced elements
of Tycho Brahe's astronomy into the Jewish rabbinical tradition. In England, John Hamon was apparently out of touch, but figures like Francis
Bacon and Henry Savile, with whom Tycho had only indirect ties, were
reshaping what remained of a Tychonic legacy as King James of Scotland
and his consort, Anne of Denmark, brought their memories of Uraniborg
to the English throne in 1603.[10] When the telescope was devised in 1608,
former *famuli* like Cort Aslakssøn, David Fabricius, Johannes Kepler,
Christian Longomontanus, and Simon Marius were quick to use it, and
Jacob Metius of Alkmaar, a brother of Adriaan Metius, was one of the in-
ventors, but the early telescopes were used individually and not as part of
large-scale team research.

From around 1620, Tycho's posthumous works attracted the attention
of Jesuit astronomers in Germany and Italy. Tycho had been in contact
with the Jesuits since 1596 through Christopher Hjort, Jens Aagesen, and
later, through his correspondence with Christopher Clavius in Rome. The
Jesuits made the empirical astronomy of Tycho Brahe their own but recast
it in an anti-Copernican, anti-Galilean tradition while preserving its irenic
overtones from the ancient theology.[11] In this form, they dispersed it far
and wide during the seventeenth century through the highly centralized
structures of their order. In China after 1669, Jesuit astronomers and Chi-
nese craftsmen rebuilt the imperial observatory as a virtual reproduction
of Tycho's island, perched on the walls of Beijing [Fig. 48]. There it stands
to this day with its dragon-festooned instruments incorporating all the features described in Tycho Brahe's instrument book.[12]

The earliest European observatories of Jesuits and others also had Tychonic associations. The first ones were two university observatories: **Snel**
and **Blaeu** provided the first instrument for the Leiden Observatory of
1632, and **Longomontanus** helped to establish a "Royal Stjerneborg" at the
University of Copenhagen in 1637–42.[13] The Paris Observatory followed,
and the very first scientific expedition from the Académie Royale des Sciences came to Denmark in 1671 to redetermine the latitude and longitude
of Tycho's island. Its leader, Jean Picard, came back to Paris with Tycho
Brahe's original observational manuscripts (which were later returned to
Denmark) and with a brilliant young Danish astronomer, Olaus Roemer
(Ole Christensen Rømer), who was elected to the French scientific academy and eventually discovered the velocity of light.[14] When Royal Green-
wich Observatory was founded in 1675 by a great-grandson of King Frederick II, it too was equipped with instruments patterned on those that had

been used on Hven. In the work of these and many other observatories, the legacy of Tycho Brahe was institutionalized in the mainstream of seventeenth-century science. The mode of operation was transformed, however, because seventeenth-century observatories used big apparatus mainly for small-scale, individual research.

Even as Uraniborg and Stjerneborg fell into disrepair, the kingdom of Denmark-Norway remained a center of Tychonic traditions where over half the former *famuli* pursued their careers. Again, they did so without the integrating influence of a single, active patron and leader like Tycho Brahe, though many of Tycho's kinsmen remained patrons of learning in the conventional sense.[15] Some of the *famuli* were artisans and printers, a couple became teachers in minor posts, and eight were parish clergymen, including Jens Wandal, who built an observatory at his parsonage and trained his sons as astronomers. Others held important administrative posts: Holger Rosenkrantz as state councillor, Jørgen Brockenhuus as a crown vassal in Norway, Johannes Stephanius and Paul Colding as heads of elite academies, and three as bishops – Niels Arctander, Christian Riber, and Iver Hemmet. Two, Johannes Pontanus and Johannes Stephanius, became Royal Historiographers of Denmark. Two others, Jacob Fabricius and Gellius Sascerides, became physicians, as had Peter Flemløse. Finally, five became professors at the University of Copenhagen and infused that institution with traditions from Tycho's island. Two had long tenures: Cort Aslakssøn, twenty-four years, and Christian Longomontanus, forty-two years. In the crucial period 1605–8, when a frontal attack was launched upon the Philippist tradition in Denmark, five of a total of fourteen endowed professors in the university and two Danish bishops were former *famuli* of Tycho Brahe.

The traditions of Tycho's island were perpetuated in various ways. Circles of Tycho's former friends and associates had autographed copies of his works, manuscripts, recollections, and favorite anecdotes about Tycho. Christian Longomontanus and Cort Aslakssøn at the university, Johannes Stephanius in Sorø Academy, and Paul Colding at Herlufsholm Academy transmitted these traditions to their students, colleagues, and kinsmen. Stephanius married a daughter of Bishop Winstrup, whose sister was married to Professor Claus Plum, a nephew of Johannes Franciscus. After the death of his first wife, Plum married Cort Aslakssøn's daughter, Christina. Caspar Bartholin, professor of medicine, was Longomontanus's brother-in-law, and he and his fellow physician, Professor Olaus Worm, married daughters of Thomas Fincke. Thus the families of Tycho's former enemies merged with those of his disciples, creating new channels for the transmis-

Fig. 48. Tycho's legacy in China: (a) Beijing
Observatory ca. 1925. (b) Tychonic equato-
rial armillary, erected ca. 1674 by Ferdinand
Verbies at the Beijing Observatory. (Both
from Needham & Wang 1959; courtesy
Cambridge University Press)

(b)

(a)

sion of Tychonic traditions. In social gatherings over many a cleared table, the congenial Longomontanus spiced the conversation with recollections of Tycho Brahe, and the younger generation of Denmark's intellectual elite grew up on these tales. They also learned their astronomy in Longomontanus's classes, using his Tychonic textbook, *Astronomia Danica*. Ties of kinship and friendship reinforced academic patronage networks, such as those that can be glimpsed in the copybooks of Johannes Stephanius, packed with letters and recommendations for clients and students at the university and Sorø Academy.[16]

The former *famuli* did everything that learned men could do in early seventeenth-century Denmark. Many of Tycho's former *famuli* maintained libraries, archives, botanical gardens, and laboratories like those of Uraniborg, while a few maintained observatories or learned households with schools. Several prepared and dispensed chemical medicaments. Those who traveled regularly, including bishops on visitation, compiled topographic and epigraphic data, measured the latitudes of places, and drew maps. They found ways to apply empirical approaches in virtually every field. **Cort Aslakssøn** seems to have been the first to tabulate historical data, using the model of astronomical tables, and he also worked to develop a theology founded upon empirical evidence from the realm of nature, combined with the evidence of Scripture.[17] The legacy of Tycho Brahe was both pervasive and diffuse in the early years of the new century.

When that legacy came under attack, all the Danish *famuli* closed ranks to defend it. They united to defend Lutheran Philippism, with its strong interest in the whole *orbis litterarum* including science, against the orthodoxy of the Gnesio-Lutherans. The Resen controversy of 1606–14 in Denmark ran parallel to contemporary battles throughout Europe that pitted orthodoxy against a more dynamic conception of science and religion, and in Denmark as everywhere else, orthodoxy emerged triumphant.[18] The Renaissance idea that nature was a book of theology was declared to be heresy.

Science remained vigorous at the University of Copenhagen after the battles with Resen, but it was forced to become a new kind of science, stripped of its speculative dynamism, unitary vision, and spiritual longings.[19] Younger Danish scientists like Caspar Bartholin, Olaus Worm, Simon Paulli, and later, Thomas Bartholin, Erasmus Bartholin, Nicolaus Steno, and Olaus Roemer achieved European renown. They had learned to honor the memory of Tycho Brahe for his thoroughgoing empiricism, and not for the scope, unity, and innovative energy of his activities.[20]

The Danish intellectual historian Alex Wittendorff saw these changes of the early seventeenth century in a broad European context. In 1597, three years before the execution of Giordano Bruno, Tycho Brahe had become the first victim of the Danish attack upon Philippist tolerance, which, in Wittendorff's view, led inexorably to the Danish witchcraft trials of 1617 and paralleled the attacks upon Galileo in 1616. Wittendorff emphasized the similarity between the holistic natural religion of witches, who conceived of the world as an organism permeated by vital power and subject to magical manipulation, and the Hermetic Neoplatonism of Tycho Brahe, who saw nature as the creative, living abode of God and believed that a natural philosopher could ultimately unlock nature's secrets to exercise a semidivine control over natural processes. However, Tycho developed rules and norms for research and theory formation that were far different from those of witchcraft. At the heart of his approach was an epistemology that insisted upon "experience," derived from controlled experimentation and observation, to produce quantified results with a known margin of error as the basis for theoretical explanation. Although Tycho's Neoplatonism seemed exotic and heretical to neo-orthodox theologians, his methodology was so persuasive that it could not be abandoned; however, they did insist that theories derived from experiments and observation of natural phenomena must be limited to natural explanations, and they demanded that all speculation about spiritual forces be left exclusively to theologians. Under Gnesio-Lutheran attack, scientists of the generation after Tycho Brahe, including many of his former *famuli,* moved away from Hermetic Neoplatonism and toward a mechanical world view, and this allowed them to carry the empirical methods of Tycho's island into the mainstream of seventeenth-century thought. As Wittendorff saw it, the combination of empiricism with a mechanistic view of nature represented a paradigm shift that brought modern science into existence.[21]

These new rules applied to commoners, but not to princes. King Christian IV never lost interest in arcane theories of nature, and he was above the orthodoxy imposed upon his church and university. King Christian had driven Tycho Brahe from his kingdom, but he slept with Tycho's horoscope under his pillow and cultivated the arcane legacy of Tycho Brahe throughout his whole long life. When the Hermetic culture of Tycho's island came under attack in academic circles, the king simply made it a part of court culture. He built a new "museum" on the edge of Copenhagen, giving it the Rosicrucian name of Rosenborg Castle and perhaps even installing emblematic ceilings taken from Uraniborg.[22] Peter Paynck, a Para-

celsian physician from the court of Emperor Rudolf II, was put in charge
of the laboratory at Rosenborg.[23] In 1616, the king appointed a Hermetic
magus and client of the Rantzaus, Niels Heldvad, as court astrologer, astronomer, and mathematician.[24] Later, he appointed the Paracelsian Jacob
Fabricius as his personal physician, and when Tycho Brahe's great globe
was returned to Copenhagen in 1632, it was originally installed among the
arcane wonders of Rosenborg.[25] Finally, when plans began to take shape
in the 1630s for a new building – Trinity Church and the Round Tower
– in the heart of Copenhagen's university quarter, the king personally
worked on the project with Hans van Steenwinckel the Younger and composed the emblematic rebus that marks the tower to this day [Fig. 49].[26]
This building had an elaborate symbolic program combining piety (the
church) and learning (the university library in the church loft) with the
celestial philosophy of the observatory (a virtual reproduction of Stjerneborg) atop the massive tower.[27] In short, the "paradigm shift" to mechanistic empiricism never took place in the mind of the Danish sovereign,
who remained on his throne until 1648.

THE MYTH

By the middle years of the seventeenth century, the legacy of Tycho Brahe
had been reinterpreted and restructured in a variety of ways that made
sense to the new generation. Kepler's laws, the Jesuit reinterpretation, and
those of the Danish court and university were parts of this process, but
the eminent French philosopher Pierre Gassendi was about to create another enduring myth of Tycho's island.[28] Gassendi was himself an observational astronomer. In preparing for lectures on astronomy at the Collège
Royale in Paris 1645–8, Gassendi pored over Tycho's published works and
those of other great predecessors. Gassendi was a Copernican at heart, an
admirer of Galileo and Kepler, but he was also a Catholic priest who could
not openly endorse Copernicus, and he had additional reasons for his
strong interest in Tycho Brahe. Gassendi was a vigorous opponent of the
Aristotelians and of Descartes, but his great contribution to philosophy
was his thoroughgoing exposition of empiricism, grounded in skepticism
and based on rational analysis of sensory evidence, which he presented in
the context of a revived Epicurean atomic theory reconciled to Christianity. It was Gassendi's radical empiricism that drew him to Tycho Brahe,
and he set out to contact the last survivors of Tycho's *familia* in order to
come in direct contact with the Tychonic legacy. He had known Willebrord Snel and Kepler, and now he managed to reach Willem Blaeu; Kep-

Fig. 49. The Round Tower, or "Royal Stjerneborg of Copenhagen," with the rebus composed by King Christian IV (see inset). Engraving by H. A. Greyss, 1646.

ler's son, Ludwig; and Longomontanus's brother-in-law, Olaus Worm.[29] In 1654, Gassendi published a life of Tycho Brahe, which became the first full-length biography of any scientist. It established an enduring canon of Tychonic scholarship and was still influential in the twentieth century.[30] He painted a picture of Tycho as empiricist, the great observer, and he did not dwell on the fact that Tycho was also the first to develop a sophisticated *theory* of observation. "Although Tycho's works are worth their weight in gold, his observations are even more precious," wrote Gassendi. "Therefore, we must praise Tycho to the highest, because he did not deny us the results of his observations, but he is even more worthy of praise because he made and wrote down so many excellent observations, from which others could draw the same or even more important conclusions."[31] The last statement was a clear allusion to Kepler. Gassendi's biography was supplemented in the years after 1662 by the sumptuous *Atlas maior* from the Blaeu press in Amsterdam, containing eleven engraved plates of Uraniborg, Stjerneborg, Tycho Brahe's instruments, and **Willem Blaeu**'s map of Hven, with accompanying text by Johannes Pontanus.[32] Meanwhile, in 1668, Peter Hansen Resen, grandson of Bishop Winstrup and Hans Poulsen Resen, aligned himself with the Tychonic tradition by publishing the inscriptions and icons of Uraniborg and Stjerneborg with a brief life of Tycho Brahe and a collection of Tychonic poetry and letters.[33]

Resen's book, Blaeu's atlas, and especially Gassendi's biography helped to perpetuate the fame of Tycho's island, but they did so in a manner that had been reshaped to the thought patterns of a later age. Gassendi in particular took Tycho's thought out of its framework of Hermetic Platonism and put it into the context of seventeenth-century philosophical skepticism. He was influenced by Galileo's polemics against Tycho's Jesuit followers, which presented the view that Tycho was a "mere" observer who was weak on theory. The persistent myth of Tycho Brahe that Gassendi articulated, like all myths, contained a grain of truth embedded within a fable, but it did not present the whole truth. The Gassendian myth was that Tycho Brahe was a great empirical scientist and no more – unable, without the theoretical brilliance of Kepler, to make any sense of his raw data. The "teamwork" of Tycho and Kepler was seen as a prime example in the history of science of complementary geniuses. Where this myth held sway, much was forgotten: the wide range of activities on Tycho's island, its arcane Renaissance elements, Tycho's contributions to solar and lunar theory, his links to many other *famuli,* and the role of patronage in shaping a large-scale scientific research center. Under the influence of the Gassendian myth, Tycho's pioneering big science was reduced to an extravagant mechanism to generate data for Kepler's flights of insight.

A FAMILIA LANDSCAPE

Victor E. Thoren was the first historian of science to reveal the depth of Tycho Brahe's theoretical brilliance as an astronomer. The present volume has aimed to show that Tycho was a master of the patronage system who unified many strands of social and cultural life, created a new organizational model for the pursuit of science on a grand scale, and brought large teams of scholars, scientists, and technicians into the enterprise. His legacy to the seventeenth century was found in the far-flung careers of the many men and a few women who had been part of his enterprise.

Nearly one hundred *famuli* have been identified who were active on Hven during 1576–97 or in Wandsburg, Wittenberg, and Bohemia during 1597–1601. Three out of five of them were university trained, and one-third were artists or master artisans. Half were in Tycho's household for six months or less, but many of these short-time *famuli* established enduring relationships with Tycho Brahe, including Blaeu, Oddur Einarson, the Langrens, Rosenkrantz, Rothmann, and Snel. A quarter served from one to five years and became thoroughly familiar with the approaches of Uraniborg: Kepler was in this category, along with Arctander, Cort Aslakssøn, Ohr, Pontanus, Riber, Schardt, Steenwinckel, and Stephanius. An important cadre of ten were in Tycho's service for more than five years, including Crol, Flemløse, Gellius, Gemperle, Longomontanus, Morsing, and Tengnagel.

The *famuli* included men and women of six different nationalities. Five were born nobles, at least a dozen were sons of professors or clergymen, thirty-five to forty grew up in middle-class urban families, and seven or eight were upwardly mobile sons of Danish peasant families. Six of the individuals listed in the Biographical Directory below were in Tycho's service before he came to Hven in 1576. Seven arrived on the island before Uraniborg was occupied in 1580, nine came in the years 1580–4, twenty-six in 1585–9, seventeen in 1590–4, and twelve in the last years on Hven, 1595–7. At Wandsburg and Wittenberg during 1597–1600, some five new coworkers were recruited, and in Bohemia in 1600–1, around eleven joined the *familia*.

The careers of a score of the *famuli* ended before the death of Tycho Brahe in 1601, but a dozen remained active until sometime in the decade 1602–11, a baker's dozen ended their careers in the decade 1612–21, another dozen in the decade 1622–31, and the most persistent dozen were still active in the period 1632–52. Legend has it that Live Larsdatter, who died in 1693, had also served at Uraniborg in her youth.

Through these individuals, the legacy of Tycho Brahe entered the mainstream of seventeenth-century science and technology with ramifications in many directions, as the upcoming biographical sketches show. It was a fragmented and diverse legacy, scattered across the face of Europe and reshaped by the pressures of religious neo-orthodoxy upon science and learning. It never regained the unity of Tycho's island or the scale of big science in any one location, but its collective impact upon the intellectual heritage of the modern world was tremendous, and that impact continues to reverberate at the end of the twentieth century.

Tycho's island survived. In the last decade of the twentieth century, the ramparts of Uraniborg were partially reconstructed, and one quarter of the garden that surrounded the mansion was replanted with fruit trees, herbs, and medicinal flowers in boxwood parterres, meticuously researched to ensure authenticity.[34] The ruins of Stjerneborg were well preserved and visible to the public, and the little Tycho Brahe Museum had new exhibits of the fragments remaining from Uraniborg and the papermill.[35] Hven remained a beautiful place in summer, well worth a trip across the Sound by sailboat or ferry, but the visitor could not expect to find more than a faint reflection of the glory that once abounded on Tycho's island.

PART TWO

TYCHO BRAHE'S COWORKERS

BIOGRAPHICAL DIRECTORY

of

Tycho Brahe's

Artisans, Assistants, Clients, Students, Coworkers

and

Other Famuli *and Associates*

Note: Short entries indicate the name (birth and death), nationality, occupation in Tycho's service and later; and place of service to Tycho Brahe. Full biographical entries on all of these individuals are found in the unabridged edition of this book, together with bibliographies.

ARCTANDER, NIELS LAURIDSEN   was a *famulus* for about a year on Hven ca. 1580–1 and later became a leading Danish Philippist bishop. Born 1561 into a prominent Norwegian clerical family, Niels Arctander grew up in Trondheim, where his father was archdeacon and his maternal uncle later became parish pastor of Nidaros Cathedral. He entered the University of Copenhagen in 1578 and came to Tycho's island around 1580 (he was number nine on a contemporary list of Tycho's disciples). In 1584–5, he studied at the University of Marburg, supported by the patronage of Christian Friis of Borreby, then governor of Trondheim. Upon his return to Denmark, he married Maria Stratmann, became Anders Sørensen Vedel's successor as preacher in the chapel of Copenhagen Castle, and served as confessor to King Frederick II until the king's death in 1588. In these years, Niels Arctander established his lifelong reputation as an outstanding preacher. He developed close ties to the family of Professor Nicolaus Scavenius, marrying Scavenius's daughter Karine (died 1615) in 1587, after the death of his first wife. (Claus Lyschander married another daughter of Scavenius; his son, Laurids Clausen Scavenius, was bishop of Stavanger, Norway, 1605–26.) In 1595, Niels Arctander was named bishop of Viborg, and he remained a leading figure in the Danish church until his death. A staunch Philippist Lutheran, Bishop Arctander did not waver when Danish Philippists came under sharp attack from their Gnesio-Lutheran opponents in 1613–14; rather, he emerged as spokesman for the Danish bishops. Through his vigorous defense of the tradition of Philippism, which in those years was closely linked to the cause of free scientific investigation, he helped to protect the careers of several of Tycho Brahe's former disciples. Bishop Arctander died 1 November 1616 in Viborg and was buried in Viborg cathedral.

ASLAKSSØN, CORT   was a leading *famulus* on **Hven 1590–3**, a faithful client throughout Tycho's life and, later, a professor of theology in Copenhagen. Born 28 June 1564 in Bergen, Norway, Cort Aslakssøn was orphaned at an early age. He came into the household of Bishop Jens Schjelderup of Bergen, who sent him in 1578 to the famous Latin school in Malmö, near Copenhagen, where he remained until 1584. On 27 April 1584, he matriculated in the University of Copenhagen and received his baccalaureate that same year. He entered the service of Tycho Brahe on 26 October 1590, apparently promising to serve for thirty months. In return, Tycho brokered his application for the *stipendium regium,* a five-year grant to study abroad. Less than a month later, the Regency Council promised the stipend to Cort Aslakssøn when his service on Hven was over.

In Tycho's service, Aslakssøn received instruction in mathematics, observational astronomy, cosmology, and related subjects. He worked as an assistant in the Hven observatory and became familiar with Tycho's methods, theories of planetary motion, and views on the relationship between science and religion. Cort Aslakssøn frequently traveled around Denmark on Tycho's behalf. In 1591, he helped him entertain the new bishop of Bergen, Anders Foss, and his wife and daughters. In 1592, he helped to arrange the visit of King Christian IV to Hven. By the time Aslakssøn left Tycho's service on 29 April 1593, he had established a warm personal relationship with the Brahe family that endured for many years.

Cort Aslakssøn returned to the University of Copenhagen, took his M.A. on 26 May 1593, and departed for six years abroad on the *stipendium regium,* also serving as preceptor to two young Danish nobles, Jens Sparre and Tycho's nephew, Otte Steensen Brahe. They studied in Philippist Rostock and the Calvinist universities of Herborn-Siegen, Heidelberg, Basel, and Geneva. He and Otte Brahe also visited Lausanne, Orleans, Paris, the English universities, and Scotland before returning to Denmark in late 1599. Meanwhile, in 1597 Cort Aslakssøn published in Siegen a small cosmological work that was well received throughout Europe.

He was appointed professor of Latin at the University of Copenhagen in 1600, advanced to the chair of Greek in 1602, married Barbara Olufsdatter (died 1628), a burgomaster's daughter, in 1603, and also taught Hebrew from 1606 on. In the controversy of 1606 over the baptismal ritual of exorcism, Cort Aslakssøn stood with his colleagues Johannes Stephanius and Jon Jacobsen Venusin in urging its abolition; Bishops **Arctander** and Hegelund were among those who argued that it should be retained. When Jørgen Dybvad was dismissed from the university in 1607, Cort Aslakssøn was created Doctor of Theology and appointed to succeed him as professor of theology – a position he retained until his death.

A Philippist or even crypto-Calvinist in his theology, Cort Aslakssøn was also strongly influenced by the ideas of modern thinkers like Ramus, Copernicus, and Thomas Mouffet, as well as by his fellow Dane Petrus Severinus and his master, Tycho Brahe. In 1613, he published a major theological work that attempted to synthesize Tychonic natural philosophy with biblical theology. His Gnesio-Lutheran colleague Hans Poulsen Resen criticized the theology of Cort Aslakssøn in a private letter, but Aslakssøn, a dangerously subtle dialectician, forced him to back off without going public.

In 1614, Resen's own writings came under the scrutiny of a royal commission chaired by Aslakssøn, and Resen was forced to revise some of his theological statements. It proved, however, to be a pyrrhic victory for the Tychonic disciples on the commission: Crypto-Calvinism and Philippism were on the wane in Denmark. After 1614, Cort Aslakssøn, the finest theologian among Tycho Brahe's disciples, abandoned his bold attempts to achieve a synthesis of natural philosophy and theology. He died 7 February 1624 in Copenhagen, Denmark, and was buried in Our Lady's Cathedral in Copenhagen. One of his daughters, Christina Cortsdatter, was married in 1638 to Professor Claus Plum, and they founded a learned family Plum that has flourished until the present.

BENEDICHT, LORENTZ   (died ca. 1604) Danish printer; Copenhagen 1573, on Hven 1595.

BERNSSØN, JOHANNES   Danish courier; Hven 1585.

BLAEU, WILLEM JANSZOON   [Fig. 50] was a young Dutch merchant who spent six months on **Hven 1595–6** and went on to become a leading European publisher of maps and scientific literature. He was born 1571 in Alkmaar (or nearby Uitgeest) to a prosperous merchant family and was apprenticed in his youth to a kinsman, Cornelis Hooft, later burgomaster of Amsterdam and father to the great Dutch poet Pieter Corneliszoon Hooft. While learning the herring trade, Blaeu developed broad interests that included navigation, astronomy, and cartography. He was twenty-four when he left his bride, Maertgen Cornilisdochter, in Alkmaar toward the end of 1595 and departed for Tycho's island.

Those six months at Uraniborg changed his life. Blaeu arrived with a blank celestial globe. He got on exceedingly well with Tycho and won his confidence. With Tycho's permission, Blaeu marked the positions of a thousand stars on his globe and possibly also copied the iconography of constellations from Tycho's great globe. He learned Tycho's methods of astronomical and geodetic observation, surveying, and instrument construction. Blaeu was ap-

Fig. 50. Willem Janszoon Blaeu. Engraving by unknown artist.
(From Koeman 1970)

parently the one who prepared the map of Hven based on Tycho's earlier survey by triangulation. He departed Hven on 24 May 1596 and returned to Alkmaar, where his firstborn son and eventual successor, Joan Willemszoon Blaeu, was born on 23 September.

Willem Blaeu left the lucrative herring trade and set up shop as a publisher in Alkmaar, where in 1598 he published his first celestial globe, elegantly engraved by Jan Pieterszoon Saenredam. This globe, together with one by the Langrens and a third engraved that same year by Jodocus Hondius the Elder for Petrus Plancius, were the first published versions of Tycho Brahe's stellar catalog. Their innovative iconography set a style for Dutch celestial cartography throughout the seventeenth century.

Blaeu proudly proclaimed his direct link to the master by referring to Tycho Brahe as his "preceptor" on his celestial globes. Blaeu later constructed a Tychonic quadrant with a radius of 2.2 meters for **Willebrord Snel** and consulted with Snel on Tycho's methods of surveying by triangulation. After Snel's death, the Blaeu quadrant was used to establish the Leiden Observatory in 1632, and it is still there. Willem Blaeu also invented the Blaeu press, which remained the standard European printing press for two hundred years.

In 1599, he moved his firm to Amsterdam and soon became the greatest marine cartographer of his day. Blaeu maps, charts, globes, atlases, navigational and scientific books helped to set the standard for Europe. Willem Blaeu published works of Grotius, Guicciardini, P. C. Hooft, **Longomontanus**, Adriaan Metius, Pontanus, and Willebrord Snel, among others. **Kepler** wrote to ask for Blaeu's observations of the supernova of 1604. Later, Blaeu supplied Pierre Gassendi with many details for the first biography of Tycho Brahe. A tolerant man with broad cultural interests, Blaeu adhered to the Remonstrant or Arminian Calvinist theology, which had many similarities to Philippist Lutheranism.

Blaeu died a rich man on 21 October 1638 in Amsterdam. Thirty years later, his son published the famous "Blaeu Atlas" in many volumes and languages, with Willem Blaeu's map of Hven, plates of Uraniborg, Stjerneborg, and Tycho Brahe's instruments, plus a lengthy description of Hven by Pontanus in the first volume. Willem Janszoon Blaeu was one of Tycho Brahe's most distinguished disciples and, with Kepler and Snel, he was a key figure in the European diffusion of the Tychonic legacy.

BLOTIUS, HUGO   (1533–1608) Dutch; assistant, Basel, ca. 1568–9 [Fig. 51] patronage broker, Vienna, 1599–1600.

BORUSSUS, SEBASTIANUS   German student, possibly translator; Hven ca. 1589–90.

BRAHE, SOPHIE   ("Urania") [Fig. 52], was Tycho Brahe's youngest sister and was close to him throughout his lifetime, especially on **Hven 1588–97**. He considered her to be one of the most learned women of her day. Sophie Brahe was born at Knutstorp Castle, probably on 24 August 1559. Her father died when she was eleven, and she was educated at home. She learned German and began to study Latin. Tycho taught her some astronomy, and she helped him to observe a lunar eclipse in 1573, when she was fourteen.

In 1579, when she was twenty, Sophie Brahe entered an arranged marriage with a rich nobleman, Otte Thott (1543–88), bringing a substantial inheritance and dowry into the match. They lived in grand style at Eriksholm Castle (now Trolleholm) in Skåne, where Sophie gave birth to their only son, Tage Thott, on 27 May 1580.

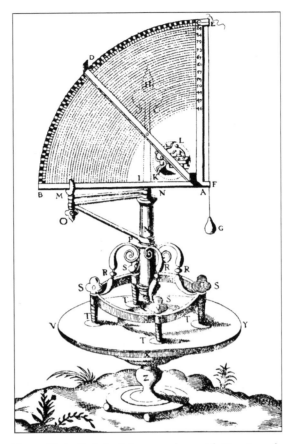

Fig. 51. Tycho Brahe's gilt brass quadrant with Nonnian sub-
divisions and an emblem of life and death, probably built ca.
1568–9 with Blotius in Basel. (From *Astronomiæ instavratæ mechanica*)

One evening around 1582–4, while visiting Uraniborg, Sophie and a young
cousin, Lisbet Jensdatter Bille (1572–1633), went outside the portal to observe
the planets and were chased by Tycho's bailiff. This inspired Tycho to compose a mock-heroic poem in Danish about Pan, influenced by Mars and Saturn to harass Urania and a little girl until the faithful Davus (a classic slave
name in Roman comedies) came to their rescue.

On 23 August 1584, Sophie's brother Knud Brahe married Margrethe Lange,
and the families came together to celebrate the wedding at Stjernholm in
Jutland. A month later, the bride's brother, **ERIK LANGE**, visited Uraniborg
for two weeks with a retinue including Nicolaus Reymers Bär, later called
Ursus. Erik Lange and Tycho Brahe hit it off exceptionally well, except that

Fig. 52. Sophie Brahe. Oil portrait in Gavnø Castle, artist unknown.
(Courtesy Gavnø Castle, Næstved, and Det Nationalhistoriske Museum på Frederiksborg, Hillerød)

Erik was an alchemist and Tycho was not. (Tycho wrote a Latin poem to re-
direct Erik's interest from alchemical transmutation to iatrochemistry, though
without success.) Sophie Brahe was apparently not present.

Two years later, in August of 1586, she was on Tycho's island a few days
before the visit of Queen Sophie with her parents from Mecklenburg, and
she may have helped to host the royal guests.

Sophie's husband, Otte Thott, sixteen years her senior, died on 23 March
1588. Tycho wrote that the young widow sought solace in cultural activities:
establishing the finest Renaissance garden in Scandinavia at Eriksholm; constructing a chemical laboratory within the garden, where she prepared Paracelsian medicaments for friends and poor folk; and, finally, turning to astrology and the casting of horoscopes. Tycho instructed her in chemistry, but

urged her to moderate her interest in astrology. Impressed with her learning, Tycho later planned to publish one of her astrological letters, though printing learned letters from women was highly unusual. He compared Sophie to the famous Italian female scholar, Fulvia Olympia Morata.

During these years of widowhood, Sophie Brahe probably also composed a Danish epistle on the philosopher's stone, addressed from Urania to Diana, with citations from Paracelsus, Arnold of Villanova, and the Emerald Table of Hermes Trismegistus. She became a frequent visitor to Tycho's island, arriving with her son, Tage Thott, or with her sister or other noble relatives. During the autumn and winter of 1589, she often met Erik Lange at Uraniborg. He had studied in Tübingen, Strasbourg, Wittenberg, and Leipzig, traveling much of the time with a fellow Danish lord, Falk Gøye. They had survived France's St. Bartholomew's Day massacre in 1572 by fleeing through the sewers of Paris. Erik Lange returned to Denmark in 1573 to take over the paternal estate of Engelsholm, which he owned jointly with his sister, Margrethe. In 1583, he sold his maternal inheritance to the crown in exchange for lifetime tenure as governor of Bygholm Castle. He began to consume immense sums with alchemy and other vain experiments. Sophie Brahe fell under the spell of this turbulent gallant, whose fief of Bygholm was falling into disarray and whose fortune was vanishing from his obsession with alchemy.

Hans Brix ascribed to Sophie Brahe a well-known Danish love ballad full of alchemical references that must have been written around 1589. Later authors, Brask and Dal, accepted the theory that Sophie Brahe wrote it, but Zeeberg called for more solid evidence.

In any case, Sophie and Erik were betrothed, early in the year of 1590, despite the opposition of every member of her family except Tycho Brahe. Soon after, in early March, Erik's creditors demanded that he be put under house arrest, though he was soon released. He and Sophie Brahe were both present when King James VI of Scotland visited Tycho's island on 14 March 1590. In a desperate attempt to stave off his creditors, Erik Lange sold his share of Engelsholm to Knud Brahe and Margrethe Lange; he then departed for Hamburg, where he apparently hoped to achieve the alchemical transmutation that would solve all his problems. Sophie remained behind in Denmark.

In August of 1590, Tycho brought **Christopher Rothmann** to visit Eriksholm, where the guest was greatly impressed by Sophie's Renaissance gardens. Tycho was redesigning his own gardens around that time, according to an elaborate microcosmic plan. The example of Sophie's gardens at Eriksholm must have had some influence upon the plan. A servant of hers came to Uraniborg in March and April, and he may have been her head gardener, bringing seeds and root stock. Sophie Brahe undoubtedly supplied plants and advice for the new gardens at Uraniborg.

Meanwhile, Erik Lange returned during the summer of 1591 and was at Uraniborg with Sophie in August. He visited several times that autumn and winter, but early in 1592, with Tycho's help, he fled abroad to avoid creditors.

Around 1594, Erik Lange was in Kassel when Tycho wrote his great Latin poem "Urania to Titan," in the form of a letter from Sophie to Erik. The model was Ovid's *Heroides* – verse letters to their beloved ones from mythological heroines in hopeless love situations – but the content and imagery were chemical and astrological. It pictured Sophie Brahe as a confident, talented woman in a moment of crisis, and it served as an admonition to Erik Lange, to whom it must actually have been sent, denouncing his deception of himself and others. The fact that a poem so rich in classical and arcane allusions could actually have been transmitted as a letter was testimony to the high level of culture at Uraniborg, in Zeeberg's view, and he compared the spirit of Uraniborg with Castiglione's Urbino or the Platonic Academy of Florence. Sophie Brahe helped to create that sophisticated milieu.

That same year of 1594, around Michaelmas (29 September), negotiations began for a match between Tycho's eldest daughter, Magdalene, and Gellius Sascerides. Mogens Bertelsen Dallin was the spokesman for Dr. Gellius; Sophie Brahe became the negotiator for Magdalene's party. These negotiations ran into difficulties during the last three months of 1594. The Consistory of the University of Copenhagen intervened in January of 1595 to mediate a new marriage contract, but Sophie continued to consult with Tycho and Magdalene. In January of 1596, she drew up a statement summarizing her view of the failed negotiations.

Sophie and Erik's friend Falk Gøye had died at the beginning of 1594, and his widow now brought action to redeem a costly diamond that Erik Lange had pawned in Lübeck. She wanted to sell it and regain the 25,000 dalers Erik had owed her husband, and the court allowed her to do so. Perhaps the diamond had once belonged to Sophie Brahe.

Through all of this, Sophie lived at Eriksholm, supervising her son's education, administering her own property, corresponding far and wide, and pursuing her interests in botanical and ornamental gardening, chemistry, astrology, and genealogical research. She carried her book of horoscopes and often showed it to friends. In her laboratory hung a portrait of Erik Lange, although he himself remained in exile. For a time in 1596, he seems to have been under house arrest in Kassel for debts incurred there. His fief of Bygholm Castle was repossessed by the crown in September of 1596, despite Knud Brahe's best efforts on Erik's behalf.

Sophie continued to visit Uraniborg with her son, now sixteen and a great admirer of his famous uncle. After Tycho departed for Rostock in June of 1597, Sophie kept in touch by correspondence. In the autumn of 1598, Tycho met with Erik Lange in Magdeburg and required him to sign a sworn statement concerning the visit of Ursus to Uraniborg, for which Tycho held him responsible.

By 1599, Sophie Brahe's longing could wait no more: She accompanied her son on his way to study abroad and, in Braunschweig, was able to meet briefly with her beloved Erik Lange. She returned to Skåne, completed her immense genealogical manuscript on Scandinavian noble families in 1600, and then came back to Holstein, longing to be with Erik. During the summer of 1601, she stayed for ten weeks in Hamburg with the family of a Dutch physician who treated her for melancholy.

Tycho tried to persuade her to come to Prague. She wanted to take the healing waters of Pfefferbad near Basel, recommended by Paracelsus, but Tycho offered to meet her in Karlsbad instead. He invited her to attend the wedding of his daughter, Elisabeth, and **Frans Tengnagel**, but that meeting did not transpire either. Toward the end of 1601, Sophie learned that her beloved brother Tycho had died in Prague on 24 October.

She took lodgings in the small Holstein town of Eckernförde, and there, at long last, Erik Lange came to her. They were married on 21 March 1602 and celebrated with a few nobles in town, then traveled into Denmark to visit relatives. By this time, Sophie Brahe's own debts had reached astonishing heights. She and Erik lived in dire poverty, paying pawnbrokers 150 percent interest. When Lisbet Jensdatter Bille visited Sophie in Eckernförde, she and her sisters were so appalled with the situation that they gave Sophie their own fine clothing and jewelry. Erik and Sophie later pawned Lisbet Bille's gloves, golden chain, and pearl-embroidered cap for 600 dalers, a quarter of their worth, and never got them back. Meanwhile, two noblemen from Hesse-Kassel moved in and stayed all winter, because Erik owed them money.

For Sophie Brahe, marriage was a life of humiliation, abundant tears, and alienation from family and friends. She tried to advise Erik and separate him from shady companions, but he was seldom home. Sophie sought peace of mind in correspondence, writing, studying, and working in her laboratory. She received inheritances in 1604 and 1605, but it all slipped away to creditors. Her son, Tage Thott, spent two years in Heidelberg and Basel, then journied through France, the Low Countries, England, and Scotland on a Grand Tour, for he was the heir to Eriksholm and as rich as his mother was poor. In October of 1606, his marriage to a daughter of Viceroy Breide Rantzau took place in high style in Copenhagen, and Sophie Brahe must have been present, perhaps Erik Lange as well. In March of 1608, Erik turned up in Prague, where he died around 1613.

Sophie Brahe lived the remainder of her life in Denmark. She took up residence in Elsinore, where she kept a garden, corresponded widely, and pursued her interests in iatrochemistry and genealogical research. Sophie developed keen methods of source criticism and used them to complete by 1626 a new version of her vast manuscript of the genealogies of sixty Danish noble families. In 1629, she sent rare plants from her garden – red and yellow tulips, crown imperial, and white Alexandrine lily – to a Swedish nobleman who had kindly supplied her with a family tree. In 1630, when Tycho Brahe's grandson, Rudolf Tycho Tengnagel, needed to prove his nobility, Sophie Brahe sent a testimonial to Prague on his behalf, written in German. In 1634, the secretary of the French ambassador to Denmark reported that they had seen the sister of the famous Tycho Brahe in Elsinore. She was still writing long genealogical letters to younger relatives and also corresponded with Holger Rosenkrantz, who visited her in 1639.

Active to the end of her long life, Sophie Brahe died in Elsinore in 1643 and was buried next to her first husband in the Thott family chapel at Torrlösa near Eriksholm. By then, her immensely rich son was one of the leading noblemen of his generation and a prominent state councillor. Their descendants include the present Count Thott and Barons Thott in Sweden, and their Danish heirs flourish as the Barons Reedtz-Thott.

BRENDER, OLUF CHRISTOPHERSEN (died 1600) Danish student; Hven 1592 and 1596.

BRENNER, STEFFEN (died ca. 1603) Danish clock and instrument maker; Copenhagen 1573.

BROCKENHUUS, JØRGEN (ca. 1570–1634) Danish nobleman and amateur astronomer; Hven 1590-1.

BUCK, HANS KNUDSEN (died 1586) Danish student; Hven 1585.

CASPAR German papermaker; Hven ca. 1589–90.

COLDING, NIELS BERTELSEN (died 1593) Danish student, educator; Hven 1588–90.

COLDING, PAUL JENSEN (1581–1640) Danish student, educator; Prague 1601.

CORNELIUS, LAMBERT Dutch engraver; Amsterdam 1595.

CORONENSIS, MARTIN INGELLI Danish student; Hven 1589–90.

CROL, HANS   was Tycho Brahe's goldsmith and instrument maker on Hven ca. 1584–91 and an unusually sharp-eyed observer. He was apparently not related to the learned Marburg family of Crollius but was a native of Westphalia.

Crol's principal task on Hven was to build, under Tycho's direct supervision, the famous instruments for the Stjerneborg observatory. He was first mentioned in the meteorological journal on 9 June 1585, less than a week after the departure of Hans Buck, but he may have arrived earlier in order to construct the equatorial armillary of Stjerneborg. In November of 1585, he observed a comet with **Elias Olsen Morsing**, and Tycho praised him as a keen observer. Crol could write Latin and frequently kept the observational journals in the years 1586–90. Norlind thought he was the one who drew up the list of Tycho's disciples around 1588.

Hans Crol's betrothal was celebrated at Uraniborg on 12 March 1587. Three years later, their young son died at three in the morning on 14 May 1590; four days later, they departed for Copenhagen, possibly to bury their son. Crol returned to Tycho's island, but whether his wife ever did is unclear. On 9 January 1591, Crol's sharp eyesight served him well: Mästlin and **Kepler** in Tübingen thought they saw a total occultation of Jupiter by Mars, but on Hven, Hans Crol observed it "on the way to the lecture" at six in the morning and could see that Jupiter was not completely hidden.

Frobenius recorded in the spring of 1591 that Hans Crol warned him against Tycho, complaining that he had served at Uraniborg for six years without receiving any wages except clothing, and that he could not get away. Crol did go to Copenhagen shortly after Frobenius departed, returning sick on 17 July 1591. He died at Uraniborg on 30 November 1591 and was buried on 4 December. In a letter of 20 February 1592 to the landgrave of Hesse-Kassel, Tycho praised the memory of "my goldsmith, named Johannes Crol, who lived with me here in the country and took care of my instruments, of which he had made many with his own hands, and served me in this diligently and faithfully for many years"; he went on to request the assistance of Joost Bürgi in finding a successor.

Works. All the innovative instruments built for Stjerneborg in the years after 1584 were probably the work of Hans Crol, possibly assisted by others including Rudolphus Groningensis 1585–8. Crol built the great equatorial armillary, the large (155 cm) revolving azimuth quadrant, the improved trigonal sextant, and the large (233 cm) portable brass semicircle that hung in the warming room of Stjerneborg. He remodeled the wobbly quadrant atop the south rotunda of Uraniborg into the largest (194 cm) azimuth quadrant of Stjerneborg. Gingerich and Voelkel pointed out that the great azimuth semicircle was mounted on this quadrant's original base, and Crol probably built the semicircle. They also argued that the revolving azimuth quadrant was built from an earlier quadrant mounted on a ball and socket. Crol may also have made the smaller, steel equatorial armillary.

EBELTOFT, CHRISTIAN   Danish chemist; Hven ca. 1587.

EINARSON, ODDUR   was a *famulus* briefly on **Hven ca. 1585** and later became a distinguished Icelandic churchman who maintained his ties with Tycho Brahe. Oddur Einarson was born 31 August 1559 at Möðruvellir Priory, Iceland, the descendant of a long line of priests, priors, and abbots (the clergy had never been celibate in Iceland). From Hólar Cathedral School, he came to the University of Copenhagen in 1583, took his baccalaureate degree in 1584, and was named chaplain of *Communitatis Regiæ,* the subsidized boarding hall for students. Oddur Einarson came to Hven around 1585, together with eight others, including Anders Jacobsen Lemvig and Joannes Varde. He was described as an average linguist but otherwise well-informed. He and the others remained on Hven for a month or more and departed with Tycho's permission.

By 1586, Oddur Einarson was back in Iceland as rector of Hólar Cathedral School. The bishop of Hólar 1571–1627 was Guðbrandur Þorláksson (1542–1627), a scholar, mathematician, patron of learning, and the outstanding Icelandic geographer of his century. The bishop had the only printing press in Iceland, cut his own woodcut illustrations, and eventually supplied a map of Iceland for the Ortelius atlas. In 1575, he made a celestial globe for the Danish governor-general of Iceland, Johan Bockholt, who observed a lunar eclipse in 1580 and sent his observations to Uraniborg.

After two years at Hólar, Oddur Einarson was named bishop of Skálholt, Iceland, in 1588. Because Iceland was a Danish possession, he returned to Denmark to be consecrated to the office on 25 March 1589, and Tycho Brahe probably attended the ceremony. Oddur Einarson visited Uraniborg for a week in April of 1589, then Tycho accompanied him to Elsinore and was gone from Hven for ten days. The night after he returned, Tycho tried out a new "Icelandic quadrant," presented to him by Oddur Einarson but undoubtedly made by Bishop Guðbrandur Þorláksson.

Oddur Einarson served as bishop of Skálholt for over forty years. In 1591, he married Helga Jónsdóttir (1567–1662), the daughter of a district governor. A scholar and patron of learning as well as a churchman, Einarson transplanted the Tychonic tradition to Iceland. He determined the latitude and longitude of Skálholt and sent his results to Tycho Brahe. He collected a large library, acquired a reputation for great learning in mathematics and astronomy, and compiled extensive topographical and chronological notes on his travels throughout Iceland, apparently planning to write a description of the island. Bishop Oddur Einarson died 28 December 1630 at Skálholt. Both of his sons carried on the Tychonic tradition in Iceland. The eldest, ARNI ODDSSON (1592–1665), translated Cort Aslakssøn's *En nyttig undervisning* (Copenhagen 1612) into Icelandic. His second son and successor, Bishop GISLI ODDSSON (1593–1638), wrote two works on Icelandic ethnography and natural history, based in part on his father's manuscripts.

ERIKSEN, JOHANNES   German student, secretary; Wittenberg 1599 and Bohemia 1599–1601.

FABRICIUS, DAVID   was an independent astronomer and cartographer who joined Tycho's *familia* briefly in **Wandsburg 1598** and **Prague 1601**. Kepler considered him the finest observational astronomer in Europe after the death of Tycho Brahe. The son of a smith, David Fabricius was born 9 March 1564 in Esens, East Frisia. He entered the University of Helmstedt in 1583 to study theology. In 1584, he settled into the life of a Lutheran clergyman at Resterhave near Dorum, East Frisia, where he married a young widow and began to pursue his interests in astronomy, astrology, and cartography on a highly sophisticated level. Six of his eight children were born in Resterhave between 1585 and 1602.

In 1589, he made the first good map of the flat, marshy territory of East Frisia, dedicated to the Count of East Frisia, as was a second map of East Frisia by Fabricius dated 1592. Such maps were powerful icons of statecraft and of considerable practical value to rulers, soldiers, merchants, and scholars. Fabricius mapped the neighboring principality of Oldenburg (1591) and prepared detailed maps of small enclaves in East Frisia, Gödens (1600) and Bederkesa Amt (1614). Fabricius was frequently summoned to Aurich to advise Count Enno III on astrological and cartographic matters, but he continued to live in the village of Resterhave.

David Fabricus began to keep a meteorological and astronomical journal in 1590. From 1593, he corresponded with astronomers and instrument designers across Europe, beginning with Joost Bürgi, who was building iron and brass instruments incorporating many of Tycho Brahe's innovations. Bürgi instructed Fabricius by letter, and in 1593, Fabricius built two highly reliable instruments, an iron quadrant and a sextant. Fabricius discovered the variable star Mira Ceti in 1596 and wrote to Tycho Brahe about it, comparing it to the supernova of 1572. Tycho entered the observations in his own journal under Fabricius's name, and **Kepler** in 1604 published the discovery.

Having begun a correspondence with Tycho Brahe, Fabricius regularly sent him his observations, and he briefly joined Tycho's *familia* on two occasions. The first was when Tycho invited him to visit Wandsburg: Fabricius arrived in late May of 1598; they got on very well, and Tycho presented Fabricius with gift copies of all his published works. Fabricius became thoroughly familiar with Tycho's methods and continued to send letters full of observational data, notes, scrawled comments, and precise corrections to Tycho's printed works – so frequently in the years 1597–1601 that his well-equipped observatory in Resterhave became a virtual affiliate to Tycho's own observatories in Wandsburg, Wittenberg, and Bohemia.

From October of 1599 onward, Tycho began to invite David Fabricius to join him as private chaplain and astronomer. Count Enno decided to help by sending Fabricius on a diplomatic mission to the imperial court. Fabricius arrived in Prague on 28 May, where he lived in the *familia* of Tycho Brahe for around three weeks. Once again, the two hit it off very well, and Tycho surely would have liked him to stay – if he had, he might have become Tycho's successor – but by 3 July, Fabricius was back home with his wife and family in Resterhave.

Kepler was in Graz that summer, so Fabricius did not meet him, but they soon became friends by correspondence. Fabricius dispatched a flood of letters to Kepler, who often replied at considerable length. In the years to come, Kepler tried out some of his most innovative ideas on Fabricius, whose frank, incisive, scrawling letters glowed with humanist wit and elegance. Fabricius also kept in touch with other Tychonic *famili*. On 10 March 1602, he received a visit in Resterhave from **Frans Tengnagel** and Johannes Eriksen. Matthias Seiffart also visited him as a courier from Kepler.

In October of 1602, Dutch troops plundered in the vicinity of Resterhave. David Fabricius fled with his family, instruments, and library to the walled city of Esens, where he remained until 1604. For a time, he served as court preacher in Aurich. In 1603, the count arranged for him to assume the call of Osteel, and the Fabricius family moved there in 1604.

After moving to Osteel, Fabricius reestablished his observatory and continued his avid correspondence. It was in Osteel that he became the first to observe the supernova of 1604, "like a burning light" near a conjunction of Jupiter and Mars. He published three German and Latin pamphlets on the supernova. Fabricius also published a long series of astrological almanacs.

In February of 1609, Fabricius was amazed by the reappearance of Mira Ceti and wrote about it in his last letters to Kepler.

On 7 May 1617, as he took his evening walk in the churchyard of Osteel, Fabricius was confronted by Frerik Hoyer, a peasant whom he had admonished from the pulpit in a matter of stolen geese. Hoyer struck him a blow to the head with a turf spade and ended the life of an outstanding observational astronomer. David Fabricius was buried in Osteel churchyard.

David's son **JOHANNES FABRICIUS** was born 8 January 1587 in Resterhave parsonage. He was sent to the University of Helmstedt in 1604, where he studied medicine as well as mathematics and astronomy. He went on to Wittenberg in 1606 and Leiden in 1609, still studying medicine. In 1610, he came home to Osteel with some telescopes acquired in Leiden, with which he and his father commenced to observe. Johannes Fabricius soon discovered sunspots, observed them at length with the telescopes and a *camera obscura,* and went on to discover the rotation of the sun. Johannes Fabricius returned to Wittenberg, where he published his discoveries in June of 1611. Because he was the first to publish these findings, and possibly the first to observe them, he was hailed as the discoverer of sunspots; however, Galileo and Christoph Scheiner later challenged his priority, touching off a learned controversy in which Kepler and the other former *famuli* of Tycho Brahe supported young Fabricius. Johannes Fabricius died on 19 March 1616, barely twenty-nine years of age, and was buried in Osteel churchyard.

FABRICIUS, JACOB   (1576–1652) German student, physician; Hven 1592–6.

FELS, DANIEL   German courier; Benátky 1600.

FLEMLØSE, PETER JACOBSEN   was Tycho Brahe's leading assistant for eleven years on **Hven 1577–88** and later served Tycho's friend and kinsman, Viceroy Axel Gyldenstierne, in Norway. Born around 1554 in the parish of Flemløse on the Danish island of Fyn, he was a student at the University of Copenhagen when Tycho Brahe lectured on astronomy 1574–5. Besides attending the lectures, Flemløse was undoubtedly one of the students who was tutored in observational astronomy by Tycho, whose patronage he sought by dedicating to him a Latin pamphlet on the solar eclipse of 1574.

Tycho began construction of Uraniborg in 1576. Flemløse joined him there in 1577, and from January of 1578, the entries in the observational journals were in his hand, including a sketch of an astronomer (presumably himself) in ancient Roman garb, peering at the starry heavens through a Tychonic sextant [Fig. 53]. In October of 1580, "Peter and Paul" (Flemløse and **Wittich**) observed a comet from Hven with a quadrant while Tycho observed it from Helsingborg with a radius. In 1581, Flemløse directed a program of observations of the fixed stars using radius, sextant, and quadrant. In the years to come, Tycho's journals were full of observations in Flemløse's hand. He worked with Tycho and **Elias Olsen Morsing** to prepare a short commentary on the comet of 1585 published under Morsing's name. In short, he was one of Tycho's most trusted assistants in astronomy, iatrochemistry, and meteorology, helping to train many others.

In 1586, Peter Jacobsen Flemløse visited the court of Kassel on Tycho's behalf and traveled on to Frankfort, returning to Uraniborg after an absence of six months. His mission helped to promote cordial, fruitful relations between Kassel and Uraniborg. In 1588, King Frederick II requested a handbook of

Fig. 53. Flemløse's self-portrait in Roman garb, showing the steel sextant in use. (From Brahe 1913–29, 10: 67)

weather prognostication, and Tycho assigned the task of composing it to Flemløse. The handbook was printed on the Uraniborg press in 1591 by Hans Gaschitz in Danish and German editions.

Tycho Brahe brokered the patronage that ensured Flemløse's successful future career. Flemløse first received a royal letter of expectance dated 13 June 1579 (renewed 25 July 1588 by the Regency), promising him the next vacant canonry in Roskilde Cathedral when he had completed his mathematical studies with Tycho Brahe.

The canonry was not available when Flemløse departed Hven on 14 August 1588 to enter the service of Tycho's kinsman, Axel Gyldenstierne, who had just been named Viceroy of Norway. Flemløse served as Gyldenstierne's physician, assisted in astrology, iatrochemistry, and other learned matters. He became an important figure in the circle of Oslo humanists around Bishop Frans Berg. In addition, after a brief visit to Uraniborg in 1589, Flemløse began a series of observations to determine the latitude of various places in Norway. That winter of 1589–90, Axel Gyldenstierne and his household, including Flemløse, were kept busy entertaining Princess Anne of Denmark, who arrived in September, and arranging for her marriage in Oslo on 23 November 1589 to King James VI of Scotland (later James I of England), with a month of celebrations thereafter.

In the summer of 1590, Flemløse was named to his long-awaited canonry in Roskilde. He received the income of the canonry but remained in Norway until 1596, where he continued to observe latitudes and serve as Tycho's intermediary with Norwegian learned circles. In 1595, he accompanied Bishop Anders Foss on several visits to Hven.

Peter Jacobsen Flemløse never married. He was on the verge of departing for Basel to receive his M.D. degree when he died suddenly in 1598.

FROBENIUS, GEORG LUDWIG   (1566–1645) German printer; Hven 1591.

GANS, DAVID BEN SOLOMON   visited **Benátky 1600** and established a link between Tycho Brahe and the Jewish rabbinical tradition. David Gans was born 1541 in Westphalia, possibly in Lippstadt, the son of Solomon ben Seligmann Gans. He studied in Cracow under the renowned Rabbi Moses Isserles. Noah J. Efron noted that the academy of Isserles was one of only two Ashkenazi Jewish academies in early modern Europe where natural philosophy was taught.

Around 1564, Gans came to Prague, where the Jewish community was rapidly recovering from a nearly total expulsion and was establishing many ties with the imperial court. He studied with Rabbi Sinai ben Bezalel and later joined the circle of his brother, Rabbi Judah Loew ben Bezalel, called Maharal, a Cabalist and mystic who became one of the most famous rabbis of the Renaissance era and was summoned by Emperor Rudolf II in 1592 for confidential discussions of esoteric matters. In that same year, Gans published a Hebrew chronicle, *Zemah David* ("The Offshoot of David") in two volumes, one dealing with the history of the Jews and the other with the nations of the world.

David Gans was a prolific author on chronology, mathematics, geography, astronomy, the use of a quadrant, and related subjects. Efron asserted that the breadth of his writings "was unprecedented and would remain unparalleled in Ashkenazi culture for centuries." His works aimed to explain to a Hebrew audience how the secular world worked in its physical as well as its political aspects. Efron noted that Gans saw natural philosophy, including astronomy, as a theologically neutral field that could bring Jews, Roman Catholics, Lutherans, and other Protestants to agreement. In other words, a profoundly irenic purpose underlay Gans's life work. This struck a responsive chord in Tycho Brahe and also harmonized with the prevailing cultural mood at the court of Emperor Rudolf II.

According to his own account, David Gans visited Tycho Brahe three times at Benátky Castle around 1600 and stayed for five days each time. He described Tycho's observatory at Benátky, with twelve learned assistants including **Johannes Kepler** and Johannes Müller, and twelve rooms in a row, each containing a large astronomical instrument. His description showed a tendency to transform personal observations into dramatic mythic images.

David Gans found Tycho Brahe to be a generous host who was willing to converse at length. They discussed the great Jewish and Gentile astronomers of the past. From the Brno manuscript of Gans's textbook of astronomy, *Magen David* ("The Star [lit. Shield] of David"), Alter quoted a passage in which Tycho described Ptolemy as a kind of Hermetic magus who ruled the heavens and was capable of ascending through the spheres by the power of his mind. Tycho and Gans discussed the medieval astronomers of Spain, and Gans translated Hebrew astronomical tables into German as a gift to Tycho Brahe. Certainly they also discussed Copernicus, whom they both admired.

Gans was writing *Magen David* at the time, and he put many questions directly to Tycho, Kepler, and Müller. One related to a problem that had arisen within *halakhah,* or Jewish ritual law: Where do days start? Jews in scattered locations needed to celebrate the Sabbath on the same day, and the global diffusion of peoples in the sixteenth century gave the matter a new urgency. Efron pointed out that Gans, true to his irenic principles, sought a resolution of this problem, not from any of the renowned rabbis of Prague, but from the Christian scholars of Benátky Castle. Tycho and his coworkers, however, were forced to admit, after several days of consideration, that they did not have a satisfactory answer. Three centuries later, the problem was finally resolved with the establishment of the International Date Line in 1884.

Gans included his description of Tycho Brahe at Benátky in *Magen David,* which he completed by 1612; it remained unpublished until 1743, when it appeared under the title *Nechmad ve-Naim* ("Pleasant and Agreeable"). Rabbi David Gans died 22 August 1613 in Prague and was buried in Prague Jewish cemetery.

GASCHITZ, HANS German printer; Hven ca. 1591–2.

GEK, PER Danish jester; Hven 1596.

GEMPERLE, TOBIAS was a painter in Tycho Brahe's *familia* on **Hven** 1577–87. During his travels abroad in 1575, Tycho, both as Frederick II's broker of cultural patronage and on his own behalf, contacted a number of artists and artisans including **Labenwolf, Schardt, Schissler,** and Tobias Gemperle. Gemperle, who had been trained by the Dutch artist Abraham del Hele (1534–98), was in Denmark by 23 April 1577, when King Frederick II granted him the privileged status of court artist.

His first known painting in Denmark is an 1578 oil portrait of Anders Sørensen Vedel, seated at a table, surrounded by emblematic objects and mottoes [see Fig. 8]. The direct gaze and living realism of the face and hands still captivate the viewer. This combination of arcane symbolism and striking realism appealed strongly to Tycho Brahe.

The most ambitious extant work by Tobias Gemperle is the altar of St. Ibb's Church on Hven, also from 1578. The main panels show the Crucifixion and Resurrection. The cross is at an oblique angle in the Crucifixion, with large foreground figures, including Mary Magdalene in an immense red cloak, against a chiaroscuro sky of light and dark grays. In the Resurrection panel, the figure of Christ hovers in an aura of bright light over a dark scene of sprawling humans. This altar represents the dramatic mannerist style that Tycho's artists, Gemperle and Schardt, introduced to Scandinavia.

In 1579, Gemperle made the first medical illustration in a Danish book: an inverted woodcut copy of a human skeleton taken from the 1568 Plantin edition of Vesalius.

Around 1580, Tycho built a large wooden equatorial armillary decorated with full-length representations of Ptolemy, Al-Battani, Copernicus, and himself; these were probably painted by Gemperle. During the 1580s, Gemperle was engaged in decorating Tycho's new "celestial palace" of Uraniborg, and perhaps also cutting woodcut illustrations for works from the Uraniborg press, including the emblems of Astronomy and Alchemy designed by Schardt. He may have painted the lost portraits of King Frederick II and Queen Sophie, as well as the emblematic portraits of Hipparchus, Ptolemy, Al-Battani, and seven Italian philosophers that hung in the museum. After 1584, the *hypocaustrum* ("warming room") of Stjerneborg was decorated with portraits illustrating the genealogy of astronomical wisdom from Timocharis, Hipparchus, Ptolemy, Al-Battani, Alfonso the Wise, Copernicus, and Tycho Brahe to the "Tychonides" of the future; these may also have been painted by Gemperle.

Around 1586, when Tycho Brahe turned forty years of age, Gemperle was involved in a series of collaborative emblematic portraits of the astronomer. A dramatic woodcut and a small pen and chalk sketch of Tycho [see Fig. 29], both aglow with vitality, may have been done from life by Gemperle; they were the basis for prints by various artists including Gheyn and Cornelius. Closely related to the iconography of these graphic portraits was the work that Tycho Brahe considered to be Uraniborg's artistic masterpiece: the painting on the mural quadrant. The quadrant was constructed 1582 and decorated 1587 by Gemperle, Knieper, and **Steenwinkel**. Knieper did the distant landscapes, Steenwinkel the symbolic views into the various areas of Uraniborg, and Gemperle the full-length foreground figure of Tycho Brahe, of which Tycho himself wrote, "The likeness could hardly be more striking, and the height and stature of the body is rendered very realistically."

Not long after completing this portrait, Gemperle died of the plague in Copenhagen. Under Tycho's patronage, Tobias Gemperle and the artists of Hven reached new heights in the integration of art and science in the late Renaissance.

GHEYN, JACQUES DE, II   (1565–1629) Dutch engraver; Haarlem (and Amsterdam?) ca. 1586–90.

GRONINGENSIS, RUDOLPHUS   Dutch instrument maker? Hven 1585–8.

HAMMER, NIELS   Danish student; Hven 1595 and Benátky 1599–1600.

HAMON, JOHN   (died 1617) English student, physician; Hven 1587.

HEGELUND, JACOB JENSEN   (1556–1600) Danish student; clergyman; Hven ca. 1586.

HEMMET, IVER IVERSEN   (1564–1629) Danish student; bishop; Hven ca. 1588–90.

HJORT, CHRISTOPHER   (1561–1616) Norwegian student, secret Roman Catholic; Hven 1596–7.

HVÆN, PETER   (died 1638/9) Danish printer; Hven ca. 1596–7.

JACHENOW, PETER   was a maker of odometers who worked for Tycho Brahe on **Hven 1590–2** and at **Wandsburg 1598**. Odometers were among the mechanical marvels described by Vitruvius and Hero of Alexander, inspiring Renaissance thinkers from Leonardo da Vinci to Jean Fernel. Tycho had been familiar with odometers since his student days. Among his teachers and associates, Homilius, Scultetus, and **Schissler** were all interested in odometers.

During the 1570s, Peter Jachenow made an odometer for Duke Ulrich of Mecklenburg-Güstrow. During a visit to Mecklenburg, King Frederick II admired the odometer so much that, in 1580, Duke Ulrich promised to send him not only the odometer but the wagon, a good strong horse, and even Peter Jachenow himself – though he made the king promise to give Jachenow a return passport. Four years later, in 1584, King Frederick wrote to thank the duke for the horse, wagon, and odometer, which measured German miles,

Fig. 54. Odometer by Peter Jachenow, 1582. The three arms record German rods, *morgen,* and miles. (Courtesy Nationalmuseet, Copenhagen)

morgen (thirty to the mile), and rods (sixty to the *morgen*). Presumably it had
three arms on a single dial, in a style developed by Jachenow [Fig. 54].

Peter Jachenow became a frequent visitor to Hven from around 1588, and
Tycho learned to value him for his honest reliability as well as his skill. Early
in the spring of 1591, he seems to have spent two and a half months on Hven:
This was probably when he fit an odometer to Tycho's carriage.

By then, Jachenow was no longer young, but he still traveled widely in pursuit of commissions. In August of 1591, he was in Prague. Thaddeus Hagecius
wrote to ask for Tycho's recommendation, and Tycho replied with lavish praise
of the artisan, who had equipped his carriage with an ingenious odometer that
recorded distances to the sonorous striking of small bells. Tycho used this carriage to transport friends and visitors around Hven.

Jachenow returned to Uraniborg in 1592. He visited Tycho Brahe, possibly for the last time, at Wandsburg in March of 1598. The following summer, Tycho rode to his first audience with Emperor Rudolf II in a splendid carriage, equipped with one of Jachenow's odometers, which the emperor admired and wanted to duplicate for his own use. A fine extant example of Peter Jachenow's work is an odometer made 1582 in Szczecin (Stettin), now in the National Museum in Copenhagen [see Fig. 54].

JENSEN, CHRISTOPHER   (1572–1611) Norwegian student, clergyman; Hven 1593–4.

JEPPE   Danish jester; Hven 1590s.

JOACHIM   German printer; Hven ca. 1584–5.

JÖSTELIUS, MELCHIOR   (1559–1611) German mathematician; Wittenberg 1598–1600.

KANDESTØBER, EVERT SIVERTSEN   Danish pewter and brass founder; Hven 1592.

KEPLER, JOHANNES   worked with Tycho Brahe for three months at **Benátky 1600** and for ten months in **Prague 1600–1**; together, they established one of the most famous collaborative enterprises in the history of science. Kepler [Fig. 55] was born into a dysfunctional family on 27 December 1571 in Weil-der-Stadt, Swabia, the son of a mercenary soldier and a mother who practiced herbal folk medicine. After early training in excellent Lutheran schools at Adelberg and Maulbronn monasteries, he entered the University of Tübingen in 1589, where Michael Mästlin aroused his interest in Copernican astronomy, though his major focus of study was theology. He was a brilliant student but broke off his studies in 1594 to accept an appointment as provincial mathematician of Styria and teacher of mathematics in the Lutheran school of Graz, where he issued an annual astrological calendar during 1595–1600 and published his first major book, *Mysterium cosmographicum* (1596), an emphatically Copernican work that searched for divine harmonies to reveal the hand of God in nature. In 1597, he married a prosperous widow of Graz, Barbara Müller.

Tycho's relationship with Kepler began early in March of 1598, when a courier from Helmstedt arrived at Wandsburg with a letter from Kepler to Tycho and a gift copy of *Mysterium cosmographicum*. Tycho read the book with great interest. Unfortunately, the same courier brought a copy of Ursus's vicious attack on Tycho, in a book that also contained fulsome praise of Ursus by Kepler. Tycho replied to Kepler on 1 April 1598, complimenting him on his book, inviting him to visit and discuss astronomical matters, but also condemning the scurrilous work of Ursus. Later that month, Tycho took the trouble to write to Mästlin about Kepler.

Tycho entered the service of Emperor Rudolf II in Prague in the spring of 1599. By late summer of 1599, he had taken residence in the castle of Benátky nad Jizérou ("Venice on the Iser") and was transforming it into a new Uraniborg. Meanwhile, Kepler was having serious problems in Graz, where militant Roman Catholics were moving to drive all Protestants from the province of Styria. When Kepler learned that Tycho was in the imperial service, he determined to seek him out. Before he could act, Tycho wrote a long, cordial letter, late in 1599, inviting Kepler to join him and offering to broker patronage on his behalf.

Kepler came to Prague and rode with Tycho Tygesen Brahe to Benátky Castle on 4 February 1600 to join the *familia* of Tycho Brahe. It was a momentous day in the history of science. At the castle, Kepler was a bit overwhelmed by the large, bustling household, but Tycho was overjoyed with Kepler, whose genius was immediately apparent to him.

Each coworker had assigned tasks. Kepler worked with Longomontanus on astronomical projects, where his performance could be measured against Tycho's ablest coworker, and where Longomontanus could keep an eye on him. Kepler was a bit too enthusiastic about heliocentrism, and moreover, Tycho harbored suspicions that he might still be an ally of Ursus. The two coworkers were assigned to reply to criticisms of Tycho: Longomontanus to Craig, and Kepler to Ursus.

Meanwhile, Longomontanus had been working on the theory of Mars, and Kepler assisted him. When Longomontanus was reassigned to resume his earlier task of revising the lunar theory, Kepler continued to work with the Mars data.

These two *famuli* were a study in contrasts. Christian Longomontanus was affable, warmhearted, sociable, faithful and loyal, beloved by Tycho and his whole family, an exacting mathematician, and an observational astronomer with eagle eyes. Johannes Kepler was near-sighted, surly and gauche, out of sympathy with Tycho's Hermetic philosophy but full of eccentric theories, and prone to reject the Tychonic system in favor of the Copernican. Nonetheless, Tycho was irresistably attracted by Kepler's intellectual brilliance and daring.

Unfortunately, Kepler remained ill at ease and unsatisfied. Tycho's household gathered for meals and discussed numerous matters over the table. Everyone competed for Tycho's attention, but Kepler felt awkward and neglected in the convivial company, and he seldom spoke with Tycho except at these meals. As the wealth of Tychonic data was gradually revealed to him, Kepler became aware that he could not finish his work in a short period of time. This realization, and his uncertain situation in Graz, had made him amenable to

Fig. 55. Johannes Kepler. Oil portrait by Hans von Aachen, ca. 1603–4. (Courtesy Zámecká Obrazárna, Rychnov nad Kněžnou, Czech Republic)

negotiations for a one- or two-year tenure in Tycho's service. The strain of the negotiations, conducted through various intermediaries including Longomontanus and Johannes Eriksen, drove Kepler toward the breaking point. "With pedantic caution," as Norlind put it, Kepler drew up numerous counterproposals to Tycho's proposals. Longomontanus suggested that he moderate his demands, and a memorandum with a modified set of conditions was entrusted to Jessenius, now in Prague, to negotiate with Tycho on Kepler's behalf. On Wednesday, 5 April 1600, Tycho, Jessenius, and Kepler met to reach an agreement, but Kepler was extremely tense, and a sharp conflict resulted. Later that day, there was another outburst, apparently at table in the presence of the whole household, after Kepler had drunk a good deal of wine, and this provoked Tycho to reply in a rage. The next day, Kepler left Benátky Castle with Jessenius and wrote a vitriolic letter from Prague, where he contacted Ursus, but then, in a fit of remorse and shame, he sent a repentant letter to

Tycho. Despite the violence of Kepler's conduct, the long-suffering Tycho was soon reconciled, and toward the end of April 1600, Kepler made his way back to the castle of Benátky. Tycho promised to broker a two-year imperial summons of Kepler.

This agreed, Kepler left for home on 1 June 1600 but soon discovered that he could not remain in Graz unless he renounced his Lutheran faith and became a Roman Catholic. This he refused to do, and he was ordered to leave the province within forty-five days. He informed Tycho Brahe, who immediately promised to do everything in his power. Kepler departed Graz on 30 September 1600 with his wife, stepdaughter, and two wagons full of household goods. On 19 October 1600, they arrived in Prague and found lodging with Baron Johann Friedrich Hoffmann, the Viceroy of Styria.

Kepler was seriously ill, racked with fever, coughing badly, and devoid of money and income. Tycho Brahe was the only one who could assist him. During the previous winter, the imperial court had removed to Plzen (Pilsen), but Emperor Rudolf II had returned to Prague in June of 1600 and immediately summoned Tycho to the city. Tycho moved to the house of the Golden Griffin on Hradčany Hill near the imperial palace. By the time Kepler and his family arrived, Tycho's instruments and library were installed in the Belvedere Palace nearby. In February of 1601, Tycho moved into the palace of the late Vice-Chancellor Kurz von Senftenau on Loreta Plaza, near the palace of Baron Hoffmann, and began to transform it into his New Imperial Uraniborg.

Kepler reentered the service of Tycho Brahe. He was assigned to work on the refutation of Ursus's plagiarism and slander of Tycho. He also spent some time with the theories pertaining to Mercury, Venus, and Mars, and he discussed these matters with Tycho.

In late April of 1601, Kepler returned to Graz to claim a portion of the estate of his rich father-in-law, who had just died. During this absence, he shot off another vituperative letter to Tycho after his wife sent a complaint from Prague. He also wrote to Giovanni Antonio Magini, trying to enlist his help in finding a Copernican solution to the orbits of Mars and the earth.

Kepler was in good health when he returned to Prague at the beginning of September 1601. Tycho brokered an audience with Emperor Rudolf II, who congratulated Kepler on the recovery of his health and commissioned him to cooperate with Tycho in compiling new planetary tables. Tycho requested permission to call them the *Rudolphine Tables,* and the request was graciously granted. Without realizing it, Tycho was virtually nominating Kepler to be the next imperial mathematician.

Kepler was not an easy person to work with, and he rejected the Tychonic planetary system in favor of the Copernican. James R. Voelkel commented that Tycho would have preferred Tengnagel, Longomontanus, or Johannes Müller as his successor, perhaps even David Fabricius or Simon Marius; but the fact is that Kepler was present when Tycho's life suddenly came to an end. Robert Westman pointed out that Kepler had come to Tycho seeking his data and patronage but found as well a role model. Tycho Brahe showed him the role of astronomer as active reformer and natural philosopher, supported by princely patrons and free from the pressures of university teaching.

Soon after the imperial audience, on 13 October 1601, Tycho Brahe fell gravely ill. Except perhaps for Matthias Seiffert, Kepler was the only mathematician and astronomer in the *familia* at the time. On his deathbed, Tycho begged Kepler to complete the *Rudolphine Tables* as quickly as possible and demonstrate their theory in terms of the Tychonic system, not the Copernican. Again and again, he expressed the hope that he had not lived in vain. Tycho Brahe died in the New Imperial Uraniborg on 24 October 1601 and was buried with great pomp on 4 November 1601 in the Týn Church of Prague.

Two days after Tycho's death, the imperial secretary, Johannes Barvitius, informed Kepler that the emperor had appointed him imperial mathematician with responsibility for Tycho Brahe's instruments and manuscripts and for the completion of his unfinished work.

His possession of Tycho's observational protocols caused problems because the instruments and manuscripts still belonged to Tycho's heirs, although the emperor wanted to purchase them. At the time of Tycho's death, his eldest son and his son-in-law, **Frans Tengnagel,** were absent, and Tycho's widow was distraught with grief, so the rightful claims of the heirs were without defense. Kepler took charge of the observational manuscripts: His work could not progress without the data they contained.

When Tengnagel came back, nearly a year later, he demanded the return of the manuscripts. Kepler resisted but eventually handed them over, though he surreptitiously held back the manuscripts of Mars. Kepler was working toward his second law and could neither formulate nor demonstrate it without Tycho's observational data. Tengnagel reclaimed the authority to work on Tycho's posthumous publications. He and Kepler squabbled constantly, but together they managed to publish Tycho Brahe's massive *Introduction to the Instauration of Astronomy* (*Astronomiæ instavratæ progymnasmata*) in 1602 and *On the Most Recent Phenomena of the Ætherial World* (*De mvndi ætherei recentioribvs phænomenis*) in 1603. Tengnagel next turned to the task of the *Rudolphine Tables* and discovered, in the spring of 1603, that some protocols of Mars observations were missing. He demanded their return, and Kepler reluctantly acceded. Without Tycho's manuscripts, Kepler needed something else to justify his position as imperial mathematician, so he turned to a study of optics.

Meanwhile, Tengnagel was becoming deeply involved in Habsburg politics and came to realize that he could not complete the *Rudolphine Tables* without help. Through the mediation of the imperial confessor, Johannes Pistorius, Tengnagel and Kepler reached an agreement on 8 July 1604 that allowed Kepler to use some of Tycho's observational journals in return for a promise to complete the *Rudolphine Tables* to Tengnagel's satisfaction and, in the meantime, to get the approval of Tengnagel before publishing anything based on Tycho's manuscripts. Later that year, a supernova appeared in dramatic array with Saturn, Jupiter, and Mars, and it was observed in the Tychonic manner by a team comprising Tengnagel, Kepler, and Joost Bürgi. Four years later, telescopes captured the attention of the learned world, including Kepler. By then, with the Tychonic data in hand, Kepler was completing his epochal *Astronomia nova,* containing his first and second planetary laws, which he published in 1609 with Tengnagel's approval. Voelkel has shown how the rhetorical shape of this work grew out of Kepler's clashes with Tycho, Tengnagel, and Longomontanus.

Relations with the Brahe heirs improved when Tycho's younger son, Jørgen or Georg Brahe, as he was called in Bohemia, replaced Tengnagel as their representative in negotiations with Kepler. Georg Brahe and Kepler established an amicable working relationship based on mutual trust. The result was an enduring agreement of 1612 that made Kepler the custodian of all the observational protocols. These were the basis of his life's work, and he never failed to honor the legacy of Tycho Brahe in every one of his publications based upon them, including his *Harmonices mundi* (1618), which contained the third law.

Kepler's wife died in 1611, and he remarried and raised a new family. After 1618, though, endemic warfare created chaos throughout central Europe and disrupted the life of the Kepler family. Jesuit astronomers tried to gain possession of Tycho Brahe's instruments, library, and manuscripts. With imperial support, they got the library, some surviving instruments, and many manuscripts, but Kepler clung tenaciously to the observational manuscripts, refusing to give them up to anybody but the Brahe heirs.

Kepler finally published the *Rudolphine Tables* in 1627, with an image of Tycho Brahe on the title page. Then he signed a contract with the Brahe heirs to prepare Tycho's observations for publication. This task was still unfinished when Johannes Kepler died on 15 November 1630 in the imperial city of Regensburg. In 1662, Kepler's son, Dr. Ludwig Kepler, sold Tycho Brahe's observational manuscripts to King Frederick III of Denmark, and they came to the Royal Library in Copenhagen, where they repose to this day.

KNIEPER, HANS   (died 1587) Flemish painter, designer; Hven 1587.

LABENWOLF, GEORG   was a Nuremberg art-bronze founder who came to **Hven ca. 1577** and constructed the fountain of Uraniborg. The landgrave of Hesse-Kassel, for whom he had built fountains and waterworks around 1570–2, may have recommended him, and Tycho Brahe met Labenwolf in Nuremberg in 1575. In his role as cultural broker, Tycho recommended Lab-

Fig. 56. Labenwolf's Kronborg fountain. Engraving 1730 of a drawing by W. J. von Stromer, 1583. (Courtesy Det Kongelige Bibliotek, Copenhagen)

enwolf to King Frederick II, who summoned him to Denmark toward the end of 1576.

In 1577, Labenwolf signed contracts for an elaborate bronze and marble fountain in the courtyard of Kronborg Castle and a hydraulic system for Uraniborg. It took several years and a tremendous amount of money to complete these projects. On 15 August 1579, when Vedel was about to depart for Germany, Tycho asked him to ensure that Labenwolf send the materials for Tycho's system before winter. King Frederick grew very impatient over the delays with the Kronborg fountain, and this led the Nuremberg town council in 1581 to appoint a patrician humanist, Joachim Pömer, to supervise the project. Labenwolf subcontracted parts of it with numerous sculptors, goldsmiths, and other artisans, but the Kronborg fountain was erected and given a trial run in Nuremberg later that year, as Pömer reported to Tycho.

Pömer enclosed a gift for Tycho of two Latin poems by the humanist laureate, Paulus Melissus (Schede), together with three motets with Latin and German texts, praising the royal Danish couple and the magnificent fountain. Tycho Brahe transmitted the motets to the king, who rewarded Pömer with a golden chain and a royal portrait medallion. Kongsted thought that Orlando di Lasso might have written the motets, but Kirnbauer showed that they were composed by Leonhard Lechner. Kirnbauer saw them as a cultural product of humanist friendship among Tycho, Pömer, Melissus, and Lechner. They were undoubtedly performed at the dedication of the Kronborg fountain [Fig. 56], which finally arrived in 1583, accompanied by Labenwolf's son and two journeymen to set it up. Two years later, Joachim Pömer visited Hven on 13 June 1585.

Besides brokering this major commission for the crown, Tycho Brahe engaged Georg Labenwolf to design the hydraulic system of Uraniborg. From a well in the cellar, pipes in the walls carried water throughout the manor house. Labenwolf's fountain of Hippocrene, the source of wisdom, was erected in a rotunda at the crossing of the main-floor corridors, under Pegasus on the spire far above. The Vitruvian spirit and rich iconography of the fountain indicate that Tycho Brahe must have played a role in its design. The central figure was a bronze Aquarius, "brought alive" and rotated by jets of water. Four heads at the cardinal points of the basin represented the four winds, reinforcing the cosmic orientation and symbolizing universal wisdom as they sprayed into the basin. Uraniborg's internal waterworks were without parallel in sixteenth-century northern Europe, but only two elegant brass faucets survive, in the shape of dolphins and with a handle representing Arion playing his violin.

The Kronborg fountain was Georg Labenwolf's masterpiece, and Uraniborg was perhaps his most innovative project. He died in the year 1585 in Nuremberg.

LANGREN, ARNOLD FLORIS VAN   (1571–after 1630) Flemish-Dutch cartographer; Hven 1590.

LANGREN, HENDRIK FLORIS VAN   (died ca. 1598) Flemish-Dutch cartographer; Hven 1593.

Fig. 57. Live Larsdatter. Oil painting by Pieter van der Hult, 1691. The legend on the painting notes that she was born 6 August 1575 in Kynstrup parish and baptized in Kynstrup Church. She died 9 July 1698 in Copenhagen. (Courtesy Det Nationalhistoriske Museum på Frederiksborg, Hillerød)

LARSDATTER, LIVE   Danish housekeeper, chemist; possibly Hven 1590s [Fig. 57].

LEMVIG, ANDERS JACOBSEN   Danish clergyman; Hven ca. 1585.

LEMVIG, JACOB MIKKELSEN   (ca. 1556–1618) Danish clergyman; Hven 1584–6.

LONGOMONTANUS, CHRISTIAN SØRENSEN   (Christian Severini, Christianus Severini Lymvicus, Christianus Longicampianus, Christen Lomborg, Chresten Lemvig, Christiern Langberg) served Tycho Brahe for eight years on **Hven 1589–97** and two years in **Bohemia 1600–1**. He was the son of peasants and was born 4 October 1562 on the farm of Lomborgberg, near Lemvig in western Jutland. He was instructed as a boy by the pastor of Lomborg and in 1577 he entered Viborg Cathedral School. His academic progress was slowed by poverty, and he did not enter the University of Copenhagen until 1588, where he became known as Longomontanus, a Latinized version of the name of his birthplace. The very next year, his professors recommended him to the service of Tycho Brahe, but the date of his arrival on Hven is not recorded.

Longomontanus was primarily an astronomer. **Elias Olsen Morsing** may have been his tutor in Tycho's methods of observation, though **Flemløse** also became a good friend and worked with him. Longomontanus was exacting and precise, had sharp eyesight, and worked well with others. These traits helped him to become an outstanding observational astronomer. He worked on the great catalog of 777 stars in his first three years on Hven.

Tycho became very fond of him and occasionally used him as his personal secretary. Around 1591, Tycho dictated the preface of Flemløse's book on weather prognostication to Longomontanus. As Tycho's sons were growing up, Longomontanus sometimes escorted them around Skåne, where their father's and mother's families lived, and he may have been the tutor who prepared them for Sorø Academy.

Early in 1597, in the last few months before leaving Hven, Tycho rushed to expand his star catalog and great globe from 777 to 1000 stars. He put Longomontanus in charge of the project, but it was carried out hastily and imprecisely in February and March of 1597: Others were busily taking inventory and moving; it was a time of great confusion and uncertainty for Tycho and his staff. On 11 April 1597, Tycho established his new "temple of the muses" in Copenhagen, but it did not last long. Less than two months later, he gave Longomontanus a letter of recommendation for eight years' faithful service and sent him on his way. The very next day, 2 June 1597, Tycho and his household took their leave of Copenhagen on the road to voluntary exile.

While Tycho moved from Copenhagen to Rostock and on to Wandsburg, Longomontanus was traveling and studying in Wrocław (Breslau), Leipzig, and Rostock. Tycho tried repeatedly to draw him back into his service. In the spring of 1598, Longomontanus received his M.A. in Rostock and came for a short summer visit to Wandsburg to discuss the Ursus affair, but then he left to continue his studies. In September, he returned, and when Tycho's carriages and wagons rumbled out of Wandsburg, Longomontanus accompanied them to Magdeburg, then took Kirsten Jørgensdatter and her daughters back to Wandsburg.

By then, Longomontanus was looking for a permanent academic position. Tycho wrote to him from Wittenberg at the end of December 1598 and offered to broker his patronage outside Denmark. Longomontanus received this letter in Ribe, where he must have been discussing his future prospects with Anders Vedel, Bishop Hegelund, and other learned power brokers. On 12 January 1599, he left Ribe for his boyhood home in Lomborg parish. There on the edge of the North Sea, he observed the lunar eclipse that caused such embarassment to Jöstelius and Tycho in Wittenberg. Longomontanus was thirty-six years of age. He longed for a permanent position so he could settle down, marry, and begin a scholarly career of his own; but to get one, he needed patrons, and his was Tycho Brahe.

On 11 January 1599, Tycho wrote and urged him to come to Wittenberg, offering to pay his travel expenses, and asking whether he wanted Ursus's job as imperial mathematician; if not, Tycho would arrange for a Wittenberg professorship. Tycho wrote once more on 21 March 1599, saying that he wanted to discuss revisions of the lunar theory with Longomontanus. Finally, on 22 September 1599, he gave his son, Tycho Tygesen Brahe, a letter to deliver to Longomontanus in Rostock on his way to retrieve the instruments left on Hven. The letter urged Longomontanus to join young Tycho and then come and help to set up the instruments at Benátky.

Longomontanus arrived at Benátky Castle by 17 January 1600, and by the time **Kepler** arrived on 4 February 1600, he was hard at work on the theory of Mars. Kepler was assigned to assist him and was allowed to stay with Mars – after promising in writing never to reveal Tycho's secrets – when Longomontanus was reassigned to revise the lunar theory. Besides these parallel theoretical assignments, each of them was given the task of refuting one of Tycho's critics: Longomontanus got John Craig, and Kepler got Ursus. Neither enjoyed the assignment, and neither ever published the polemics.

In Tycho's household, Longomontanus had some clear advantages over Kepler: He was a skilled observational astronomer with a warm, convivial personality, and he had developed close personal ties to Tycho and his family over many years. He served as one of Tycho's intermediaries in negotiating Kepler's terms of service.

Longomontanus continued to work on the lunar theory and was almost finished when a solar eclipse of 10 July 1600 showed the need for further revision of Tycho's solar theory. Longomontanus took up that task during July. On 18 July 1600, Longomontanus was planning his final departure from Tycho's service and gave Johannes Eriksen his manuscript polemic against Craig and a copy of Copernicus's *Commentariolus*. On 4 August 1600, Tycho Brahe gave him a final letter of recommendation, and Longomontanus, thirty-eight years of age, set off in the direction of Denmark.

He traveled widely in the following years, always well received by Tycho's far-flung friends and former associates. When Tycho died in Prague on 24

October 1601, Kepler was present and Longomontanus was not. Kepler was named imperial mathematician, though James R. Voelkel is certainly right in asserting that Tycho would have preferred Longomontanus as his successor.

Longomontanus now returned to Denmark, and in 1603 became rector of his alma mater, Viborg Cathedral School.

By 1605, Longomontanus had found a new patron in Royal Chancellor Christian Friis of Borreby, who commended him to the University of Copenhagen and offered personally to pay his salary as Professor Extraordinary of Mathematics; but when Laurids Scavenius left the university shortly thereafter, Longomontanus was appointed to the vacant chair of Latin literature. In 1607, **Cort Aslakssøn** was promoted to a chair of theology, Christian Hansen Riber succeeded him as professor of Greek, and Longomontanus followed Riber in the chair of mathematics, which he held until the end of his long life, forty years later. Besides these three old *famuli* from Hven, Johannes Stephanius and Gellius Sascerides also held university chairs in Copenhagen, and there was a solidarity among the former *famuli* (except for Gellius) that made them a powerful force in university affairs.

In 1607, Christian Longomontanus married Dorthe Bartholin (1590–1637) and entered a family circle with great influence in Danish academic life. Her brother, Professor Caspar Bartholin, was married to a daughter of Professor Thomas Fincke, and Professor Olaus Worm would soon become another Fincke son-in-law. The influence of this academic dynasty survived throughout the seventeenth century. Two of Caspar Bartholin's sons, Thomas and Erasmus Bartholin, became well-known professors in Copenhagen, and Erasmus's daughter married the astronomer Ole Rømer (Olaus Roemer). The genial Longomontanus soon came to play an important role within this family circle. Forty-five years of age, he had finally arrived.

Around 1608, he wrote to inform Tycho's family of his marriage and professorship and received a cordial reply from Magdalene Brahe, Tycho's daughter. On 21 June 1610, Longomontanus received a grant from the university to build a telescope with some lenses obtained by Cort Aslakssøn. He was still corresponding occasionally with Kepler.

Longomontanus was no theologian and played a passive role in the 1614 Resen hearings chaired by Cort Aslakssøn. In 1621, his chair was transformed into an extraordinary chair of astronomy, which he occupied until his death. In 1632, Tycho Brahe's great celestial globe was brought back to Denmark as war booty, and Longomontanus composed an inscription to commemorate its return.

Five years later, construction began on a university observatory. King Christian IV had conceived a plan for a cosmic building to unite heaven and earth, faith and learning, by combining the university church, library, and obser-

Fig. 58. Christian Longomontanus. Engraving by Simon de Pas, 1644.
(Courtesy Det Kongelige Bibliotek, Copenhagen)

vatory in a single structure with a rich spatial and emblematic framework. The
building was completed in the year 1642 and still stands in the Latin quarter
of Copenhagen as Trinity Church and the Round Tower. At the west end was
the massive observatory tower, symbolic of heavenly wisdom, bearing a rebus
composed by the king himself [see Fig. 49], and containing the great celestial globe of Tycho Brahe. Atop this tower was the observatory named Royal
Stjerneborg of Copenhagen, a modernized version of Tycho Brahe's observatory, transmuted from underground to the heights above the city. It was the
second university observatory in Europe, following that of Leiden. Longomontanus [Fig. 58] was director of the Copenhagen observatory until his
death on 8 October 1647 in Copenhagen, where he was buried in Our Lady's
Cathedral.

Fig. 59. Simon Marius. Woodcut by unknown artist, 1614. (From his *Mundus jovialis*)

MARIUS, SIMON   (1573–1647) [Fig. 59] German astronomer; Prague 1601.

METIUS, ADRIAAN   (1571–1635) Dutch astronomer, educator; Hven 1594–5.

MIKKELSEN, NIELS   Danish student; Herrevad 1572.

MORSING, ELIAS OLSEN   was a leading assistant to Tycho Brahe for seven years on **Hven 1583–90** and was sent on an expedition to Copernicus's observatory in **Frombork 1584.** He was born around 1550 on the island of Mors in northern Jutland, from which he took his surname of Morsing. Anders Lemvig, Jacob Lemvig, and **Christian Longomontanus** later came from the same vicinity, and Morsing may have recommended one or more of them to Tycho's service.

Morsing was undoubtedly trained in Tycho's methods of observation by **Peter Jacobsen Flemløse**, as well as by Tycho himself. Morsing in turn trained a number of coworkers including Rudolphus Groningensis and **Hans Crol**, who assisted him in observing the comet of 1585. On the basis of handwriting analysis, Friis in 1876 determined that Morsing kept the meteorological diary on Hven for most of the time from 13 April 1583 until 5 April 1589.

One of the first scientific expeditions in history was Elias Olsen Morsing's 1584 journey to the former residence of Copernicus in Frombork (Frauenburg). Tycho Brahe needed to redetermine the latitude of Frombork in order to make sense of Copernicus's data. Early in 1584, a Prussian envoy arrived at the Danish court, and when King Frederick II, later that spring, outfitted two ships to take the envoy home, Tycho Brahe arranged for Morsing to receive passage as far as Gdańsk (Danzig). By 13 May 1584, Morsing was in Frombork, where he stayed for three weeks of overcast, rainy weather, broken by a few clear nights when he was able to observe the stars. Morsing made friends with the canons of the cathedral chapter. The dean, Johannes Hanovius, presented him with a triquetrum (parallactic instrument) of pine, eight feet long, made and used by Copernicus himself. A portrait of Copernicus, said to be a self-portrait, may also have been presented to him during his stay in Frombork. Morsing also visited Kaliningrad (Königsberg), seat of the Prussian government, to determine its latitude for the administrator of the duchy, Margrave George Frederick of Brandenburg-Ansbach. He arrived back on Hven on 23 July 1584. Tycho was so excited to receive Copernicus's triquetrum that he immediately composed a Latin poem. The instrument and portrait of Copernicus were esconced in places of honor at Uraniborg, and Tycho built an improved triquetrum of the same type.

In 1585, Morsing was assigned by Tycho Brahe to prepare an astrological calendar for 1586, which Tycho wanted to print in some two thousand copies on his new press. The manuscript was already in press when a comet was sighted on 10 October 1585. It was observed carefully, and an appendix on this comet, with an astrological prognostication, written in part by Tycho, was added to the calendar. The work came out in 1586 under the name of Morsing. It was the first bound product of the Uraniborg press, and Tycho sent copies to his learned friends around Europe.

Three years later, in the spring of 1589, Anders Sørensen Vedel discussed with Tycho Brahe his plans for expeditions throughout Denmark to gather historical and chorographical materials. He needed maps to accompany these materials, so geodesy and cartography became part of Vedel's plan, and Tycho assigned both Flemløse and Morsing to get involved. In 1589, Morsing determined the latitudes of Viborg, Nykøbing on Mors, and Aalborg, undoubtedly using the portable sextant. Early in June, the participants all gathered again at Uraniborg. On 11 June 1589, Vedel and Morsing sailed to Skåne, where Morsing observed latitudes of several locations and angular distances between points around the horizon. This seems to have been the first actual use of the method of geodetic surveying combined with triangulation. Morsing returned briefly to Hven, then continued his geodetic observations on Sjælland and Jutland.

Elias Olsen Morsing died on Tycho's island at 11:30 P.M. on 11 March 1590 in the arms of his friend, Christian Longomontanus. A brother of Elias Olsen Morsing, name unknown, visited Hven on 22 October 1591.

MULE, CLAUS   (ca. 1575–1647) Danish nobleman, courier; Wandsburg, Wittenberg, Bohemia 1598–1600.

MÜLLER, JOHANNES   (died after 1608) German astronomer, alchemist; Hven 1596, Wandsburg 1598, Bohemia 1601.

OHR, PHILIPP VON   German printer; Wandsburg 1597–8.

OSWALD, JOHANNES   German mathematician, physician; Wittenberg 1600.

PAESCHEN, HANS VAN   (died after 1582) Flemish or Dutch architect; possibly Hven 1576–7.

PAULSDATTER, ELISABETH   was a chemist who served in Tycho Brahe's *familia* on **Hven 1596**. There is an old story that one of Tycho Brahe's maidservants on Hven secretly labored in his laboratory at night: Scharling told it in 1857 without citing his source, and Troels-Lund later repeated it. If there is any substance to the story, it must refer to Elisabeth Paulsdatter, the only woman known to have worked among the learned members of Tycho Brahe's *familia* on Hven, except for his sister, **Sophie Brahe**. Elisabeth was the daughter of Paul Thiisen (1549–1625, also called Paul Apotheker or Paulus Pharmacopola), a prominent merchant apothecary in Elsinore, and Clara Jansdatter (died 1594), whose family came from Amsterdam. The Elsinore alchemist, "Doctoress" Barbara Thiisen, who purchased flasks, stills, and bowls from the Herrevad glassworks in 1579, was apparently Elisabeth Paulsdatter's aunt.

Paul Thiisen supplied customers with exotic goods like Seville oranges and also became an important contractor of paint, varnish, horn glue, sal ammoniac, nails, spikes, and other building supplies for Kronborg Castle. Elisabeth Paulsdatter grew up in a comfortable middle-class home in one of the most dynamic, cosmopolitan cities in Denmark. Her brother, Mattis Paulsen, herbalist, carried on the family profession.

The family visited Hven frequently and may have supplied Tycho Brahe with chemical equipment, rarities, and construction materials. Clara Jansdatter was on Hven in 1590, and her husband came at least twice in 1593.

Some time thereafter, Elisabeth Paulsdatter joined the *familia* of Uraniborg. Her date of arrival on Tycho's island is not recorded, but her departure is documented dramatically in the meteorological journal, where the entry for 22 July 1596 says, "Elisabeth, daughter of Paulus Pharmacopola, Joachimus and Theodoricus dismissed for sedition." This may be the event referred to by Scharling. Paul Thiisen visited Uraniborg a month later, perhaps to make amends for his seditious daughter, and she returned with him on 27 October 1596. Paul Thiisen was back on Hven in 1597. Few people called at Uraniborg more regularly than he.

Some time after she left Tycho's island, Elisabeth Paulsdatter married Berent Facke (died 1620), clerk of Kronborg Castle and later clerk in the Elsinore Tollhouse. However, Elisabeth Paulsdatter did not live long to enjoy the social status of a prominent official's wife. She had died by 14 November 1598, when her father and widower met to settle her estate. The manuscript probate records give a clear impression of solid middle-class wealth, but they do not tell much about her learned interests. If she had any chemical apparatus at the time of her death, the probate clerk must have confused it with everyday household equipment, or perhaps she worked in her father's, brother's, or aunt's laboratory. The probate inventory did record, however, that Elisabeth Paulsdatter owned no less than thirty-eight books, a large library for a woman. Did she own any of Tycho Brahe's works, medical or chemical treatises, writings of Paracelsus or other German vernacular authors? Did she have books in Latin, Danish, or Dutch, as well as German? A list of titles could have told a lot about her learned interests, but the probate clerk was in a hurry and simply noted a large German Bible, a book of sermons, and "various other books of all sorts, thirty-six items." That Elisabeth Paulsdatter was a woman of learning is beyond doubt, but the specific details unfortunately escape us.

PEDERSEN, DAVID   Danish bailiff; Hven 1597–1601.

PEDERSEN, RASMUS   Danish farmer, vassal; Gundsøgaard ca. 1587–92.

PEPLER, CHRISTOPHER   Dutch-Norwegian bailiff; Nordfjord 1578–86, 1589–91.

PLOW, JENS PEDERSEN   Danish student, clergyman; Hven ca. 1577.

Fig. 60. Johannes Isaksen Pontanus. Engraving by J. van der Velde, 1630, after a painting by Isaac Isaacsz, the son of Pieter Isaacsz. (Courtesy Det Kongelige Bibliotek, Copenhagen)

PONTANUS, JOHANNES ISAKSEN   (1571–1639) Dutch-Danish poet, historian, chorographer; Hven 1593–5 [Fig. 60].

RHODIUS, AMBROSIUS   (1577–1633) German mathematician; Prague 1600–1.

RIBER, CHRISTIAN HANSEN   (1567–1641) Danish student, bishop, cartographer; Hven 1586–90, Wandsburg 1598.

RICHTER, PETER   Danish student, teacher; Hven ca. 1586–7.

Fig. 61. Holger Rosenkrantz and Sophie Axelsdatter Brahe. Funerary portrait in Hornslet Church, artist unknown. (Courtesy Hornslet Church and Det Nationalhistoriske
Museum på Frederiksborg, Hillerød)

ROSENKRANTZ, HOLGER   (1574–1642) Danish noble politician, theologian, educator; Hven 1592, Wandsburg 1598 [Fig. 61].

ROTHMANN, CHRISTOPHER   was astronomer to Landgrave William
IV of Hesse-Kassel, corresponded with Tycho Brahe for many years, and visited **Hven 1590**. Tycho never succeeded in recruiting Rothmann into his *familia*, but there was a tremendous exchange of ideas and data between Uraniborg and Kassel, in addition to transfers of technology and empirical methods
from Tycho's island to the landgrave's court.

Rothmann was born 1550 in Bernburg, Anhalt, and began his university
studies in Wittenberg around 1570, where he wrote a manuscript on astron-

omy, combining Aristotelian and Copernican elements with original ideas. His studies were supported by Duke Joachim Ernst of Anhalt-Bernburg, the ruler of his native land. In 1577, two years after Tycho had visited Kassel, Rothmann was sent there to inspect the landgrave's famous astronomical instruments and automata, and he was persuaded to remain. The instruments had been constructed during the previous decade by Eberhart Baldewein. In 1579, two years after Rothmann's arrival, Joost Bürgi joined the landgrave's court as maker of clocks, automata, and instruments.

Paul Wittich visited in 1584 and showed Bürgi many of Tycho Brahe's innovations in instrument design, and he also taught Rothmann and Bürgi what he knew of Tycho's methods of observation and data analysis, including the method of *prosthaphæresis* used on Hven. Ursus visited Kassel in the spring of 1586, after he had been at Uraniborg, and **Peter Jacobsen Flemløse** came that same spring as Tycho's emissary.

Following Flemløse's visit, Tycho Brahe began a lively learned correspondence with the Kassel astronomers. Rothmann's study of the comet of 1585 led him to refute Aristotle's assertion that comets were sublunary and to support Tycho's earlier conclusions. Tycho confirmed Rothmann's rejection of the existence of solid celestial spheres and his discovery of systematic changes in stellar latitudes. Rothmann agreed with Tycho's explanation that the tail of the comet was due to the effect of sunlight passing through the comet's rarified substance. He shared Tycho's rejection of a sphere of fire and his assumptions about light and the nature of refraction, but they disagreed over the measurement of stellar longitudes and refraction, as well as what these measurements showed about the structure of the heavens. They discussed the aurora borealis and the physical nature of celestial space, and Rothmann chided Tycho for his continued belief in the existence of celestial ether. Rothmann was more willing than Tycho to explore heliostatic Copernican models, though he agreed with Tycho that mathematical demonstrations of the nature of the universe had to be grounded in evidence derived from precise observational data.

In his letters to Rothmann and Landgrave William, Tycho willingly revealed a great deal about his instruments and how he used them. The observers in Kassel produced excellent results, but Tycho's theory and methods of empirical observation remained superior, and he was much better at publishing his results. Landgrave William and Rothmann developed a strong desire to visit Tycho's island, see its marvels, and continue the discussions in person. The landgrave almost came in 1588, but his plans fell through.

That same summer of 1588, Gellius Sascerides visited Kassel and presented to Rothmann an autographed prepublication copy of Tycho Brahe's great work on the comet of 1577, *De mvndi ætherei recentioribvs phænomenis*, which contained a description of the Tychonic theory of planetary motion. Tycho now learned that Ursus had passed off an earlier version of the Tychonic theory as his own during his visit to Kassel, and that Bürgi had made a clockwork model of it.

Christopher Rothmann suffered ill health in these years, and he took leave in the hope that a change of air would be beneficial. He set out for Uraniborg and arrived on the evening of 1 August 1590, where he remained as Tycho's guest for a month. On 13 August 1590, Tycho, Rothmann, and Lauge Urne, a learned Danish nobleman, went to see the sights of Skåne, where Rothmann was particularly impressed with the herbal, medicinal, and allegorical gardens of Tycho's sister, **Sophie Brahe**. When time allowed, Tycho and Rothmann conversed, with Tycho arguing against the heliostatic theory of Copernicus and Rothmann defending it. Eventually, either Tycho's eloquence convinced Rothmann, or the latter's courtesy led him to yield to his host; in any case, he conceded the argument to Tycho, adding that he would never publish anything in support of the Copernican theory.

On Sunday, 30 August 1590, when the rainy weather turned clear and still, Christopher Rothmann joined Tycho Brahe and his assistants in observing the heavens and viewing some splendid aurora borealis. Two days later, Rothmann took his leave of Tycho's island.

Tycho thought Rothmann was returning to Kassel, but instead he traveled to his native Bernburg. Tycho did not hear from him until 1594, when Rothmann sent a short letter, complaining of continued ill health. Tycho sent a long reply, but he never received another letter from Christopher Rothmann. Tycho published their correspondence in 1596, and in 1599, he heard that Rothmann was still in Bernburg. Tycho hoped to attract him to Benátky in 1600, but Rothmann did not come. By 1608, he had died in Bernburg.

SADELER, MARCUS   Dutch printer; Haarlem ca. 1586–8.

SASCERIDES, DAVID   (died ca. 1622) Danish-Dutch student, clergyman; Hven 1585–6.

SASCERIDES, GELLIUS   (1562–1612) Danish-Dutch student, chemist, astronomer, physician; Hven ca. 1582–7, 1593–4.

SCHARDT, JOHAN GREGOR VAN DER   was an Italian-trained sculptor who played a major role in the construction of Uraniborg on **Hven ca. 1576–9**. His career has been documented and his oeuvre established by Hanne Honnens de Lichtenberg. Schardt was born about 1530/1 in Nijmegen, trained as an artist in the Netherlands, and studied in Rome and Florence; he then worked for a time in the Veneto, possibly on Palladio's villas, and in Bologna in 1568–9, whence he was recruited to serve the Holy Roman Emperor. Daniele Barbaro, the editor of Vitruvius and patron of Palladio, recommended him enthusiastically to Emperor Maximilian II as a sculptor, brass founder, outstanding architect, and a highly intelligent individual. For seven years, Schardt served the imperial court. He probably helped to build the Pheasant Garden at Vienna, with many features later found at Uraniborg. Schardt's splendid bronze miniatures influenced later imperial artists, including Bartholomäus Spranger, Hans von Aachen, and Adriaen de Vries.

Tycho Brahe met and recruited Schardt, as well as **Gemperle** and **Labenwolf**, during his travels abroad in 1575. By November of 1576, he was staying on Tycho's island, where Uraniborg was under construction. By May of 1577, he had entered the service of King Frederick II. In 1578–9, Schardt was living in Elsinore, sculpting a lively polychrome terra-cotta bust of King Frederick [see Fig. 33] and casting bronze busts of the king and queen, while also working on Tycho's island.

Honnens de Lichtenberg characterized his sculpture as combining realism, naturalness, masterly anatomical representation, harmonious balance, inner energy, and a latent movement [Fig. 62] that expressed both the present and the coming moment (e.g., Queen Sophie on the verge of smiling, her consort about to turn to action). Without abandoning the profound serenity of the Renaissance, Schardt infused his works with the dynamism and allegorical richness of the early baroque. The cornerstone of Uraniborg may have been laid before Schardt arrived, and undoubtedly the proportions of the building and its grounds had been worked out personally by Tycho Brahe in accordance with Renaissance Vitruvian theory. Schardt, however, helped to execute the plan, enriching the allegorical program of the building, and carving or casting much of the statuary. He made the statues of Astronomy and Chemistry over the entrances of Uraniborg, giving them the form of Roman river gods. The sole surviving figure from Uraniborg, a putto from one of the entrances, has been ascribed to Schardt by Honnens de Lichtenberg. Whether the gilt Pegasus on the spire and the statues of the four seasons (or ages of man) on the side cupolas were by Schardt cannot be determined. In any case, Uraniborg was Schardt's masterpiece: He and Tycho Brahe worked together to give the building and grounds their original form, while **Steenwinckel** contributed to the project primarily as master builder. Schardt probably also created the statue of Mercury/Hermes in antique style, later placed on the spire of Stjerneborg, which so intrigued the learned Duke Heinrich Julius of Braunschweig-Wolfenbüttel that he demanded to have it.

By late 1579, Schardt was back in Nuremberg, where he vanished from the records around 1581, fifty years of age. He probably died in Germany, possibly in Nuremberg, where many of his miniatures have survived.

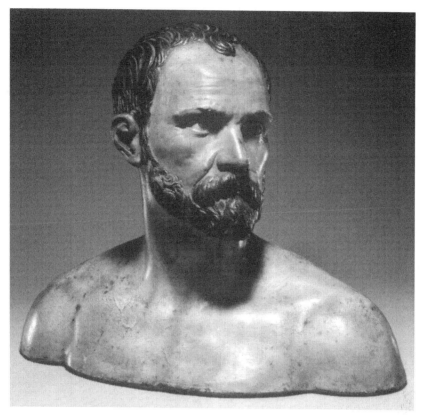

Fig. 62. Johan Gregor van der Schardt. Polychrome terra-cotta self-portrait. (Courtesy Collection of Michael Hall, Esq., New York)

SCHISSLER, CHRISTOPHER (Christoffervs Schissler senior, Christoph Schißler) was an instrument maker who constructed the basic elements of Tycho Brahe's great celestial globe in **Augsburg ca. 1570–9**. Schissler was active in Augsburg from 1546 as a fabricator of fine astronomical and geometrical instruments. Tycho had spent some time in Augsburg after concluding his university studies, and in 1570, he commissioned Schissler to construct the globe. Schissler was at the height of his career.

He made the globe out of many pieces of wood, fit together like a great puzzle, with an overall diameter of five Tychonic feet (ca. 1.5 meters). The large size made it difficult to transport, and it remained in Augsburg. There, Tycho saw it for the first time in 1575 – cracked, warped, and not at all up to his standards – and set Schissler to work on repairing it. The cracks were filled, and the surface was covered with hundreds of parchment patches until it was perfectly spherical. Then it was observed for two years to determine that it did not warp with changes in the weather.

When Tycho was confident that his globe was perfectly spherical, he had it covered with thin brass sheets, so carefully joined that it looked like a solid sphere – precise work that may have been done by **Johan Gregor van der Schardt**. By December of 1580, the globe was ready for engraving. One of Tycho's artists, possibly **Tobias Gemperle**, devised a whole new iconography of the constellations that was engraved on the globe, and each star was engraved in place and gilded as its position was determined with satisfactory accuracy. The globe functioned as a great analog computer of stellar positions and became one of the most famous of Tycho's instruments [see Fig. 23b]. It was the only commission completed by Schissler for Tycho Brahe.

Christopher Schissler remained active until 1605. Many of his splendid instruments survive in contemporary collections. In 1596, he published a book in German on the geometrical quadrat. He died in his native Augsburg on 14 September 1609. His son, Christopher Schissler the Younger, was trained in his father's shop but worked in Prague from 1591, where he became the Imperial Court Clockmaker in 1610.

SEIFFART, MATTHIAS   German mathematician, astronomer; Benátky and Prague 1600-1.

SEVERINUS   Danish student; Hven ca. 1583.

SIGURDSSØN, FINN   (1567–1627) Norwegian student, clergyman; Prague 1600.

SNEL, WILLEBRORD   [Fig. 63], was a *famulus* of Tycho Brahe in **Prague 1600** and went on to become one of the leading mathematicians of the early seventeenth century. Snel was born 1580 in Leiden, where his father, Rudolphus Snellius (Rudolph Snel van Royen, 1546–1613), an avowed Ramist, was professor of mathematics for thirty years. Tycho Brahe, having great respect for Rudolphus Snellius, proposed in 1595 to send his son, Tycho Tygesen, to be raised and educated by him. In return, he offered to take Willebrord Snel into his own household as a foster son and pupil. Nothing came of this proposal, however, and Willebrord Snel grew up in the learned atmosphere of Leiden when that new university was experiencing a golden age.

He matriculated in 1590 at the age of ten and at nineteen received permission to lecture on mathematics and astronomy. In the summer or autumn of 1600, he set off on a grand tour of foreign centers of learning and went almost directly to Prague, where he joined the *familia* of Tycho Brahe. He stud-

Fig. 63. Willebrord Snel in 1617. Engraving by unknown artist.
(Courtesy Det Kongelige Bibliotek, Copenhagen)

ied Tycho's epistemology and methods of data reduction and learned from
experienced observers how to use Tycho's instruments. He had access to the
manuscript observational journals and later used Tycho's unpublished data in
his own publications. Willebrord Snel also learned about Tycho's theory and
methods of geodesy and went on to use those methods with striking success.
He shared Tycho's adherence to a geostatic view of the universe. Snel also met
Kepler.

After this fruitful period in Prague, Willebrord Snel continued his peram-
bulations to European centers of learning. Back home in Leiden by 1604, Snel
took his M.A. degree in 1608 and married Maria de Lange (died 1627), daughter of the burgomaster of Schoonhaven. He began lecturing the following year.
On the death of his father in 1613, Snel became his successor at Leiden and
taught mathematics there for the rest of his life.

In 1615, he launched an audacious project to determine the circumference of the globe by exact measurement of a segment of a meridian. This had not been attempted in nearly two millenia, since the time of Eratosthenes, who had used a much simpler approach. Snel proposed to use the method of triangulation, described in theory by Gemma Frisius in 1533 and applied in actual practice around 1588–91 by Tycho Brahe, with the help of coworkers including **Flemløse, Jachenow, Morsing**, and Riber. Tycho's method had been to measure a series of azimuth angles around the horizon from various high points, beginning with Uraniborg, the tower of St. Ibb's Church, and a third point on the southern end of Hven. Morsing's azimuth measurements from Lund and Malmö had been integrated into the system, as were additional azimuth observations from other points around the Sound. Exact determinations of the latitudes of the observational stations, together with precision in orienting the northern point of the azimuths, established the positive coordinates of the system. Using Jachenow's odometer, Tycho had measured a baseline from Uraniborg to St. Ibb's. From the measured length of one side and the measured angles of one triangle, he was was able to calculate the lengths of the sides of all the interlocking triangles. Tycho soon discovered, however, that the baseline on Hven was too short and the angles in the upper part of the system at Elsinore and Helsingborg were too small, resulting in a margin of error too large to satisfy him. Tycho never prepared a map of the whole Sound region, nor of Denmark and Norway, as Vedel wanted him to do, but only of the island of Hven, executed by **Willem Janszoon Blaeu** and published in 1596 [see Fig. 35b].

Willebrord Snel used precisely the same methods, twenty years later, when he set out to the measure the earth. By then, Adriaan Metius was also applying the surveying methods of triangulation around Franker. However, Blaeu's map of Hven was still the only map ever made on the basis of a triangular survey. Snel consulted with Blaeu and commissioned him to build a large quadrant with all the Tychonic innovations of design, scale, construction, alignment, sighting, and calibration. Snel used a semicircle with a diameter of about 1.10 meters for azimuth observations, and a bronze-mounted iron quadrant with a radius of about 1.75 meters for determinations of latitude. The types, scale, and iron construction of these instruments clearly indicated that they were also patterned on the instruments of Tycho's island.

On the flat, open terrain of the Netherlands, Snel established his baselines with great care. Two locations in the vicinity of Leiden were chosen and measured in 1615–16, then integrated into a series of azimuth angles measured in the general vicinity of Leiden, Delft, and the Hague. These in turn became the core of a triangulation network reaching from Alkmaar in the north to Bergen op Zoom in the south and as far eastward as Utrecht. Snel computed the length of the meridian of Alkmaar within this network and, on the basis of the latitudes covered by the meridian segment, computed the circumference of the globe. He followed Tycho's practice in taking repeated observations to establish the margin of error in his latitudes. He described his methods in *Eratosthenes Batavus* (Leiden 1617) and continued to work on the problem for the rest of his life.

Snel's work was emulated during 1669–70 by Jean Picard, who measured a baseline in France and linked it to a triangulation network in order to prepare an accurate map of France. In 1671, Picard traveled to Denmark to determine the precise latitude and longitude of Uraniborg. Assisted by the Danish astronomer, Olaus Roemer (Ole Rømer, 1644–1710), he also measured the azimuths of landmarks around the Sound and compared his results with those in Tycho's manuscripts in Copenhagen, which Roemer had been helping to edit. Roemer returned to Paris with Picard and became the first foreign member of the Académie Royale des Sciences.

Willebrord Snel died on 30 October 1626 in Leiden and was buried in Pieterskerk, where his monument may still be seen. The great quadrant made for him by Willem Janszoon Blaeu is still preserved in the Leiden Observatory.

SOMMER, LAURIDS ISAKSEN (died ca. 1613/14) Danish student, clergyman; Hven 1596.

STEENWINCKEL, HANS VAN was an architect and artist who helped to build both Uraniborg and Stjerneborg on **Hven 1578–93**. He was born around 1545 in Antwerp and received his early architectural training under his father, who built the great town hall of Emden in the Dutch Renaissance style. Hans van Steenwinckel was still working there in 1577, when Anthonis van Obbergen came to the Netherlands to recruit skilled craftsmen for work on Kronborg Castle. Steenwinckel was in Elsinore the next spring but was quickly drawn away from the Kronborg project by Tycho Brahe.

Johan Gregor van der Schardt was working with Tycho on the design, sculpture, and iconography of Uraniborg, but Tycho needed a master builder to take charge of construction. Steenwinckel had been trained as a stonemason and construction supervisor; now, Tycho Brahe and Schardt introduced him to classical and Italian architectural theory while instructing him in the disciplines of geometry and astronomy, which Vitruvius considered essential for an architect. Steenwinckel also became skilled in the theory and practice of perspective drawing. Working with Schardt and Tycho Brahe helped to clarify and simplify Steenwinckel's style; moreover, he and Tycho developed a friendship that endured for many years.

The plans of Uraniborg were well-developed by the time Hans van Steenwinckel came to Hven. He took charge of their execution in the years 1578–81 and constructed the whole splendid building, up to the cupola influenced by Bramante's proposal for St. Peter's (as reproduced in Serlio's massive book on architecture). Besides directing the construction of Uraniborg and working with Schardt, who created the sculpture and other iconographical elements of the building, Steenwinckel also helped to observe the comet of 1580.

When Uraniborg was finished, Steenwinckel entered the service of King Frederick II in 1582. He received his commission as royal master builder the following year but continued to maintain close ties with Tycho. Around 1584, he helped to plan and construct Tycho's subterranean observatory of Stjerneborg. Working hand in hand with Tycho himself, Steenwinckel designed the building and its entrance, which placed the arms of Brahe and Bille on a triumphal portal amid elaborate Hermetic iconography. In the spring of 1585, he was living on Hven and was one of two architects recommended to rebuild the Kullen lighthouse on Tycho's fief, patterned after the famous Pharos of ancient Alexandria.

After 1585, Steenwinckel worked on a variety of commissions around Denmark, including many for relatives of Tycho Brahe. When Tycho, around the time of his fortieth birthday in 1586, commissioned a series of portraits of himself, Steenwinckel was involved. First came a series of graphic portraits in woodcut and copperplate showing Tycho Brahe within a triumphal arch that bore the arms of his sixteen great-great-grandparents. The arch is reminiscent of the Stjerneborg entrance and may have been designed by Steenwinckel. This imagery audaciously transferred the emblems of the Renaissance prince to a private citizen, showing Tycho with the majesty and power of a "celestial lord." The same theme was presented in the painting executed in 1587 within the arch of the mural quadrant, where Steenwinckel painted the background, featuring views through triumphal arches into the various areas of Uraniborg packed with emblematic contents.

In the years beginning 1588, Hans van Steenwinckel worked on the fortifications of Akershus, Bohus, and Varberg but also accepted private commissions. He was on Hven in June of 1590, undoubtedly in connection with the construction of Tycho's papermill. In 1593, Steenwinckel was working for Tycho's brother, Jørgen Brahe, governor of Varberg Castle, when he sent Tycho some pen sketches of the ex libris that were later stamped on the front and back covers of Tycho's books. In 1595, Steenwinckel constructed an elaborate sepulchral monument for Tycho's brother, Steen Brahe.

Working closely with Tycho Brahe over many years, Hans van Steenwinckel helped to realize the architectural and iconographic program of the island of Hven. He died in Halmstad in 1601 and was buried there. His descendants played a leading role in Danish art and architecture throughout the seventeenth century.

STEPHANIUS, JOHANNES   (1561–1625) Danish student, educator, administrator; Hven 1582–4.

STOCKELMANN, HANS   (died 1619) Danish printer; Hven ca. 1596–7.

TENGNAGEL VAN DE CAMP, FRANS GANSNEB GENAAMD   (Franciscus Tengnaglius, Franz Tengnagel, Francis Teynagel, Francesco Tegnagel de Monstoria Vestalia, Frans Gansneb genaamd Tengnagel tot de Camp, Franz Gansneb genannt Tengnagel van Kamp, zum Campe, Ritter von Camp, Ritter von und zu Camp) served Tycho Brahe for six years at **Hven, Wandsburg, and Bohemia 1595–1601**, married Tycho's daughter, published Tycho's posthumous works, and went on to a distinguished political career in the Habsburg service. Though sometimes maligned by scholars as a politician who knew little about natural philosophy, Tengnagel played an important role in transmitting the legacy of Tycho Brahe into the mainstream of seventeenth-century science.

He descended from a noble house of Gelderland named Gansneb, later called (*genaamd*) Tengnagel, and he was born 1576 at De Camp near Neede, a seat of his family from 1553–1742. Frans spent part of his youth at Bocholt in Westphalia, where his maternal grandfather and namesake, Franz van der Wyck, was an official of the prince-bishop of Münster.

Frans Tengnagel was raised in the Reformed religion, matriculated at the University of Franker in 1593, and arrived 17 February 1595 at Uraniborg with recommendations from Janus Douza the Younger of the University of Leiden. Two years later, he was home in the Netherlands when Tycho left Denmark, but he rejoined the *familia* at Wandsburg.

In 1597–8, Tengnagel was assigned to deliver presentation copies of *Astronomiæ instavratæ mechanica* and the star catalog to Archbishop-Elector Ernest of Cologne and Prince Maurice of Orange, and he also called upon Jan van Oldenbarneveldt, Prince Maurice's rival for leadership in Dutch affairs. Tycho preferred couriers of noble birth like Tengnagel and Claus Mule whenever he sent gifts and messages to princes. In November of 1598, Tengnagel traveled to Italy as Tycho's emissary, met with Magini in Bologna, was created Knight of San Marco by the doge of Venice after conveying Tycho's gifts, spent some time at the university in Padua, and also went to Florence and Rome.

Tengnagel visited the Netherlands again in 1599 and returned to Tycho at Benátky Castle toward the end of the year. He helped to bring **Johannes Kepler** to Benátky and acted as Tycho's agent in requesting Kepler to express in writing his views on the dispute between Ursus and Tycho, which Kepler did in March 1600.

Although the Bohemian diet never formally naturalized the Brahe family, Kirsten Jørgensdatter and her children were accorded de facto noble status in

Bohemia, and Emperor Rudolf II even proposed to make Tycho a baron. This made it possible for marriage negotiations to commence between the aristocratic Tengnagel and Tycho's daughter, Elisabeth Brahe. The betrothal must have been celebrated during the Yuletide of 1600, and the couple began living together. The nuptial ceremony on 17 June 1601 was less elaborate than originally planned. **David Fabricius** was present, and Tengnagel wanted him to cast his horoscope.

The newlywed couple and their servants, including Johannes Eriksen, departed for a year in Deventer, where Tengnagel's father held a military command, and where, on 28 September 1601, Lady Elisabeth gave birth to a daughter, baptized Ida Catherina after Tengnagel's sister. Less than a month later, the infant's grandfather, Tycho Brahe, died in Prague. In 1602, Frans Tengnagel and Johannes Eriksen visited the learned Landgrave Maurice in Kassel and David Fabricius in East Frisia.

In the late summer of 1602, the Tengnagels returned to Prague, where Frans took charge of the Brahe heirs' long struggle to gain control of their inheritance. Tycho's widow and children had moved from the palace of Vice-Chancellor Kurz von Senftenau to a smaller residence in the Old Town of Prague. They had already sold the printed sheets of Tycho Brahe's astronomical correspondence to Levin Hülsius of Nuremberg, who added a new title page and published the book in 1,500 copies in 1601. They also sold the woodcuts and engravings of Tycho Brahe's instruments, which Hülsius used to bring out a scholar's edition of *Astronomiæ instavratæ mechanica* in 1602. Tycho's observational protocols had been entrusted to Johannes Kepler. Tycho's instruments were stored in the Kurz von Senftenau palace on Loreta Plaza. Emperor Rudolf II had agreed to purchase them for 20,000 thalers and pay 6 percent interest on the unpaid principal, but no money had been forthcoming. Moreover, Tycho had died with his pension in arrears, and it had not been paid to the heirs.

Tengnagel managed by 1603 to recover the sum in arrears and got a down payment of 4,000 thalers plus 1,000 thalers interest on the instruments, which allowed Lady Kirsten Jørgensdatter to complete the purchase of a dower estate, west of Prague on the road to Dresden, where she took residence with her unmarried children by 1603. Tengnagel also recovered most of the observational protocols from Kepler: ten volumes of raw observational data compiled over nearly forty years, and twenty-four volumes of careful transcriptions, arranged by year and celestial object. Kepler covertly retained Tycho's data on Mars because his work could not go on without it.

Tengnagel's goal at that time was to publish Tycho Brahe's whole *Theater of Astronomy* posthumously. He began with Tycho's unfinished volume on the supernova of 1572, including the Tychonic solar and lunar theories and stellar tables. With the assistance of Kepler (with whom he constantly squabbled),

Matthias Seiffart, and probably also Johannes Eriksen, Tengnagel published the massive tome, *Astronomiæ instavratæ progymnasmata*, with the Schumann press of Prague in 1602. The second great volume on the comet of 1577, *De mvndi ætherei recentioribvs phænomenis*, which included Tycho's description of his planetary system, was published in February of 1603. Tengnagel and the others did their work well, and the fame of Tycho Brahe soared as these post-humous tomes joined those published by Levin Hülsius to make known throughout Europe the results of thrice seven years' herculean labors on the island of Hven.

When Tengnagel turned to his next task, the *Rudolphine Tables*, he discovered that the Mars protocols were missing and demanded that Kepler return them. By this time, Tengnagel was becoming deeply involved in Habsburg politics and began to realize that he would need major assistance to complete the tables. In July of 1604, the imperial confessor, Johannes Pistorius, mediated an agreement between Kepler and the Brahe heirs that gave Kepler possession of some of Tycho's observational manuscripts in return for a promise to work on the *Rudolphine Tables*.

When Kirsten Jørgensdatter died in 1604 and was buried next to her husband in Týn Church in Prague, the Tengnagels took the unmarried Brahe daughters into their household. That same year, Frans Tengnagel undoubtedly helped to arrange the marriage of Tycho Tygesen Brahe to a young noble widow, Margaretha Vitzthum von Eckstädt, whose family connections launched young Tycho on a career in imperial politics.

Frans Tengnagel was clearly a man of action, with many irons in the fire as astronomer, astrologer, Latin poet, courtier, and diplomat. In 1605, the emperor named him Councillor of the Court of Appeals with an income from the Benátky estate to support his work in astronomy, and that same year, he was sent on a diplomatic mission to England. In London, Tengnagel and his faithful Johannes Eriksen managed to observe the solar eclipse of October 1605.

Like many aristocratic Habsburg courtiers, Frans Tengnagel and his household, including Sophie Tygesdatter Brahe, converted to Roman Catholicism around 1605, but the other Brahe heirs remained Lutherans. The Tengnagels were involved when Tycho's youngest daughter, Cecilie Brahe, became a baroness by marrying the handsome Baron Gustav Sparre of Sundby in 1608, after a betrothal of many years. He was an officer in the Polish army, the son of Lord Chancellor Erik Sparre, and nephew of the Polish ambassador in Prague, Count Erik Brahe of Visingborg. In that same year of 1608, Tengnagel was rewarded with a grant of 12,000 thalers following another diplomatic mission to England.

Diplomacy was certainly needed in those years because the Habsburg lands were a powder keg. Frans Tengnagel now entered the service of one of the firebrands of the Habsburg family, Archduke Leopold, prince-bishop of Passau and Strasbourg, the emperor's young cousin. Tengnagel was the archduke's Privy Councillor in 1609, when Bohemian magnates forced the emperor to sign a Letter of Majesty handing over much of his power to local authorities. Two days later, with Frans Tengnagel acting as intermediary, Emperor Rudolf II and Archduke Leopold formed an alliance and launched a series of political initiatives aimed at crushing the Estates and augmenting Habsburg power throughout Europe.

The last Duke of Jülich-Cleves had just died, and the strategic location of his territories near the mouths of the Rhine, where France, Germany, the United Netherlands, and Spanish Belgium all came together, meant that the ducal succession was hotly contested. With Dutch assistance, two Protestant claimants occupied Düsseldorf, the capital of the territory, in May of 1610. Archduke Leopold and his advisors developed a counterplan to seize Jülich-Cleves in the name of the emperor. In July, the archduke galloped across Germany to Jülich in disguise, accompanied by only five men, one of whom was Frans Tengnagel. When they arrived, the local commandant surrendered the keys to the castle.

From Jülich, Tengnagel was dispatched on a diplomatic mission to secure the neutrality of King Henry IV of France. He went next to the court of the Spanish Habsburgs and laid bare the secret plan to link all the Roman Catholic princes of Europe into a mighty league led by the Habsburgs.

Events unfolded far differently than anticipated, however. A major European war over Jülich-Cleves was averted when the king of France was assassinated on the eve of intervening in force, and a march on Prague by Leopold's troops turned into a bloody fiasco ending with the overthrow of Emperor Rudolf II. The troops of the emperor's brother, King Matthias of Bohemia, sent them on the run in February of 1611.

Tengnagel tried to flee Prague at the head of a caravan of booty but was captured at Velvari, twenty miles outside Prague on the road to Dresden, near the estates of his Brahe in-laws. He was cast into the dungeon of Witte tower in Prague, where he was compelled to eat his meals with the executioners. On 18 April 1611, he was hauled into the torture chamber, stripped naked, stretched on the rack more than four inches for the length of time it took to say one Our Father, and burned along his ribs with candles. Under torture, he incriminated various individuals at the imperial court and admitted to a plan to unseat King Matthias, make Leopold king of Bohemia, and crush the Bohemian Estates by force. After torture, he was thrown naked on a bed of straw. The intervention of the Spanish ambassador finally secured his release, and he hobbled off to Passau. For the rest of his life, Tengnagel suffered pain as a result of his torture.

Emperor Rudolf was deposed and died soon after. The seat of the Holy Roman Empire moved to Vienna, and Matthias became emperor. The *Rudolphine Tables* were far from Tengnagel's thoughts in these years.

Jørgen Tygesen Brahe, called Georg Brahe in Bohemia, took over as representative of the Brahe heirs in dealings with Kepler and soon established a cordial working relationship with him, based on mutual trust. All the observational manuscripts were turned over to Kepler. Meanwhile, Gottfried Tampach, a Frankfort bookseller who had purchased the stock of the late Levin Hülsius, reissued *Astronomiæ instavratæ progymnasmata* with a new title page in 1610, and bound *De mvndi ætherei recentioribvs phænomenis* together with the astronomical correspondence for a new 1610 edition.

Lady Elisabeth Brahe died in 1613 in the imperial city of Regensburg. A few years later, Tengnagel came to the Polish court as Habsburg ambassador and found his second wife, a Prussian noblewoman. He and Elisabeth Brahe had had four children: Ida Catherina, who became a lady-in-waiting at the Polish court and married a Polish nobleman named Sigmund Wybranoffsky; Vendela, who died at the age of two; Rudolf Tycho, born in Prague around 1604; and Leopold, born 1608. The sons were educated by the Jesuits in Passau.

When the wars of religion broke out in Bohemia after 1618, the Brahe heirs, who were Lutherans, fled to Saxony, where they had many connections; but Frans Tengnagel remained safe in Passau as Chancellor and Privy Councillor to Archduke Leopold. Rudolf Tycho Tengnagel matriculated at the University of Vienna in 1622. In that same year, after striking a deal with the Jesuits, Frans Tengnagel was named Imperial Councillor to Emperor Ferdinand II and granted a debenture of 25,000 florins for his services to the house of Habsburg. He remained on the winning side until his death on 1 December 1622 [Fig. 64].

A few years later, his son RUDOLF TYCHO GANSNEB GENANNT TENGNAGEL VON UND ZU CAMP was required to prove his nobility by demonstrating that his eight great-grandparents had all been of noble rank. This presented difficulty because his maternal grandmother, Kirsten Jørgensdatter, had been born a commoner, but the uncontested nobility of all his other ancestors and nearest relatives, and the illustrious service of his father and his grandfather, Tycho Brahe, finally led to his acceptance into the ranks of the Bohemian nobility. With his stocky build, thick sandy hair, and sweeping moustaches, Rudolf Tycho Tengnagel was the very image of Tycho Brahe. He became a rich landowner, Imperial Councillor, Chamberlain of the High Court, Reformation Commissioner, and in the years 1641–4, Commander of Hradčany Castle and all of Prague west of the Vltava (Moldau) River. Rudolph Tycho Tengnagel died on 25 March 1672 and was buried in St. Vitus Cathedral. Through the fifteen children born to his daughter, Elisabeth Lud-

Fig. 64. Frans Gansneb genaamd Tengnagel van de Camp. Funerary monument in the Church of Our Lady under the Chain, Prague, with sketches of the four coats of arms. (Photo by the author)

milla Baroness Klebelsberg von Lysau (died 1698), the biological heritage of Tycho Brahe was transmitted to many generations of central European aristocrats.

VALENTIN   German? contractor; Hven 1589–90.

VARDE, JOHANNES   Danish student, clergyman; Hven ca. 1585.

VIBORG, ANDERS (SCHYTTE?)   Danish student, astronomer, clergyman; Hven 1581–4.

VINGAARD, MADS   (died ca. 1600) Danish printer; Copenhagen 1572.

WALDKIRCH, HENRIK   (died 1629) German-Danish printer, publisher; Hven 1591.

WANDAL, JENS IVERSEN   (1570–1628) Danish student, clergyman, astronomer; Hven 1589–92.

WEIDA, CHRISTOPHER   German printer; Hven ca. 1586–90.

WENSØSIL, JENS JENSEN   (died after 1608) Danish clergyman; Hven ca. 1587–8, 1594–7, Rostock 1597.

WITTICH, PAUL   (Paulus Wittichius), spent three or four months in the *familia* of Uraniborg on **Hven 1580** and introduced the data-processing method of *prosthaphæresis* to Tycho's island. He was born around 1546 in Wrocław (Breslau, Wratislavia) in Silesia into a middle-class family that included a learned uncle with ties to Rheticus and Copernicus. Wittich matriculated at the University of Leipzig in 1563, where he met Tycho, who was also a student there. He went on to Wittenberg in 1566 and to Frankfort-on-the-Oder in 1576.

In Frankfort, Wittich instructed the Scots mathematician, John Craig, in his nascent method of *prosthaphæresis,* which may have built on earlier work by Rheticus. Craig later discussed the method with his countryman John Napier, who went on to invent the even more ingenious method of logarithms.

Wittich spent the years 1579–80 in Wrocław, where his circle of friends and correspondents included many with ties to Tycho Brahe, including Bartholomeus Scultetus, **Christopher Rothmann**, and Thaddeus Hagecius.

Paul Wittich arrived on Tycho's island around the end of July 1580, with a letter of recommendation from Hagecius. Uraniborg was in the last phases of construction, and Tycho was still living in temporary quarters. They got on very well: Wittich was a very able mathematician, and Tycho liked working with him. Wittich, like Kepler, was nearsighted, and he was more interested in the mathematical and theoretical aspects of astronomy than in its empirical dimensions, but Tycho and the Uraniborg staff did introduce him to the observational methods of Hven. When a comet appeared in October of 1580, Wittich helped **Peter Flemløse** to observe it with a quadrant in Tycho's absence.

Paul Wittich made two important contributions to the scientific culture of Hven. First, he introduced Tycho to *prosthaphæresis,* his method of converting multiplication and division into more simple addition and subtraction to solve trigonometric problems. It was tremendously useful in processing astronomical data, and Wittich worked with Tycho to improve the method, which Tycho continued to use until the end of his days. Handbooks of *prosthaphæresis* eventually lay on every worktable of Uraniborg for the use of the staff and were revised and augmented from time to time by Tycho and others.

Wittich's second contribution was to stimulate Tycho Brahe's thought on the nature of the solar system. Like Tycho, Wittich was a great admirer of Copernicus but had tried to find ways to transpose Copernican astronomy into geostatic models. He owned several copies of Copernicus's *De revolutionibus orbium coelestium*, full of annotations by Wittich and Erasmus Reinhold of Wittenberg. In one copy, Wittich drew numerous diagrams of planetary, solar, and lunar motion using geostatic Copernican models. Tycho had wrestled with this same approach since his 1574 lectures at the University of Copenhagen, and again when he had tried to establish the position of the comet of 1577 in planetary space. The two of them may have come up with some identical hypotheses independently, but Tycho was clearly impressed with Wittich's ingenuity in pursuing the matter. Now, in the autumn of 1580, when the new comet appeared, they quickly agreed that it was located in superlunar space. Their conversations were tremendously stimulating to Tycho and helped him to advance his own thoughts about the nature of the planetary system.

Shortly after observing the comet, however, Wittich told Tycho that his rich uncle had died in Wrocław and he needed to go home to claim his inheritance. He promised to return in seven or eight weeks. Tycho gave him a parting gift of an opulent and costly book, the *Astronomicum Cæsareum* of Petrus Apianus, perhaps in the hope that Wittich would reciprocate with one of his annotated copies of Copernicus, but he did not. Wittich took his books with him, and Tycho never heard from him again.

Paul Wittich returned to Wrocław and continued to move in humanist circles. He was temperamental and could be both boastful and secretive, muttering under his breath at times and refusing to give a straight answer, jealous of his innovations and those of others. At other times, however, he was completely open and exceedingly generous. In 1581, Henry Savile studied with Wittich, who introduced him to the geocentric adaptations of Copernican astronomy used in Lutheran universities. **DUNCAN LIDDEL** studied mathematics with Wittich around 1582 and was also allowed to copy many of Wittich's astronomical marginal notes. Liddel visited Tycho's island in June of 1587 and in 1588.

In 1584, Paul Wittich visited the court of Landgrave William IV of Hesse-Kassel, where Tycho Brahe had been in 1575. Wittich met **Christopher Rothmann** and Joost Bürgi, taught Bürgi the method and proof of *prosthaphæresis,* and showed the Kassel scholars how things were done at Uraniborg, including the technical innovations of Tycho's instruments as they had been in 1580, apparently passing off the innovations as his own. He was rewarded by the landgrave with a golden chain. Together, Wittich and the Kassel scholars observed a lunar eclipse in November 1584.

Paul Wittich did not have long to live when he took his leave of Kassel. He died around the age of forty in Vienna on 9 January 1586. When Flemløse came to Kassel as Tycho's envoy, later that year, he discovered that the sextant, an instrument invented by Tycho, was in use there, and so was an early version of Tycho's parallel-slit sights and his transversal arc graduation, all thanks to Wittich's instruction of Bürgi. This technology transfer disturbed Tycho because he had not yet published his innovations and considered them to be his intellectual property. However, Flemløse reported that the landgrave and his courtiers had acted in good faith, and cordial relations were soon established between Kassel and Hven.

Tycho felt remorse when he heard of the death of Wittich and regretted that he had ever questioned his integrity. He later made strenuous efforts to acquire Wittich's library and manuscripts, through Hagecius and the Barons Liechtenstein and Herberstein in 1589, through Hagecius and Jacob Monaw in 1595, and finally, through **Longomontanus** and Monaw in 1598, when he was successful. Tycho must have considered them to be of considerable value. Clearly a brilliant mathematician, Paul Wittich left no published works.

GLOSSARY OF TECHNICAL TERMS

ALCHEMY   search for the chemical processes producing the *philosopher's stone* (also called the *elixir*), which was thought to have the power to prolong human life and transmute base metals into gold; also, a religious or philosophical system growing out of this search; disavowed by Tycho Brahe in favor of the *spagyric art*

ARC   segment of a circle (360°) on the graduated scale of an instrument, each *degree* of arc containing 60 *minutes* of arc; also, instrument used to measure arc (e.g., bipartite arc, transversal arc)

ARMILLARY   (also *armillary sphere*) astronomical model in which the sky is represented by a skeletal framework of solid rings (*armilla*), used to measure star positions; e.g., the elaborate *zodiacal armillary* [see Fig. 5b], aligned to the ecliptic, known since the time of the ancient Greeks; the simplified *equatorial armillary* [see Figs. 13c, 16, 25b], invented by Tycho Brahe, aligned to the earth's equator

ASTRONOMICAL CLOCK   elaborate clock incorporating various astronomical phenomena, such as phases of the moon, sunrise and sunset, month and day of the year

AURORA BOREALIS   luminous displays of changing colored light (also called *northern lights*) generated in the night skies of northern regions when solar wind interacts with the earth's magnetic field to energize atoms and molecules in the upper atmosphere

AZIMUTH   angle measured clockwise (as with the graduated azimuth circle on several of Tycho's instruments) from the northern point to the horizontal point that intersects with a vertical circle through a given celestial body

CHOROGRAPHY   description of the natural and cultural features of places and events associated with them; intended to be used in conjunction with maps

CROSS-STAFF   *See* radius

ECLIPTIC   observed annual solar path in a great circle through the sky

EPHEMERIS   (pl: *ephemerides*) table giving the coordinates of a celestial body at intervals during a given period of time

EQUATORIAL ARMILLARY   *See* armillary

GEOCENTRISM   theory that the earth is the center of our planetary system

GEODESY   branch of applied mathematics that determines the latitude and longitude of points on the earth's curved surface and links them together

GEODETIC SURVEYING   (or *geodesic surveying*) method of geodesy

GEOHELIOCENTRISM   theory that the planets revolve around the sun while the sun revolves around the earth

GEOSTATISM   assumption that the earth is in a fixed position

HELIOCENTRISM   theory that the sun is the center of our planetary system

HELIOSTATISM   assumption that the sun is in a fixed position

IATROCHEMISTRY   Paracelsian chemistry aimed at producing medicaments (medicines)

MEDICAMENT   *See* iatrochemistry

OCCULTATION   one celestial body hiding another from view, as when the moon moves between the earth and the sun to cause a solar eclipse

PARALLACTIC INSTRUMENT   *See* triquetrum

PINNULE   sighting device; configured on Tycho Brahe's instruments to provide four parallel lines of sight, for greater accuracy [see Fig. 13b]

PROSTHAPHÆRESIS   method of simplifying trigonometric calculations by converting multiplication and division into addition and subtraction; invented by Paul Wittich, enhanced by Tycho Brahe and Melchior Jöstelius

QUADRANT   instrument with a 90° graduated arc for measuring altitudes of celestial bodies; so named because its arc is one-quarter of a circle [see Figs. 11b, 15a, 25a, 31, 34, 39a]

QUADRAT   square plate, generally of wood or brass, graduated on two sides and used to measure altitudes and distances; so named because it is quadratic (square) in shape

RADIUS   (or *radius astronomicus,* also called *cross-staff*) instrument with a long shaft and a sliding crossbar bearing pinnules at each end, used to measure the angle between two points

SEMICIRCLE   instrument with a 180° graduated arc for measuring altitudes of celestial bodies [see Fig. 39b]

SEXTANT   instrument with a 60° graduated arc (one-sixth of a circle), invented by Tycho Brahe for measuring altitudes and angles between two celestial bodies [see Figs. 5a, 17, 53]

SPAGYRIC ART   (also *pyronomic art*) the Paracelsian philosophy of harmonious health by means of cosmic spiritual forces and iatrochemical medicaments

TRIANGULATION   division of part of the earth's surface into a network of interconnected triangles based on a line (*baseline*) of known length constituting the side of one triangle; allows distances and directions of all points within the system to be determined accurately

TRIGONAL   shaped like an ancient Greek triangular harp

TRIQUETRUM   (also *parallactic instrument*) instrument with three arms connected by hinges to form an adjustable triangle; used to measure altitudes [see Fig. 15c]

VARIABLE STAR   star whose brightness varies because of either internal changes or periodic eclipsing by mutually revolving stars

ZODIAC   band following the ecliptic across the celestial sphere that contains the apparent paths of the sun, moon, and planets; it is divided into twelve equal parts or *signs,* each named after a contiguous constellation

ZODIACAL ARMILLARY   *See* armillary

NOTES

CHAPTER I. IN KING FREDERICK'S SERVICE, 1575–1576

1. Tycho's travels in 1575 are described in Thoren 1990, 92–102.
2. S. Heiberg 1984. His beard was forked and longer until around 1576, when he trimmed it to a Van Dyke.
3. These fiefs, as well as Ravnsborg Castle, Halsted Abbey, and Roskildegaard, were all awarded to new vassals around this time, see Erslev 1879.
4. Brahe 1913–29, 7: 25. Zeeberg 1991a,b deal with Tycho's ambivalent and changing attitude toward aristocratic life; cf. Thoren 1990, 62–5.
5. Wolter 1982, 10, 43–55. An alternative path of education for Danish magnates' sons began with service as a page and then squire in a great noble household, which was less expensive but equally exclusive because it depended on aristocratic connections.
6. Fabritius 1950, 14; Thoren 1990, 22–4.
7. The document is in Brahe 1913–29, 14: 3. On Tycho's visit home in 1567, see Brahe 1913–29, 7: 3, and Thoren 1990, 26–9. Kettering 1986, 3–11, describes the role of broker in patron–client relationships.
8. Canon Morten Pedersen of Roskilde was already involved with astronomy; his observations of the 1572 nova and the comets of 1577 and 1580 are noted in Norlind 1970, 46–7, 104, 129–30.
9. Thoren 1990, 37–9. The underlying principle was one of equal inheritance: Each daughter would also, upon marriage, receive property – dowry, betrothal gift, and morning gift (i.e., gift from the husband on the morning after consummation of the marriage)– equal to her inheritance. In lieu of such wedding disbursements, each son was therefore given a doubled share of inheritance initially.
10. Karker 1964, 267. On aristocratic honor and patronage, Biagioli 1993, 80–1. On the importance of ensuring that the value of the gift was commensurate with the status accorded its recipient, see Findlen 1991, 19–21.
11. Boldfaced names have entries in the Biographical Directory.
12. Brahe 1913–29, 14: 299–300; Christianson 1972b, 238.
13. One cannot be certain of Kirsten's parentage, but strong circumstantial evidence points to the pastor and his wife. Carlquist 1948–85, 8: 444–8; Fabritius 1961, 125–6; Thoren 1990, 45–6.

14. On women's attire in the late sixteenth century, see Troels-Lund [1879–1901] 1968–9, 2: 344–63.
15. On balladry at Uraniborg and in the family circle of Tycho Brahe, see Brix 1933–55; Dal 1962, 222–32, 286–7; Piø & Pedersen 1984, 6–11.
16. Brahe 1913–29, 9: 174–5.
17. Thoren 1990, 46–8. Kirstine, born earlier, had died in infancy.
18. Wad 1893 gives a full account.
19. Troels-Lund 1910–12, 2: 294–312; Wad 1893, 2: 142–3.
20. On King Frederick's wedding, see Thoren 1990, 54–5.
21. Bech 1979–84, 4: 525; F. Jensen 1978, 29–30. See also F. Jensen 1982. Victor Hermansen demonstrated in 1956 that King Frederick was dyslexic, but F. Jensen 1978, 28–32, made the fact more widely known among historians.
22. Shackelford 1991, 87–95. His dyslexia was undoubtedly why King Frederick II remained more of what Moran 1991a described as a prince-collector than a prince-savant or a prince-practitioner like his brother-in-law, Elector August of Saxony; his sister, Electress Anna; and his father-in-law, Duke Ulrich of Mecklenburg. See also Moran 1981.
23. Hartfelder [1899] 1964. Westman 1975a showed how this created scientific structures or "academies" within the Lutheran universities, rather than externally.
24. Stybe 1979, 34–8.
25. Norvin 1937–40, 1: 144–56.
26. Ibid., 1: 162–3.
27. Ibid., 1: 285–97.
28. Ibid., 1: 304–5, 2: 271–2.
29. Christianson 1964 and Thoren 1990, 48–54, on Herrevad as a model for Hven.
30. Brahe 1913–29, 1: 27.
31. Brahe 1913–29, 1: 30–4.
32. Moesgaard 1972a, 31–3; Rørdam 1863–74, 2: 170–5; Stybe 1979, 38; Thoren 1990, 78–85. There were only a dozen professors and around 150–200 students, despite the fact that Copenhagen was the only university in all of Denmark and Norway.
33. Christianson 1979, 112–13; Krstović & Lazzari 1999, 27–30; Moesgaard 1972a, 31–3; Thoren 1990, 85–92.
34. Westman 1980 showed how Copernicus merged the roles of astronomer and natural philosopher, whereas Tycho transferred these merged roles to an aristocratic context that promoted innovation. Biagioli 1993, 293–4, saw the life-style of "aristocratic natural philosopher" as an alternative to that of courtier for seventeenth-century noblemen, but Tycho found a way to combine those roles.
35. Christianson 1979, 116.
36. On Tycho's income, see Christianson 1973a; Thoren 1990, 36–9.
37. Kongsted 1990, 66, thought that Landgrave William IV was the probable link between Tycho and Labenwolf.
38. Brahe 1913–29, 7: 25–9; Norlind 1926, 21–30.
39. Brahe 1913–29, 7: 26; Norlind 1926, 23–4.
40. Brahe 1913–29, 7: 26; Norlind 1926, 24.
41. Brahe 1913–29, 7: 26–7; Norlind 1926, 24–5.
42. Brahe 1913–29, 7: 27; Norlind 1926, 25.

43. Brahe 1913–29, 7: 27; Norlind 1926, 25–6. Tycho capitalized the words in italics.
44. Brahe 1913–29, 7: 27–8; Norlind 1926, 26–7.
45. Brahe 1913–29, 7: 28; Norlind 1926, 27.
46. Brahe 1913–29, 7: 28; Norlind 1926, 27.
47. Pratensis's letter, 15 February 1576, Brahe 1913–29, 7: 30–1; Thoren 1990, 103–4.
48. Biagioli 1993, 90–1, on patronage as a productive system.
49. The pension was almost discontinued in 1581, but Tycho continued to receive it until 1597. Biagioli 1993, 49–52, asserted that an absolute prince, who theoretically lacked nothing, could not receive "gifts" but only "favors," and that he responded with *gratia*. Such views were not universally accepted, even at absolutist courts, see P. Christianson 1991. For his part, King Frederick II was well aware of his constitutional limits but still used a standard formula of *Gunst* (grace, favor, good will, *gratia*) in all his acts and proclamations.
50 Brahe 1913–29, 14: 4–5.
51. Ibid., 10: 42.
52. Ibid., 14: 5.
53. Ibid., 14: 5–6.
54. Ibid., 14: 6–9, 12–14, 21–9, 49–50; Christianson 1972b, 231–5, 241–2; Thoren 1990, 133. The eleven farms were merged with the Kullen fief in 1581, and Tycho held Kullen until his death.
55. Brahe 1913–29, 14: 7–11, 14–17, 19, 32, 34–8, 48–9, 54, 62–3, 67, 97–9, 130–9; Christianson 1972b, 243; Thoren 1990, 133. In 1586, Tycho was granted an annual pension of 300 dalers from the Sound Toll (raised six months later to 400) in return for the fief of Nordfjord. Nordfjord was granted to him again 1589–96, but the pension was not canceled until 1592.
56. Brahe 1913–29, 14: 3–4, 10, 12–17, 19, 24–5, 35–6, 55–60, 63–6, 68–71, 91 et seq., 106; Christianson 1972b, 240–1; Thoren 1990, 142–3. Tycho held the Roskilde canonry of the Three Holy Kings (i.e., the Magi) until 1597.
57. The documents were published by J. O. Arhnung and Svend Gissel.
58. On patronage at the Danish court, see Christianson 1973a; Shackelford 1991. On courtly patronage of science, see Biagioli 1993; Eamon 1991; Moran 1981; Moran 1982; Westfall 1985; Westman 1980. On patronage in general, see P. Christianson 1976; Findlen 1991; Kettering 1986; Kettering 1989; Sarasohn 1993.
59. Biagioli 1993, 84–90, citing Alain Viala.
60. Biagioli 1993, 37 (but see 39, 41, where the transitional, cashlike value of golden chains as gifts is discussed); Findlen 1991, 5.
61. Biagioli 1993, 23, comments on the role of informers.
62. Much of this analysis is based on Kettering 1986 and Kettering 1989, but see also P. Christianson 1976. All of these services were rendered to King Frederick by members of Tycho's closest family circle.
63. Svend Cedergreen Bech in Danstrup & Koch 1962–6, 6: 342–6, 406–8, 424–8, 449–50. F. Jensen 1982, 194–203. Ladewig Petersen 1980a, 213, 290–3.
64. G. Ilsøe 1980; K. Jespersen 1980a; Ladewig Petersen 1980b; Wolter 1982.
65. Biagioli 1993, 90–1.
66. The phrase "socioprofessional identity" is from Biagioli 1993, 87, see also 90 and 293–4.

CHAPTER 2. JUNKER AND PEASANTS, 1576–1581

1. The parsonage had acres in only two fields.
2. Frandsen 1988, 2: 29–30, discusses Danish rotation systems.
3. The waters of the Sound usually stayed open all winter, though in very harsh years they could freeze solid.
4. Brahe 1913–26, 5: 151, and Brahe 1996, 158, mentioned about forty farmhouses; Gissel 1956, 189, recorded forty-six tithe-paying households in 1567. The demographic carrying capacity of the traditional island economy must have been around two hundred people, roughly equal to the number living there in 1576.
5. The classic account of Danish farm buildings of this era is Troels-Lund [1879–1901] 1968–9, 1: 205–72.
6. The Danish word *toft* could mean both toft and croft; when they needed to be distinguished, the croft was called the *tofteager* ("toft acre"), cf. Hanawalt 1988, 31–44. On Danish peasant gardens, see Skougaard, Hansen, and Rasmussen 1984.
7. Thoren 1990, 106, mentions a constable (*sognefoged*), but I have not found evidence of such an official on Hven before 1576.
8. Norlind 1970, 390.
9. Ven 1980. Brahe 1913–29, 5: 293, reproduces Tycho's map of Hven in the 1580s.
10. Brahe 1913–29, 14: 5.
11. The German word *Grundherrschaft* means "authority over the land." The Danish social and agrarian contexts are described in Bolin [1941] 1966; and Maarbjerg 1995.
12. Fussing [1942] 1973.
13. O. T. Brahe 1570. Otte Brahe had inherited 143 farms, 22 cottages, and 4 mills of this estate, while 33 of the farms were Beate Bille's inheritance; together, they had established 8 farms and purchased 105 farms, 6 cottages, and 3 mills; 33 farms and 1 cottage were held in 1570 as surety for loans. The widow could claim dower rights to her own inheritance and half the property acquired during her marriage, as well as the property held as surety.
14. The German word *Gutswirtschaft* means "estate production."
15. Maarbjerg 1995 described the Danish cattle trade.
16. Frandsen 1988, 2: 13–18.
17. The tithe amounted to a tax of every thirtieth sheaf of grain and the increase of livestock and bee swarms. One-third of the tithe, the *fabrica ecclesiæ,* was earmarked for maintaining the church, and another third, the *mensa presbyteri,* supported the pastor; the remaining third from St. Ibb's was dedicated to the hospital in Elsinore.
18. Fussing [1942] 1973, 193–277; Olsen [1957] 1975, 148–53. Cf. Thoren 1990, 112–13.
19. Brahe 1913–29, 14: 12–13, refers specifically to firewood.
20. Frandsen 1988, 2: 79–87, 123–9.
21. Brahe 1996, 147–8; Norlind 1970, 73–4; Thoren 1990, 114.
22. Jern 1976, 20. The outbuildings stood on the present site of Kungsgård manor, just south of Uraniborg.
23. Brahe 1913–29, 5: 293. Jern 1976, 20, thought the tower may have served as Tycho's temporary observatory, but Tycho and his assistants could just as well have observed from an open field. Cf. Thoren 1990, 114–15.

24. *Vornedskab* grew out of the late-medieval labor shortage in Denmark. By 1578, it was interpreted to mean that peasants could not leave their native estate. See Skrubbeltrang 1978, 18–19, 24–6, 175–8.

25. Brahe 1913–29, 14: 7; Thoren 1990, 132–3.

26. Jern 1976, 14–24; Thoren 1990, 144–65.

27. Gunnarsson 1993; Ödman 1996.

28. Brahe 1913–29, 14: 15–16.

29. Ibid., 14: 17–19.

30. Frandsen 1988, 2: 89.

CHAPTER 3. AMONG FRIENDS, 1570–1576

1. B. Andersen 1971, 18–26, 103–12; Fabritius 1950, 15–18, for Tycho and his siblings. Aymard 1989 saw masculine networks of friendship emerging in early modern Europe as family networks weakened and "spiritual kinship" declined; he mentioned fosterage on 484–5. The aim of fosterage was to reinforce wider clan solidarity, and several of Tycho's siblings were fostered by members of their mother's family; but Tycho's fosterage began with violence and seemed permanently to strain his relationship with his birth parents.

2. Nehamas & Woodruff 1989, xiii–xiv.

3. Brahe 1913–29, 1: 15.

4. Brask et al. [1984] 1990, 400–4. See also Christianson 1998, 474–6, and Zeeberg 1993. Zeeberg 1988 analyzed an *amor* poem written 1584 by Tycho Brahe. On the obligation to repay gifts and the function of gift exchange in a premarket economy, see Mauss [1925] 1967; see also Findlen 1991.

5. The preferred meter was a distich of one hexameter and one pentameter line, which Tycho Brahe used in virtually all of his poetry, and the preferred classical models were the Augustan poets, see Brask et al. [1984] 1990, 386–412.

6. Zeeberg 1994b, 9–16, showed that half of all Latin poetry in Denmark was written by young academics eager to demonstrate their mastery of Latin learning and win a patron. See also Zeeberg 1993.

7. Brahe 1913–29, 1: 1–72. Thoren 1990, 65–73.

8. *Amores*, III, 1, see Friis-Jensen and Skafte Jensen in Brask et al. [1984] 1990, 404–7; see also Zeeberg 1991a,b. Tycho wrote this poem in the sixteenth-century genre of "parody," where a specific classical poem was reshaped with the same structure but a new and often contrasting meaning. See also Thoren 1990, 71–2.

9. Tycho had helped his uncle to import Venetian glassmakers and German papermakers, set up a chemical laboratory, and establish an astronomical observatory, and the smiths of Herrevad, who had a water-powered trip hammer, helped to construct some of Tycho Brahe's instruments. Herrevad Abbey became a prototype of Uraniborg, see Christianson 1964; Thoren 1990, 48–54.

10. Skov & Bastholm 1979, 49.

11. Heavenly Eros, as distinguished from Sensual Eros, which binds the soul to temporal things and drags it downward, see Nygren 1953, 160–77.

12. See *Symposium* 180D–180E and 187D–187E in Nehamas & Woodruff 1989, 13, 22.

13. Nygren 1953, 173.

14. Ibid., 174–81, see *Symposium* 203B–212C in Nehamas & Woodruff 1989, 48–60.
15. Nygren 1953, 183–6.
16. Ibid., 193.
17. Copenhaver 1990; Yates [1964] 1969, 20–61.
18. Nygren 1953, 449–558.
19. Ibid., 691–741.
20. Nelson 1958; Robb [1935] 1968, 31–86; Yates [1964] 1969, 20–83.
21. *Heptaplus,* see Robb [1935] 1968, 64–7. Yates [1964] 1969, 84–116, discusses Pico's use of Cabalist magic.
22. Robb [1935] 1968, 72.
23. Ibid.; see also Copenhaver 1990.
24. Nelson 1958, 75–84; Robb [1935] 1968, 74–86; Zeeberg 1988, 170–2.
25. Brahe 1913–29, 1: 18. On Tycho and the Paracelsian tradition, see Shackelford 1991, 95–105; Thoren 1990, 24, 52–4, 59–61, 66, 80–7. On Tycho's integration of the Paracelsian, Philippist, Copernican, and Ptolemaic traditions, see Thoren 1990, 11, 17–19, 30–3, 80–92. On the Aristotelian tradition and Tycho's opposition to it during the 1570s, see Thoren 1990, 56–7, 67–8, 125, 137. On Tycho's differences with Ramus, see Thoren 1990, 33–5. On Tycho and the authority of the Bible, see Blair 1990, 362–4. On Tycho's world view and that of Danish Philippists, see Christianson 1979.
26. Garstein 1953, 173–4.
27. Ashworth 1990 described an "emblematic world view" that became an ever more important part of the *mentalité* of Uraniborg.
28. Robb [1935] 1968, 73.
29. Brahe 1913–29, 5: 130; 7: 34. On Pratensis, see Shackelford 1991, 86–95; Thoren 1990, 43, 62–5, 102–4, 113–14.
30. On Dançay, see Christianson 1979, 111–14; Thoren 1990, 36, 42, 62, 79, 84–5, 102, 114.
31. On Steen Bille, see Christianson 1964, 27–35; Thoren 1990, 49–50, 102–3, 267.
32. On Vedel, see Thoren 1990, 10, 13–14, 51, 70–1, 140, 206–7.
33. On Aalborg, see Bech 1979–84, 16: 233.
34. On Severinus, see Gillispie 1970–80, 12: 334–6; Shackelford 1991; Thoren 1990, 43–5, 52, 113–17, 127.
35. Shackelford 1991, 86–92, 105.
36. On Johannes Franciscus, see Shackelford 1991, 94–5; Thoren 1990, 43, 85, 114.
37. On Lætus, see Bech 1979–84, 9: 275–6.
38. Tycho would hardly have refrained from mentioning him had he attended. On Kaas, see Thoren 1990, 43, 101, 118.
39. On Lange, see Bech 1979–84, 8: 479–80.
40. On Gyldenstierne, see Bech 1979–84, 5: 411–12.
41. Tycho was on good terms with his brothers, his sister, Margrete Brahe, and his brothers-in-law, but only his youngest sister, Sophie Brahe, and to some degree his youngest brother, Knud Brahe, shared his learned interests. Other friends from Tycho's student days who cannot have attended the 1576 dedication because they were not in the vicinity included Anders Foss, Peter Hegelund, and Jens Nilssøn.
42. Brahe 1913–29, 1: 131; 10: 38.

CHAPTER 4. FOUNDING THE FAMILIA, 1584–1587

1. Herlihy 1991 explores the various classical and medieval meanings of *familia,* which derives from *famulus,* "slave," and originally meant a "band of slaves." The old Danish vernacular word *slegt* in Tycho's social class meant all direct, noble descendants of a common progenitor; they bore the same arms and often had the same surname. As commoners, Tycho's children were not part of "the Brahes' *slegt.*" There was no common sixteenth-century Danish expression for "nuclear family," and no common generic term for "cousin" or "grandparent." Instead, Tycho and his aristocratic contemporaries used a wide range of concrete terms such as mother, father, son, daughter, sister, brother, maternal grandmother (*mormor*), paternal grandmother (*farmor*), father's sister (*farster*), mother's sister (*morster*), mother's sister's son (*morstersøn*), and the like. Tycho recognized second and third cousins as kinsmen and referred to distant relations as "kinsman" (*frænde*) or "family and relative" (*slegt og byrd*). Kin by marriage were generally addressed as "brother-in-law" (*svoger*), or "sister-in-law" (*svogerske*), even when the relationship was more distant. In writing to Danish nobles who were not kin, Tycho addressed them as "especially good friend."

2. Vammen 1977, 169–71.

3. Hansen 1964; Ladewig Petersen 1980a.

4. Hanawalt 1986, 205–19, presented a model of "partnership marriage" in medieval England that applies well to late sixteenth-century Denmark.

5. Among the factors that determined a farm's labor requirements and set a lower limit to the size of the household were dues and services owed to the lord, the size of fields, and the number of livestock.

6. Ladewig Petersen 1980a, 86–98. The demographic pattern of the nobility was different, see Hansen 1964, 242–76.

7. Cf. Frederiksen 1976 on the island of Sejerø in the seventeenth century.

8. Tycho Brahe and Kirsten Jørgensdatter had lived together since around 1571–2, so they were legally married by 1576 according to Jutish Law, the common law applied to cases involving noblemen; Thoren 1990, 45–8.

9. The surnames of Kirstine Tygesdatter and Magdalene Tygesdatter were patronymics based on the Danish form of their father's given name. They could not bear the name of Brahe because their mother was a commoner. Their brother was named Claus Tygessøn or Tygesen.

10. Thoren 1990, 115–16. Tycho held farms at Ornakärr, Ry, Ingelstäde, and Svanebäck in Väsby parish.

11. There is much literature on the early modern family, but comparatively little on households and precious little on Scandinavia, see Berkner 1973, Hareven 1991, and Vammen 1977.

12. Honnens de Lichtenberg 1985; Honnens de Lichtenberg 1991. Cf. Thoren 1990, 106–13.

13. Christensen & Beckett 1921, 13, make this point.

14. On Kassel, see Moran 1981; Moran 1982. On Prague, see Evans 1973; see also Moran 1991a. King Frederick's Paracelsian practitioners were scattered among the court, the University of Copenhagen, and Hven, see Shackelford 1991.

15. Brahe 1913–29, 1: 179–207; Thoren 1990, 119–22; Wittendorff 1994, 122–5. On astrology and free will, see Christianson 1979, 113–14. S. Heiberg 1988, 17–18, saw an image of the ideal Renaissance prince in the nativity.

16. Jern 1976, 93.

17. Brahe 1913–29, 4: 5; Christianson 1979, 120–1; Thoren 1990, 123–7.

18. Brahe 1913–29, 5: 12–19, 52–5, 76–87, 97, 108; Brahe 1996, 11–20, 57–61, 83–95, 104–5, 119; Thoren 1973, 25–9, 33. Thoren dated the zodiacal armillary to 1581, but Steen Bille wrote Tycho about its construction in 1577, see Brahe 1913–29, 7: 45.

19. Brahe 1913–29, 7: 47.

20. Brahe 1913–29, 4: 381–96; Christianson 1979, 127–40; Jarrell 1989, 24–6; Thoren 1990, 124–32, 135–40, 236–50. See Westman 1972 on the influence of this comet upon Kepler's thought.

21. Brahe 1913–29, 1: 209–50.

22. A third axis of technological recruitment ran from Copenhagen to Hven, and a fourth from Herrevad Abbey to Hven. Craftsmen at Herrevad and in Copenhagen helped to build some of Tycho's early instruments before he had his own instrument makers on Hven. See Brahe 1913–29, 7: 45; Thoren 1990, 123, 149.

23. Brahe 1913–29, 5: 88–91; 7: 5–7, 13–21, 35–6, 46–53, 57–8, 83–7; Thoren 1990, 92, 117, 133–5, 139–41, 216, 313; Tycho's friends and regular correspondents in Augsburg included Johannes Major, Hieronymus Wolf, and the brothers Paul and Johann Baptist Hainzel.

24. Copenhaver 1990, 273–5; Yates [1964] 1969, 37.

25. Zeeberg 1994c, 11–12, 254. Cf. Thoren 1990, 75, 140–1.

26. Brask et al. [1984] 1990, 423–4.

27. In 1582, Aalborg would become the official bookseller of the University of Copenhagen.

28. Christianson 1979, 116; Thoren 1990, 118.

29. Gingerich & Westman 1988, 5–41. Wittich eventually acquired two copies each of the 1543 and 1566 editions of Copernicus.

30. Gingerich & Westman 1988, 77–140 et passim. The earth was not seen as a planet.

31. Christianson 1979, 120–7; Thoren 1990, 85–92.

32. Victor Thoren conjectured a "hybrid theory for the superior planets" for Tycho around 1574 that turned out to be identical with one of Wittich's 1578 diagrams. See Gingerich & Westman 1988, 46.

33. Gingerich & Westman 1988, 11–12. At the time, Napier had not yet come up with his own method of logarithms.

34. Gingerich & Westman 1988, 62; Thoren 1990, 236–8, 280–3, 407.

35. Brahe 1913–29, 76–9, 154–5; Brahe 1996, 165; see also 83–6, 163–5.

36. Thoren 1973, 39, noted on the basis of Tycho's observational journals that it was ready for use in the summer of 1584, but Tycho's portrait on the instrument bears the date of 1580, which may be the date of its construction. See Brahe 1913–29, 5: 56–9; Brahe 1996, 62–6.

37. In 1581, Wittich shared his models with the itinerant Oxford scholar, Henry Savile, see Goulding 1995.

38. Thoren 1973, 30–1, was uncertain whether the instrument named as Q. maius in the observational journal for 21 March 1581 was the same as the one later called Q. max. Gingerich & Voelkel 1998, notes 19 and 36, argued persuasively that it was

the instrument later remounted in Stjerneborg as the revolving azimuth quadrant, *Quadrans volubilis azimuthalis.* On this latter instrument, see Brahe 1913–29, 5: 32–5, and Brahe 1996, 36–9.

39. This was the instrument called *Q. max.* in the observational journals, see Brahe 1913–29, 5: 92–3; Brahe 1996, 100–2.

40. The trigonal sextant was also adapted for simultaneous observation of two stars. Brahe 1913–29, 5: 20–31, 48–51, 68–75, 92–3, 96; Brahe 1996, 21–35, 53–6, 75–82, 100–3; Gingerich & Voelkel 1998, 5; Thoren 1973, 30–7.

41. Gingerich & Voelkel 1998, 1. See also Blair 1990, 364–6.

42. Ibid., 1–5.

43. Brahe 1913–29, 14: 25.

44. This tradition flourished with sixteenth-century prelates like Tycho's friend, Abbot Iver Bertelsen of Sorø, and his great-uncle, Archbishop Torbern Bille of Lund.

45. Robb [1935] 1968, 57–89. On Tycho Brahe's contact in 1575 with academies in the Venetian Republic, see Norlind 1970, 66–8; Thoren 1990, 97.

46. Burke 1995 identified several copies of Castiglione owned by friends and kinsmen of Tycho Brahe.

47. Ozment 1983; Schwiebert 1950, 224–31, 594–9.

48. Moran 1982; Norlind 1970, 65–7; Thoren 1990, 93–6.

49. Troels-Lund [1879–1901] 1968–9, 3: 9–256, 478–82, deals with sixteenth-century Danish food and dining customs in detail.

50. Brahe 1913–29, 5: 142–5; Brahe 1996, 145–51; Thoren 1990, 146. *Vinterstue* ("Winter Room") was the common Danish name for a combined living room, dining room, and master bedroom. In Latin, Tycho called this room the *coenaculum.*

51. In summer, they sometimes ate in one of the garden pavilions or in the larger Summer Room upstairs, with its splendid views of the Sound.

52. Troels-Lund [1879–1901] 1968–9, 3: 26–30.

53. Ibid., 3: 223.

54. Ibid., 3: 37–8.

55. This was Jeppe, whose prescience gave rise to many legends.

56. These holidays were noted in the meteorological diary. The ordinance of 1539 also included Palm Saturday, the Visitation of Mary, and All Saints Day.

57. Regular meetings of top staff are still an essential part of big science.

58. Brahe 1913–29, 9: 98, 125, 140, 143, 145.

59. J. Andersen 1896, 112–21.

60. S. A. Brahe 1955, 21–2, 47, 147–8, 185.

61. Thoren 1985; Thoren 1990, 200–1.

62. Fussing [1942] 1973, 355–70, dealt with the roles of bailiffs and stewards.

CHAPTER 5. BREAKTHROUGH, 1584–1587

1. Thoren 1990, 149–80.

2. Panofsky 1992, 131–4, focused on the origins of the Stanford Linear Accelerator Center (SLAC) in 1957–66 but considered the nine-year SLAC startup fairly normal for most late twentieth-century big science. See also R. W. Smith 1992 on the interaction of lobbying, funding, and instrument design in the startup phase of the Hubble Space Telescope.

3. Tycho produced a third nativity when Prince Hans was born in 1583.

4. Moran 1981, 259–61, noted the political importance of maps and surveying in the late sixteenth century.

5. L. Nielsen 1946, 11; Thoren 1990, 185–6.

6. Eamon 1990, 338–40, discusses the emergence of intellectual property rights; see also Boas Hall 1975 and Gingerich & Westman 1988, 19. Tycho's work brought its rewards in the patronage "economy of honor," not the marketplace of commerce.

7. Regiomontanus had ambitious plans for a scientific press in Nuremberg during the 1470s but achieved rather modest results. Eisenstein 1979, 2: 575–635, discussed printing and the Copernican revolution.

8. L. Nielsen 1946, 11–13.

9. Brahe 1913–29, 5: 44–7; Brahe 1996, 48–52; Dreyer [1890] 1977, 123–5.

10. Brahe 1876, 37–42; Brahe 1913–29, 5: 24–7, 72–5; 9: 25–8; 10: 345–8; Brahe 1946, 24–7, 72–5; Brahe 1996, 25–9, 79–82; Thoren 1973, 39.

11. Pierre Belon (1517–64), the Wittenberg-trained founder of comparative anatomy, undertook extensive scientific journeys 1546–50 through Greece, the Levant, and Arabia, publishing his findings in 1553, see Gillispie 1970–80, 1: 595–6.

12. Blair 1990, 368–9; Moesgaard 1972a, 34–6; Thoren 1990, 194–6; Van Helden 1989, 115.

13. Westman 1990, 185.

14. Brahe 1913–29, 6: 265–7. Copernicus used three instruments, including the triquetrum. See Zinner 1956, 282–4.

15. Friis 1890–1901, 4, discusses the monograms of unidentified graphic artists working for Tycho Brahe.

16. Brahe 1913–29, 9: 319.

17. Brahe 1913–29, 7: 387; translated in Rosen 1986, 39.

18. Rosen 1986, 251.

19. Dreyer [1890] 1977, 183–5, 273–6; Thoren 1990, 255–6.

20. Brahe 1913–29, 7: 387–8; translated in Rosen 1986, 40.

21. Translated in Rosen 1986, 251–2.

22. Discontent among junior staff members has been common in twentieth-century big science, due in part to the hiearchical nature of the enterprise, see Hounshell 1992, 239–41; Panofsky 1992; Seidel 1992, 28–31; Traweek 1992.

23. Gingerich & Westman 1988, 17–18, 58. See Goulding 1995, 174–7, on the boastful and secretive sides of Wittich's character.

24. Moran 1982, 90–7. On Bürgi and his instruments, see Zinner 1956, 268–76.

25. Zeeberg 1994c, 74–5.

26. Brahe 1913–29, 9: 193–207. Zeeberg 1988 demonstrated that the poem revealed much about Tycho Brahe's scientific and cultural program around 1584.

27. Zeeberg 1988, 170–4.

28. Brahe 1913–29, 6: 144–6; Shackelford 1991, 97–8; Thoren 1990, 213. These emblems had undoubtedly been designed by Schardt, see Honnens de Lichtenberg 1991, 39–41.

29. Zeeberg 1991b, 835–8.

30. Brahe 1913–29, 9: 180–7.

31. Zeeberg 1991b, 838.

32. Christensen & Beckett 1921, 7–10.

33. Brahe 1913–29, 4: 375; 5: 16.
34. Brahe 1913–29, 7: 89–90; L. Nielsen 1946, 123.
35. Brahe 1913–29, 9: 187–90; Thoren 1990, 266.
36. Zeeberg 1988, 176.
37. L. Nielsen 1946, 26.
38. Brahe 1913–29, 386–7; Christensen & Beckett 1921, 8–9; Thoren 1990, 208–10.
39. Brahe 1913–29, 5: 64–7; Brahe 1996, 71–4.
40. Gingerich & Voelker 1998, 13–16; Thoren 1973, 40.
41. This did not occur in sixteenth-century universities, but it did at a few courts, see Moran 1991a.
42. Hammond & Scullard 1970, 704–5; Pfeiffer 1968–76, 1: 95–104; Sandys 1921, 1: 104–44; Sarton [1952–9] 1964–5, 2: 3–34, 141–57.
43. Quoted in Sandys 1921, 1: 107.
44. Haskell 1993, 43–51; Zimmermann 1995. See Findlen 1989 and Findlen 1994, 48–50, on meanings of "museum."
45. Findlen 1989, 60–1, 66–7.
46. Brahe 1913–29, 6: 104.
47. William IV built a similar complex, the Marstall, at his castle in Kassel. See Schepelern 1971, 76–7.
48. Brahe 1913–29, 6: 250–2, 262–4; Norlind 1926, 71–4; Thoren 1990, 146–7.
49. Cf. Hannaway 1986 and Shackelford 1993.
50. The grounds of Uraniborg were excavated by archaeologists in 1988–92 and the ramparts and gardens partially reconstructed. See Christianson 1998.
51. Yates [1964] 1969, 73–6.
52. See Evans 1973, 162–95, 243–74, on Emperor Rudolf II's collections as objects of contemplation and magical manipulation.
53. Schepelern 1971, 29–30.
54. Brahe 1913–29, 4: 512–13; Thoren 1990, 186.
55. Hellman 1944; Thoren 1990, 138–40.
56. Eisenstein 1979, 2: 575–623.
57. Gingerich 1986 comments on the surprisingly wide readership of the 1543 and 1566 editions of *De revolutionibus,* each of which he estimates to have been printed in four or five hundred copies.
58. Christianson 1972a.
59. Brahe 1913–29, 4: 1–378.
60. Dreyer [1890] 1977, 163–72; Norlind 1970, 108–26.
61. Brahe 1913–29, 4: 155–79. Boas & Hall 1959; Rosen 1986, 21–2; Thoren 1990, 258–62.
62. Brahe 1913–29, 6: 25–40; Moran 1982; Thoren 1990, 152–7, 269–71.
63. Christianson 1979, 111. See Bech 1979–84, 13: 151–62, on Queen Sophie; 2: 98 on Tycho Brahe's mother, Beate Bille.
64. Brahe 1913–29, 9: 45.
65. Ashworth 1990 defined the "emblematic world view" as a view of the natural world that combined Renaissance hieroglyphic, antiquarian, mythological, and emblematic traditions with adages, epigrams, and Aesop's fables.
66. Brahe 1913–29, 6: 255–62, 271–88, 292–4; Norlind 1926, 82–108, 110–12.
67. See Zeeberg 1994a, 41–3, and Zeeberg 1995 on the symbolic program of Uraniborg and Stjerneborg. The varieties of trees in Tycho's orchard are unknown, but Gun-

narsson 1993, 50–60, included the ones named here as likely possibilities. See also Christianson 1998.

68. Dreyer [1890] 1977, 138–9; Piø & Pedersen 1984, 7.
69. Norlind 1970, 73, 390.
70. Brahe 1913–29, 6: 64.
71. Findlen 1991, 22. On "representation of the self," see Martin 1997.
72. Friis 1890–1901, 216–18.
73. Bagge & Mykland 1987, 8, reproduced the portrait of Christian II. See also Geisberg 1974, 1: 394; 2: 604; 4: 1231.
74. Westergaard 1930–3, 1: no. 1376.
75. Ibid., 1: no. 1377.
76. Bille Brahe 1985, 14–27, 79–81.
77. Westergaard 1930–3, 1: no. 1394.
78. Brahe 1996, 33. Cf. Brahe 1946, 30.
79. Brahe 1913–29, 9: 53–8. On Liddel, see Schofield 1989, 34.
80. Gingerich & Voelkel 1998, 17–21.
81. Ibid., 16–23.
82. Thoren 1989, 12.
83. Gingerich & Westman 1988, 42–5, 69–76. Gingerich & Voelkel 1998 argued that the "central, driving theme" of Tycho's observations in the whole period 1582–95 was the attempt to determine the distance to Mars, and that this campaign helped to motivate his building ever larger and more innovative instruments, as well as fostering his pioneering investigations of refraction.
84. Schofield 1989; Thoren 1989, 6–9; Thoren 1990, 250.
85. Thoren 1990, 236–64, detailed the long and difficult struggle to come up with the Tychonic system. See also Gingerich & Voelkel 1998; Gingerich & Westman 1988; Moesgaard 1972a.
86. Brahe 1913–29, 4: 1–378, 491–511. Cf. L. Nielsen 1946, 28; Thoren 1990, 259.
87. Brahe 1913–29, 7: 132; Dreyer [1890] 1977, 162–4, 182–3; Norlind 1970, 106–8, 169–70, 215–17, 267; Thoren 1990, 312.

CHAPTER 6. THE PROBLEM OF CONTINUITY, 1580–1591

1. On leadership in twentieth-century big science, see Galison, Hevley, & Lowen 1992; Pestre & Krige 1992, 93; Seidel 1992.
2. For example, Laursen 1885–, 7 May 1577, 10 January 1579, 21 January 1582, 29 November 1582.
3. Rørdam 1883–9, 2: 310–12.
4. Brahe 1913–29, 14: 105, asserted in 1597 that Tycho Brahe had not received Holy Eucharist for eighteen years.
5. Rørdam 1883–9, 2: 337–9.
6. Every one of Tycho Brahe's children received a traditional Bille or Brahe name, though none was named after his parents.
7. Brahe 1913–29, 14: 26–7; Thoren 1990, 353.
8. Karker 1964 concluded that he died of a lung infection.
9. Officers of the realm were chosen from members of the State Council. In descending order of rank, they were Lord Steward, Royal Chancellor, Marshal of the Realm,

Admiral of the Realm, and Chancellor of the Realm. The Treasurer was a household officer. See Tamm 1987, 48.

10. S. Heiberg 1988, 30–49. Cf. Thoren 1990, 340.

11. Brahe 1913–29, 9: 63–5; Wegener 1846, 159–62.

12. Brahe 1913–29, 14: 39–40. The amount was paid in full before the end of the year see ibid., 14: 40, 43, 44, and Christianson 1972b, 236–7.

13. Wegener 1846, 160–2.

14. Ladewig Petersen 1980b.

15. Brahe 1913–29, 14: 41–3; Thoren 1990, 344.

16. Rosen 1986, 238–40. How long Méhérenc remained at Uraniborg is unknown; he later visited Kepler in Graz and was in Tycho's *familia* in Bohemia toward the end of 1599.

17. Ibid., 46–50; Jardine 1984, 30–4.

18. Wegener 1846, 168–76.

19. Ibid., 146–51.

20. Moran 1981, 260

21. Sello 1896, 33.

22. Brahe 1913–29, 14: 30; Friis 1889, 7–8; Thoren 1990, 207, 210.

23. Haasbroek 1968, 29–58; cf. Thoren 1990, 210.

24. Brahe 1913–29, 7: 222; L. Nielsen 1946, 32–3.

25. Hall 1983, 214–15, 230–2; Norlind 1970, 224–5.

26. Brahe 1913–29, 6: 198; Møller Nicolaisen 1946, 14–15.

27. Møller Nicolaisen 1946; Zeeberg 1994c, 21.

28. L. Nielsen 1946, 32–3, cf. Nordstrand 1961.

29. Møller Nicolaisen 1946, 19–20.

30. Brahe 1913–29, 14: 45–8; Thoren 1990, 344. Tycho promised in return to erect a good dwelling and workshop near the water for the master dyer.

31. S. Heiberg 1988, 39; Thoren 1990, 334–5.

32. Zeeberg 1995, 264.

33. Braheana [1589–92], fol. [8–9]; Zeeberg 1995, 264.

34. Brahe 1913–29, 7: 307, 331; Zeeberg 1994c, 47–8; Zeeberg 1995, 264.

35. Brahe 1913–29, 2: 12, 9: 83; Dreyer [1890] 1977, 202–4.

36. Norlind 1970, 146, 148, 165, 186, 398–9; Thoren 1990, 315–16, 335. Dançay, Hagecius, and other courtier friends helped Tycho obtain gifts of copyrights from the kings of France and Spain, Emperor Rudolf II, and of course the King of Denmark-Norway.

37. Brahe 1913–29, 6: 261, 287; 9: 84. Willem Blaeu, who was not present, described this visit to Gassendi in melodramatic terms, see Gassendi 1655, 119–20, 196, and Norlind 1951, 115, 180. Cf. Norlind 1970, 91–2; Thoren 1990, 335–6.

38. Brahe 1913–29, 9: 113; Hübner 1737, tab. 98.

39. Brahe 1913–29, 14: 49.

40. Ibid., 14: 51–2.

41. Bech 1979–84, 15: 252–5; S. Heiberg 1988, 39–40.

42. Alver 1971, 104.

43. Brahe 1913–29, 9: 89–91; Moesgaard 1972a, 51–3; Thoren 1990, 336. His brother, Johannes Rothmann, visited Hven in June of 1589, see Norlind 1970, 213–14.

44. Brahe 1913–29, 9: 101–2; Carlquist 1948–85, 8: 444–8; Fabritius 1961, 125–6.

45. Christianson 1968, 313; Zeeberg 1994c, 80–5.
46. Brahe 1913–29, 9: 100.
47. Gascoigne 1990 found no similar relationship between any university and a scientific academy, though he did note that some eighteenth-century academies became "virtual extensions of the university community" in their localities, see 252.
48. Christianson 1979, 119–20, 124–7, 130; Krstović & Lazzari 1999, 31–5; Moesgaard 1972b, 117–19. On the relationship among universities, academies, and courts as loci of patronage in sixteenth-century Germany, see Moran 1991a.
49. Rørdam 1863–74, 2: 621–6.
50. Ibid., 2: 653–9.
51. Brahæana [1589–92]. This manuscript was discovered by Wilhelm Norlind, see Norlind 1970, 101–2, 168–70; see also Zeeberg 1995, 263–4.

CHAPTER 7. THE SCHOOL OF EUROPE, 1591–1593

1. Traweek 1992, 101.
2. Frobenius 1587–99; Butler & Christianson 1989 is a Danish translation.
3. Brahe 1913–29, 9: 88, 119; Thoren 1990, 354.
4. Honnens de Lichtenberg 1994, 92.
5. Brahe 1913–29, 9: 100; Jern 1976, 21–2; Lundquist 1993; Lundquist 1995; Lundquist 1998.
6. Butler & Christianson 1989, 51.
7. Rørdam 1863–74, 4: 369–70. Garstein 1953, 52–5, 63–7.
8. Schama 1988 examined the emerging Dutch sense of national identity.
9. Christianson 1979, 114–16, treated the religious consensus of Danish Philippist Lutheranism under fire from abroad.
10. Except for Rasmus Pedersen and Christopher Pepler, this paragraph is based on Brahe 1913–29, 9: 69–116.
11. Christianson 1972b, 238–9; Norlind 1970, 368–9; Thoren 1990, 354–5.
12. B. Andersen 1971, 88; Friis 1905; J. Heiberg 1846; Zeeberg 1994c.
13. Bech 1979–84, 8: 479– 80; Friis 1899–1900; Friis 1905, 7–21.
14. Brahe 1913–29, 14: 68; Bech 1979–84, 2: 428–9; Fabritius 1950, 16–17.
15. Brahe 1913–29, 7: 327; Norlind 1926, 113–14.
16. Brahe 1913–29, 9: 112. Troels-Lund 1910–12, 1: 157–94.
17. Biagioli 1993, 39, 41.
18. Bjørnbo & Petersen 1908, 9–10, 26–30.
19. Helk 1987, 411 (see also 368 and 379 for Skeel and Gellius); Tamm 1987, 25–35.
20. Rosen 1986, 98–100.
21. Norlind 1970, 191–8, 374–5. Gassendi 1655, 147, mentioned this plan, see also Norlind 1951, 137, 293–4, 334. Tycho on 1 December 1590 wrote regarding this plan to Magini and also to Kurz at the imperial court, see Brahe 1913–29, 7: 286–8, 298–9; Thoren 1990, 405.

CHAPTER 8. MAGDALENE AND CALUMNY, 1593–1597

1. Venusin, see Bech 1979–84, 15: 394.
2. Dallin, see Bricka 1887–1905, 4: 163–4; Helk 1987, 198.
3. Niels Krag, see Bech 1979–84, 8: 235–8; Helk 1987, 284.

4. Anders Krag, see Bech 1979–84, 8: 226–7; Helk 1987, 283.

5. [Thiset] 1904. The brothers Krag were sons of Ribe patricians.

6. Troels-Lund [1879–1901] 1968–9, 5: 30–49, 70–106; for the view that love was a fruit of marriage and not a precedent to it, see 49–54.

7. Brahe 1913–29, 14: 82.

8. Troels-Lund [1879–1901] 1968–9, 5: 132–9.

9. Brahe 1913–29, 9: 126.

10. Ibid., 5: 139–47.

11. Cf. Fichtner 1976 on negotiations of Habsburg marriages.

12. Troels-Lund [1879–1901] 1968–9, 5: 276–306.

13. Troels-Lund [1879–1901] 1968–9, 5: 169–78. Sometimes, an aristocratic suitor added a promissory note for land or a very large sum of money, to be given to the bride at the time of his death.

14. Ibid., 5: 295–314.

15. Ibid., 5: 320–473, 6: 19–359.

16. Ibid., 5: 471–3, 420–38, and 438–68.

17. Brahe 1913–29, 9: 126; 12: 318–19.

18. Troels-Lund [1879–1901] 1968–9, 5: 150–1.

19. Ibid., 5: 187.

20. Rørdam 1883–9, 2: 341.

21. Ibid., 2: 340.

22. Troels-Lund [1879–1901] 1968–9, 5: 385–90, 404–68; 6: 11–39, 56–8, 75–84, 94–184, 200–6, 217–33, 238–50, 295–314, 355–9.

23. Brahe 1913–29, 9: 126; 14: 66, 68–9; Christianson 1972b, 240.

24. Brahe 1913–29, 14: 81.

25. Ibid., 14: 82.

26. Ibid., 14: 81.

27. Ibid., 9: 127–8; 14: 84–5. This raises the question of whether Magdalene was already betrothed to Gellius. If so, they were already man and wife: Betrothal occurred when the couple pledged themselves to one another, followed by physical sexual union. When Venusin later said that Gellius had pledged himself to Magdalene before God and verified his pledge with hand and mouth, Gellius did not deny it, see Brahe 1913–29, 14: 88, 90–1. This seems to indicate that they were betrothed, though when the betrothal occurred is difficult to say. It might have been during Gellius's visit to Hven in mid-October 1594. A year later, Tycho referred to 15 December as the date set for the wedding, but the most likely explanation is that it was the betrothal date.

28. See false rumors two and six, Brahe 1913–29, 14: 78.

29. Brahe 1913–29, 14: 82–3, 89–90, 94–5.

30. Ibid., 14: 89–90.

31. Ibid., 9: 128; 14: 82. Lisbet Jensdatter Bille (1572–1633).

32. Brahe 1913–29, 14: 90, 94–5.

33. Ibid., 14: 86–90, 94–5. The date of these events was in dispute. Tycho later said that they occurred on 6 December, but Anders Krag said 8 December, ibid., 14: 86, 88. Gellius later claimed that he had been drinking, spoke in jest, and had heard a rumor that Magdalene no longer wanted him, but all the witnesses denied each of these claims.

34. Ibid., 14: 93.
35. Ibid., 14: 82.
36. Hegelund [1578] 1972.
37. Brask et al. [1984] 1990, 337–9.
38. Brahe 1913–29, 14: 82.
39. Ibid., 14: 82.
40. Ibid., 14: 93–4.
41. Ibid., 9: 129.
42. Ibid., 14: 92.
43. Ibid., 14: 93–4.
44. Ibid., 14: 92–3.
45. Ibid., 14: 71–3, 92–4.
46. Gingerich & Voelkel 1998, 27; they argued that the distance to Mars was "a central, driving theme" of Tycho's observations throughout the whole period 1582–95 and had been an important motivation for the development of the large Stjerneborg instruments.
47. Thoren 1990, 287–300.
48. Ibid., 322.
49. Brahe 1913–29, 14: 95, verb tense altered.
50. Ibid., 14: 94–5.
51. Ibid., 14: 72.
52. Ibid., 14: 87.
53. Ibid., 9: 21; 14: 88.
54. Laursen 1885–, 24 March 1595.
55. Brahe 1913–29, 14: 79.
56. Ibid., 14: 95–6. Students promoted to the degree of Master of Arts customarily put on a feast for their professors and friends, see Norvin 1937–40, 1: 267. Hans Resen, an ally of Gellius, reluctantly verified this incident.
57. Brahe 1913–29, 9: 134; 14: 87–8.
58. Ibid., 14: 79.
59. Ibid., 9: 136; 14: 80.
60. Ibid., 14: 79. On Urne, see Bech 1979–84, 15: 188.
61. Brahe 1913–29, 7: 374–5. The letter to Holger Rosenkrantz is dated Uraniborg, 22 January 1596.
62. Ibid., 14: 73, 79.
63. Ibid., 14: 73–4.
64. Brahe 1913–29, 14: 79, 82. "Vrint" may have been Nicolaus de Frunt, citizen of Bergen in Norway, see Alver 1971, 81.
65. Brahe 1913–29, 14: 77–8. Langebek 1746 first published the transcript of the Consistory hearings but printed the list of false rumors out of context. This led to confusion on the part of Friis 1871, 216, who cited the rumors as if they were true. Dreyer [1890] 1977, 224–7, Norlind 1970, 252, and Thoren 1990, 358, did essentially the same.
66. Brahe 1913–29, 14: 79–80.
67. Ibid., 14: 80–1.
68. Ibid., 14: 84.

69. Ibid., 14: 84–5.
70. Ibid., 14: 85.
71. Ibid., 14: 89–95.
72. Ibid., 14: 94–6.
73. Weistritz 1756, 2: 275–22.
74. Brahe 1913–29, 14: 103–4.
75. Cf. Dreyer [1890] 1977, 264.
76. Brahe 1913–29, 9: 136, 138–40, for this and the following paragraphs.
77. Rørdam 1863–74, 3: 67–71.
78. S. Heiberg 1988, 46–9.
79. The others were Theodoricus and Joachimus, see Brahe 1913–29, 9: 140–2, for this and the previous paragraph.
80. Brahe 1913–29, 9: 142–3; S. Heiberg 1988, 51–67.
81. Brahe 1913–29, 9: 143.
82. Christianson 1972b, 243.
83. Brahe 1913–29, 14: 7–11, 14–17, 19, 32, 34–5, 37–8, 48–9, 62–3; Christianson 1972b, 243.
84. Brahe 1913–29, 14: 99–101; Thoren 1990, 368–70.
85. Brahe 1913–29, 14: 101–2.

CHAPTER 9. THE TEMPEST, 1597

1. Brahe 1913–29, 9: 145. On beached whales, see Schama 1988, 130–44.
2. Erslev 1885, 15, 59, 78, 84. On Sehested, see Bech 1979–84, 13: 332–3.
3. Brahe 1913–29, 7: 380; Thoren 1990, 370.
4. Brahe 1913–29, 9: 146. Earlier fires recorded in the same source were a 1587 fireplace fire, a 1590 kitchen fire, and a 1591 fire in a furnace.
5. See Weistritz 1756, 2: 11–22, and Wittendorff 1994, 235.
6. Fincke 1583.
7. All scholars in that era were polyhistors who could teach anything, and careers advanced from junior professorships to the more lucrative chairs of law, medicine, or theology.
8. Brahe 1913–29, 8: 180, 440; Wegener 1846, 205.
9. Kornerup 1928–68, 1: 271–4; Rørdam 1883–9, 1: 69; Troels-Lund [1879–1901] 1968–9, 4: 403–27. Iver Bertelsen was fired from his pastorate in 1567 for omitting exorcism and other offenses, and Jon Jacobsen Venusin was temporarily suspended for omitting exorcism in 1588.
10. Kornerup 1928–68, 1: 143–4.
11. Brahe 1913–29, 9: 146; 14: 102, 127–30; Thoren 1990, 370–1.
12. Perhaps many more volumes than that; we know only that one volume still extant was numbered 2781 during the registration, see Brahe 1913–29, 9: 146; Norlind 1970, 335.
13. Brahe 1913–29, 14: 102–3; Thoren 1990, 372.
14. Brahe 1913–29, 14: 103–4; S. Heiberg 1988, 33, 39–41, 48–9.
15. Brahe 1913–29, 9: 146. Thoren 1990, 371, says that Tycho and his family departed Hven on 29 March 1597.

16. Brahe 1913–29, 14: 105.
17. Ibid., 14: 122; Dreyer [1890] 1977, 249–50.
18. Brahe 1913–29, 14: 104.
19. Ibid., 9: 146.
20. Ibid., 14: 104–5.
21. Ibid., 14: 105; Thoren 1990, 372–3.
22. Brahe 1913–29, 8: 7.
23. Ibid., 8: 176, 14: 123, 129; Weistritz 1756, 1: 155. The king was abroad at the time.
24. Brahe 1913–29, 7: 384; Gassendi 1655, 140–1.
25. Brahe 1913–29, 14: 129.
26. Nyström 1927 argued that flaws in Tycho's character, including his singleminded ruthlessness; his hasty, violent, and unpredictable manner; and his injured vanity drove him into what he thought would be a short, temporary exile. On Tycho's personality see Christianson 1998, 476.
27. Brahe 1913–29, 9: 208–11; Brask et al. [1984] 1990, 410–12.
28. Brahe 1913–23, 14: 106; Christianson 1972b, 240–1. Walkendorf clearly wanted to break the power of Tycho Brahe and his faction, but there is some evidence that Christian Friis played a double game, carrying out the wishes of the king in ways that mitigated their harmful effect upon Tycho, in part by including in the process supporters of Tycho like Axel Brahe and Peter Munk and by going himself as a royal commissioner to Hven. This interpretation is consistent with Friis's patronage of Tycho's former *famuli,* both then and later.

CHAPTER 10. EPILOGUE: IN SEARCH OF MÆCENAS, 1597–1599

1. Brahe 1913–29, 14: 139.
2. Esge Jørgensen Bille (1552–1619), Tycho's third cousin and good friend, had been a regular visitor to Uraniborg since 1590.
3. Brahe 1913–29, 14: 129, "over 20 people."
4. Cf. Braun & Ordway 1966, 114–18, on the transfer of the German rocket program to the United States in 1945.
5. This was the normal route; there is no explicit evidence that it was Tycho's route.
6. Troels-Lund 1910–12, 2(2): 1–358, was challenged by F. Jensen 1978, 45–82. See also F. Jensen 1982.
7. Bech 1979–84, 2: 423–4; Danstrup & Koch 1962–6, 6: 355.
8. Brahe 1913–29, 8: 3, 9; Roberts 1968, 364–75.
9. Greenblatt 1980 developed a concept of Renaissance self-fashioning which Biagiolo 1993 applied to Galileo. On the validity of self-fashioning theory as an explanation of sixteenth-century behavior, see Davis 1988; Finlay 1988; Martin 1997.
10. Tycho sometimes recast it as *Esse potius, quam haberi* ("To be rather than to seem").
11. Brahe 1913–29, 14: 108–11, translated into English in Dreyer [1890] 1977, 243–5; Thoren 1990, 377–8.
12. Greenblatt 1980, 32–73, argued that Thomas More's futile attempt to bring his two identities together led to self-cancellation on the gallows: Tycho may have had dark dreams of canceling his opposing identity of aristocratic courtier and simply being Prince of the Muses.

13. The embassy had negotiated a treaty of marriage between Anne Catherine of Brandenburg and King Christian IV.

14. Brahe 1913–29, 8: 3–4; Engelstoft & Dahl, 1933–44, 3: 581.

15. Kornerup 1928–68, 1: 64–71.

16. Brahe 1913–29, 8: 5, 14: 111–12; Thoren 1990, 379.

17. Brahe 1913–29, 14: 113–14.

18. Ibid., 14: 114–18, 136–8; Norlind 1970, 256–7.

19. Brahe 1913–29, 14: 120, "jnsula Dalan prope Rigam, vel Bornhovium, aut jnsula Werden prope Dantiscum."

20. Brahe 1913–29, 8: 5–6; Norlind 1970, 260.

21. Bech 1979–84, 11: 622–7; Evans 1984.

22. Lorenzen 1912, 74.

23. Hoffmann 1867, 6.

24. Lorenzen 1912, 63–5, pl. XIII.

25. Ibid., 84–91.

26. Norlind 1970, 261–3, thought the king hardly did more than approve a letter written by advisors, but did Walkendorf or Friis write it?

27. Brahe 1913–29, 14: 121–3; Dreyer [1890] 1977, 248–52, translated the letter into English.

28. Brahe 1913–29, 9: 208–11; Dreyer [1890] 1977, 254–5.

29. Brahe 1913–29, 5: 119–24; Evans 1973, 71, 136, 223; Thoren 1990, 273, 315–16, 390.

30. Brahe 1913–29, 7: 166. Cf. Gingerich & Westman 1988, 20.

31. Hübner 1737, tabs. 246, 247. On Prince Karl von Liechtenstein, see Evans 1973, 41, 72–4, 156–7. On Minckwitz, see Friis 1902, 15–16.

32. L. Nielsen 1946, 57–61; Norlind 1970, 267–96; Thoren 1990, 381–4.

33. Thoren 1990, 383–4. Longomontanus's work on this project in the last days on Hven had been incomplete.

34. Brahe 1913–29, 14: 124–5, 321. Norlind 1970, 263–4, wonders whether this letter was actually sent to Rantzau.

35. Brahe 1913–29, 14: 126–7; Thoren 1990, 384–5.

36. Brahe 1913–29, 14: 131–3; Thoren 1990, 385.

37. Helk 1987, 104–5, 125–34. For an example, see Kornerup 1928–68, 1: 108–9.

38. Ashworth 1990 applied his concept of an "emblematic world view" to zoology, but it also fits Tycho's world view in the 1590s.

39. Brahe 1913–29, 13: 124–6; Engelstoft & Dahl 1933–44, 20: 81–2; Norlind 1970, 179–80, 265.

40. Brahe 1913–29, 13: 117–21.

41. Biagioli 1993, 23–4, discusses the role of informers in courtly patronage networks.

42. Brahe 1913–29, 8: 24–5; 14: 205–7.

43. Brahe 1913–29, 8: 60, 14: 133; [Thiset] 1904, 329.

44. Biagioli 1993, 60–73, dealt with the role of scientific disputes in patronage; following Biagioli, one might argue that Tycho needed to crush the slander and libel of Ursus in the fearless, skillful, and heroic manner of his persona as Prince of the Muses in order to gain honor from the clash. Tycho's dispute with Ursus has been examined in detail, from various other points of view, by Gingerich & Westman 1988, Jardine 1984, and Rosen 1986.

45. Brahe 1913–29, 8: 28–31, 34–5; Gingerich & Westman 1988, 20–3.
46. Brahe 1913–29, 5: 1–162; Brahe 1946; Brahe 1996. Webster 1975 does not include Tycho in his study of the Great Instauration, but see Webster 1982.
47. Brahe 1913–29, 3: 344–77; 5: 1–162; Norlind 1970, 267–99.
48. Brahe 1913–29, 8: 80–3. Elector Ernest in 1589–90 had sponsored publication of the collected works of Paracelsus in ten volumes, see Trevor-Roper 1985, 153–4. He later patronized Kepler and gave him his first telescope, see Caspar [1948] 1962, 165, 204–10.
49. Norlind 1970, 286–93, 295–6.
50. Heinrich Julius eventually went to Prague and became one of the closest advisors to Emperor Rudolf II, but not until 1607. See Evans 1973, 73, 231.
51. Thoren 1990, 401.
52. Brahe 1913–29, 14: 141–2. On Barvitius, see Evans 1973, 73.
53. Brahe 1913–29, 14: 145–6.
54. Brahe 1913–29, 14: 156, 159; Thoren 1990, 400.
55. Brahe 1913–29, 14: 142–5; Norlind 1951, 306–8.
56. Brahe 1913–29, 8: 50, 72–3; Rosen 1986, 125–30, 347. Evans 1973, 193.
57. Brahe 1913–29, 8: 97, 181; Evans 1973, 69, 111.
58. Brahe 1913–29, 8: 128; Thoren 1990, 402–3. Debts were due on Michaelmas, and annual labor contracts began on this day, which was also a day for harvest festivals.
59. Brahe 1913–29, 8: 31, 164, 339, 366–7; Hübner 1737, tab. 189; Norlind 1970, 288; Thoren 1990, 403, 411.
60. Brahe 1913–29, 8: 52, 125, 180–1; Rosen 1986, 254.
61. Biagioli 1993, 60–73. Ursus was what Biagioli, 71, called a "dismissible outsider" who delivered wild attacks, but he was also an insider as Imperial Mathematician, which meant that he had to be taken seriously, especially since he stood directly in the patronage path leading from Tycho Brahe to Emperor Rudolf II.
62. Thoren 1990, 402.
63. Brahe 1913–29, 8: 119–20, also 118; Norlind 1970, 297–8; Thoren 1990, 404–5.
64. Norlind 1970, 288, questions whether the copy dedicated to Ferdinand de' Medici, Grand Duke of Tuscany, was ever delivered; see 374–5 for Tycho's letter to the doge, Marino Grimani, and 375–6 for Tengnagel's knighthood. Brahe 1913–29, 8: 118–25.
65. Evans 1973, 74–5, 280; Norlind 1970, 289–91, 296, 375–6; Thoren 1990, 405–6.
66. Brahe 1913–29, 14: 148–9; Hübner 1737, tabs. 159, 167; Norlind 1970, 287–8, 295.
67. See Hannaway 1986.
68. Schwiebert 1950, 223, 233–4, pls. II, XXXII, XXXIII.
69. Thoren 1990, 407.
70. Ibid., 406.
71. Warner 1979, 28–9, 203–4.
72. Thoren 1972; Thoren 1989, 12–19; Thoren 1990, 407–10.
73. Brahe 1913–29, 8: 149–51.
74. Evans 1973, 71–2.
75. Brahe 1913–29, 8: 136–7, 145; Thoren 1990, 409.
76. Brahe 1913–29, 14: 191; Norlind 1970, 300.

77. Brahe 1913–29, 8: 163–6, see also 175–9. Quoted with slight changes from the translation in Thoren 1990, 412–13.

78. Norlind 1970, 302; Thoren 1990, 413.

CHAPTER 11. LEGACY

1. Kæmpe & Thykier 1993 reported analysis at the Institute of Forensic Chemistry in Copenhagen of a fragment of Tycho Brahe's beard, which showed very high levels of lead and especially of mercury. See also Kæmpe, Thykier, & Pedersen 1993. Cf. Gotfredsen 1955.

2. Brahe 1913–29, 14: 183, 217–18, 233, 247–51, 261–3, 299, 325–6; Fabritius 1950; Thoren 1990, 471–5, 477–9, 512, 517. Tycho Tygesen Brahe in 1604 married Margaretha Vitzthum von Eckstädt, widow of a Bohemian aristocrat with influential kinsmen on the imperial council. Cecilie Brahe married Baron Gustav Sparre of Sundby, a son of Lord Chancellor Erik Sparre of Sweden and nephew of the Polish ambassador in Prague, Count Erik Brahe of Visingborg, see Elgenstierna 1925–36, 1: 555–6, 7: 409–10.

3. Brahe 1913–29, 14: 242–3, 246–9, 251–2, 254–61, 263–7, 270–9, 289–92, 295–6, 300–4; Friis 1902; Norlind 1951, 196; Thoren 1990, 472, 475.

4. Brahe 1913–29, 14: 262, 298–300; Friis 1902, 28, 37–9; Thoren 1990, 477–8.

5. Brahe 1913–29, 2: 1–435; 3: 3–330; 4: 491–7, 1–378; Caspar [1948] 1962, 144.

6. Schofield 1989.

7. Brahe 1913–29, vols. 10–13; List 1961.

8. Kepler treasured Tycho's observational manuscripts until his death, and his son later sold them to the king of Denmark, see List 1961, 23.

9. On the Snels and Metius in Leiden, see also Gascoigne 1990, 223–8.

10. Bacon was well aware of Tycho, and his House of Solomon in *The New Atlantis* reads like an idealized description of Tycho's island. On Henry Savile's ties with Wittich, see Goulding 1995 and Westman 1980, 130. Tycho's friends and correspondents Daniel Rogers, Thomas Savile, and Thomas Digges, were dead, Thomas Mouffet and John Dee elderly; but the Scot, Peter Young, was still vigorous, as were John Craig and Duncan Liddel. The brothers Sir Anthony and Robert Sherley, with whom Tycho had come into contact in Prague, were embroiled in Persian adventures.

11. Donahue 1975; Gascoigne 1990, 213–14, 222; Schofield 1989, 41–3. On Clavius and Tycho, see Norlind 1970, 176, 202–4, 325, 376–82. On mathematics in the curriculum promoted by Clavius, see Westman 1980, 131–2. See Biagioli 1993, Drake & O'Malley 1960, and Redondi 1987 for various views of the relationship between Galileo and the Jesuits. Walker 1972, 194–230, discusses the survival of the ancient theology among irenic Jesuits in France and China. On Kepler and the Jesuits, see also List 1961.

12. On the rebuilt Beijing Observatory and Tychonic astronomy in China, see Needham & Wang 1959, 266–70, 295–6, 339–59, 366, 376–80, 382–90, 444–5, 448, 449, 451–2, and figs. 157, 158, 173, 176, 190–2.

13. On Leiden, see Haasbroek 1968, 22. On Copenhagen, see Thykier 1990, 1: 43–9; 2: 203–8; 3: 331–3. Telescopes did not play a role in astrometrical measurements

until Robert Hooke came up with the idea of using telescopic sights on large instruments.

14. Taton & Wilson 1989, 150–4; Van Helden 1989, 116. On Rømer, see Dreyer [1890] 1977, 372–5; Gillispie 1970–80, I: 418–82; II: 525–7.

15. Tycho's memory was venerated by his Danish kinsmen of the seventeenth century, many of whom were highly educated and held high office; they saw him as a renowned man of learning and a champion of aristocratic constitutionalism.

16. Stephanius n.d.; Stephanius 1597–1612.

17. Garstein 1953 discussed Cort Aslakssøn's theology at length; see 318 on historical tables. On Tycho and the authority of the Bible, see Blair 1990, 362–4.

18. Conflict began 1606–7 with another struggle over exorcism. Then two Philippists were forced out of the university in 1607–8. In 1614, Gnesio-Lutherans launched an attack on Cort Aslakssøn, who forced them to back off. In the years to come, the tide turned against Philippism in Denmark. See Garstein 1953; Kornerup 1928–68.

19. The suppressed undercurrent of Renaissance Neoplatonism resurfaced in the work of Olaus Borrichius (Ole Borch) during the second half of the century.

20. Donahue 1975, 259–60, 269–73; Gascoigne 1990, 236; Gillispie 1970–80, I: 317–18, 479–83; 13: 30–5; Moesgaard 1972b; Schofield 1989, 41.

21. Wittendorff 1994, 303–15.

22. See Christianson 1979, 127.

23. Schepelern 1988, 296.

24. Gregersen 1957, 124–35.

25. Dreyer [1890] 1977, 366; Gassendi 1655, 216–18; Norlind 1951, 197; Resen 1668, 238–43, 337; Thykier 1990, 3: 457.

26. The rebus incorporates icons of a sword, heart, crown, the royal monogram, and the Hebrew name of God into a Latin text with the message, "Direct learning and justice, O Jehovah, into the heart of King Christian IV!"

27. Schepelern 1988, 297–9; Thykier 1990, 1: 43–9; 2: 203–8; 3: 331–3.

28. On Gassendi, see Gillispie 1970–80, 5: 284–90.

29. On Olaus Worm, the polyhistor and avid correspondent who was Gassendi's Danish contact, see Christianson 1970a; Schepelern 1971.

30. Gassendi 1655 was virtually identical with the first edition (Paris: Mathurinus Dupuis, 1654). Gassendi's interpretative influence was perpetuated by Dreyer [1890] 1977, who used Friis 1871 as his principal source for biographical information, and Friis had leaned heavily on Gassendi.

31. Gassendi 1654, 211–12; Norlind 1951, 192.

32. Koeman 1970.

33. Resen 1668, 310–429.

34. Lundquist 1993.

35. Jern 1986.

REFERENCES

MANUSCRIPT SOURCES

Brahe, Otte Tygesen. 1570. Privatarkiv: Jordebog [cadaster], 17 April 1570. Rigsarkivet, Copenhagen.

Braheana [1589–92]. Thott 1927b quarto. Royal Library, Copenhagen.

Christianson, John R[obert]. 1972a. Astronomy in Print, 1450–1600. Paper presented at the Sixteenth Century Studies Conference, Concordia Seminary, St. Louis, on 26 October.

Frobenius, Georg Ludwig. 1587–99. Selvbiografi 1587–99. Ny Kongelig Samling 2596 folio. Royal Library, Copenhagen.

Stephanius, Johannes. N.d. Konceptbog. Kall 513 quarto. Royal Library, Copenhagen.
_____. 1597–1612. Konceptbog 1597–1612. Ny Kongelig Samling 1390 folio. Royal Library, Copenhagen.

PUBLISHED SOURCES

Alter, George. 1956. *Two Renaissance Philosophers: David Gans, Joseph Delmedigo*. Prague: Czechoslovak Academy of Sciences.

Alver, Bente Gullveig. 1971. *Heksetro og trolldom: En studie i norsk heksevæsen*. Oslo: Universitetsforlaget.

Andersen, Birte. 1971. *Adelig opfostring: Adelsbørns opdragelse i Danmark 1536–1660*. Copenhagen: G. E. C. Gad.

Andersen, J. Oskar. 1896. *Holger Rosenkrantz den Lærde*. Copenhagen: August Bang.

Andersson, Sven-Ingvar. 1993. Uraniborgs Renässansträdgård: Restaureringsprojektets förutsättningar och hypoteser. In Lundquist 1993, 12–20.

Ashworth, William B., Jr. 1990. Natural History and the Emblematic World View. In Lindberg & Westman 1990, 303–32.

Aymard, Maurice. 1989. Friends and Neighbors. In *A History of Private Life: Passions of the Renaissance*, edited by Roger Chartier, translated by Arthur Goldhammer, 3: 447–91. Cambridge, Mass.: Belknap Press of Harvard University Press.

Bagge, Sverre, and Knut Mykland. 1987. *Norge i dansketiden*. Translated by Ole Schierbeck. Copenhagen: Politiken.

Bech, S. Cedergreen, ed. 1979–84. *Dansk biografisk leksikon*. 16 vols. Copenhagen: Gyldendal. *See also* Bricka 1887–1905; Engelstoft & Dahl 1933–43.

Berkner, Lutz K. 1973. Recent Research on the History of the Family in Western Europe. *Journal of Marriage and the Family* (August): 395–405.

Biagioli, Mario. 1993. *Galileo, Courtier: The Practice of Science in the Culture of Absolutism.* Chicago: University of Chicago Press.

Bille Brahe, J. C. 1985. *Gravmindernes vidnesbyrd.* Heraldiske studier 2. Copenhagen: Societas Heraldica Scandinavica.

Bjørnbo, Axel Anthon, and Carl S. Petersen, eds. 1908. *Anecdota cartographica septentrionalia.* Copenhagen: Societatis Regiæ Scientiarum Danicæ.

Blaeu, Joan. 1667. *Le Grand Atlas, ov Cosmographie Blaviane, en laqvelle est exactement discritte la Terre, la Mer, et le Ciel.* 13 vols. Amsterdam: Chez Jean Blaeu.

Blair, Ann. 1990. Tycho Brahe's Critique of Copernicus and the Copernican System. *Isis* 51(3): 355–77.

Boas, Marie, and A. Rupert Hall. 1959. Tycho Brahe's System of the World. *Occasional Notes of the Royal Astronomical Society* 3(21): 253–63.

Boas Hall, Marie. 1975. The Spirit of Innovation in the Sixteenth Century. In *The Nature of Scientific Discovery,* edited by Owen Gingerich, 308–34. Washington, D.C.: Smithsonian Institution.

Bolin, Sture. [1941] 1966. Medieval Agrarian Society in Its Prime: Scandinavia. In *The Cambridge Economic History of Europe.* 2d edition. Edited by M. M. Postan, 5: 633–59. Cambridge: Cambridge University Press.

Brahe, Sophie Axelsdatter. 1955. *Sophie Brahes regnskabsbog 1627–40.* Edited by Henning Paulsen. Aarhus: Jysk selskab for historie, sprog og litteratur.

Brahe, Tycho. 1876. *Tyge Brahes meteorologiske dagbog, holdt paa Uraniborg for aarene 1583–1597.* Edited by F. R. Friis, Poul la Cour, and H. F. Rørdam. Copenhagen: Det kongelige danske videnskabernes selskab.

——— 1913–29. *Tychonis Brahe Dani opera omnia.* Edited by I. L. E. Dreyer and Eiler Nyström. 15 vols. Copenhagen: Gyldendal.

——— 1946. *Tycho Brahe's Description of His Instruments and Scientific Work as Given in "Astronomiæ instauratæ mechanica" (Wandesburgi 1598).* Edited by Hans Ræder, Elis Strömgren, and Bengt Strömgren. Copenhagen: Ejnar Munksgaard.

——— 1996. *Instruments of the Renewed Astronomy.* English translation (*see* Brahe 1946) revised and commented by Alena Hadravová, Petr Hadrava, and Jole R. Shackelford. Prague: Koniasch Latin Press.

Brask, Peter, Karsten Friis-Jensen, Jens Glebe-Møller, Tage Hind, Minna Skafte Jensen, Anders Jørgensen, Sigurd Kværndrup, Jens Rahbek Rasmussen, Janne Risum, and Bjarne Thorud Thomsen. [1984] 1990. *Dansk litteratur historie.* 2d edition. Vol. 2, *Lærdom og magi 1480–1620.* Copenhagen: Gyldendal.

Braun, Wernher von, and Frederick I. Ordway. 1966. *History of Rocketry and Space Travel.* New York: Crowell.

Braunius, Georg, and Frans Hogenberg. 1572–1617. *Civitates orbis terrarum.* 6 vols. Cologne: Braunius & Hogenberg.

Bricka, C. F., ed. 1887–1905. *Dansk biografisk lexikon.* 19 vols. Copenhagen: Gyldendal. *See also* Bech 1979–84; Engelstoft & Dahl 1933–43.

Brix, Hans. 1933–55. To elskovsviser fra det 16. aarhundrede. In *Analyser og problemer: Undersøgelser i den ældre danske litteratur.* 7 vols. 4: 220–47. Copenhagen: Gyldendal – Nordisk forlag.

Burke, Peter. 1995. *The Fortunes of the Courtier: The European Reception of Castiglione's "Cortegiano."* University Park: Pennsylvania State University Press.

Butler, Bartlett R., and J. R. Christianson, eds. 1989. Georg Ludwig Frobenius hos Tycho Brahe på Hven, 15. maj–29. juni 1591. *Magasin fra Det kongelige bibliotek og Universitetsbiblioteket I* 4(2): 51–7.

Carlquist, Gunnar. 1948–85. *Lund stifts herdaminne från Reformationen till nyaste tid.* 2d series, *Biografier.* 10 vols. Lund: C. W. K. Gleerup. *See also* Fabritius 1961.

Caspar, Max. [1948] 1962. *Kepler.* Translated and edited by C. Doris Hellman. New York: Collier Books.

Christensen, Charles, and Francis Beckett. 1921. *Uraniborg og Stjærneborg.* Copenhagen: Aage Marcus, and London: Oxford University Press.

Christensen, Grethe, Karl-Erik Frandsen, Kai Hørby, Benito Scocozza, and Alex Wittendorff, eds. 1984. *Tradition og kritik: Festskrift til Svend Ellehøj den 8. september 1984.* Copenhagen: Den danske historiske forening.

Christianson, John [Robert]. 1961. The Celestial Palace of Tycho Brahe. *Scientific American* 204(2) (February): 118–28.

    1964. *Cloister and Observatory: Herrevad Abbey and Tycho Brahe's Uraniborg.* Ann Arbor: University Microfilms.

    1967. Tycho Brahe at the University of Copenhagen, 1559–1562. *Isis* 58: 198–203.

    1968. Tycho Brahe's Cosmology from the *Astrologia* of 1591. *Isis* 59: 313–18.

    1970a. Tycho Brahe's Facts of Life. *Fund og forskning i den kongelige biblioteks samlinger* 17: 21–8.

    1970b. En gådefuld Tycho Brahe forsker. *Personalhistorisk tidsskrift* 15th series, 4: 76–8.

    1972a. *See* "Manuscript Sources."

    1972b. Addenda to Tychonis Brahe opera omnia tomus xiv. *Centaurus* 1972: 231–47.

    1973a. Tycho Brahe and the Patronage of Science, 1576–1597. In *American Philosophical Society Year Book 1972,* 572–3. Philadelphia: American Philosophical Society.

    1973b. Copernicus and the Lutherans. *Sixteenth Century Journal* 4(2): 1–10.

    1979. Tycho Brahe's German Treatise on the Comet of 1577: A Study in Science and Politics. *Isis* 70: 110–40.

    1998. Tycho Brahe in Scandinavian Scholarship. *Journal for the History of Astronomy* 36: 467–84.

Christianson, Paul. 1976. The Causes of the English Revolution: A Reappraisal. *Journal of British Studies* (Spring): 40–75.

    1991. Royal and Parliamentary Voices on the Ancient Constitution c. 1604–1621. In Peck 1991, 71–93, 289–98.

Copenhaver, Brian P. 1990. Natural Magic, Hermetism, and Occultism in Early Modern Science. In Lindberg & Westman 1990, 261–301.

Dal, Erik. 1962. *Danske viser: Gamle folkeviser, skæmt, efterklang.* Copenhagen: Rosenkilde & Bagger.

Danstrup, John, and Hal Koch, eds. 1962–6. *Danmarks historie.* 14 vols. Copenhagen: Politiken.

Davis, Natalie Zemon. 1983. *The Return of Martin Guerre.* Cambridge, Mass.: Harvard University Press.

    1988. On the Lame. *American Historical Review* 93: 572–603.

Dictionary of Scientific Biography. See Gillispie 1970–80.

Dobrzycki, Jerzy, ed. 1972. *The Reception of Copernicus' Heliocentric Theory*. Boston: D. Reidel.

Donahue, William H. 1975. The Solid Planetary Spheres in Post-Copernican Natural Philosophy. In Westman 1975c, 244–75.

Drake, Stillman, and John W. O'Malley. 1960. *The Controversy on the Comets of 1618*. Philadelphia: University of Pennsylvania.

Dreyer, J. L. E. [1890] 1977. *Tycho Brahe: A Picture of Scientific Life and Work in the Sixteenth Century*. Edinburgh: Adam & Charles Black, reprinted Gloucester, Mass.: Peter Smith.

———. 1916. On Tycho Brahe's Manual of Trigonometry. *Observatory* 39: 127–31.

Eamon, William. 1990. From the Secrets of Nature to Public Knowledge. In Lindberg & Westman 1990, 333–65.

———. 1991. Court, Academy, and Printing House: Patronage and Scientific Careers in Late Renaissance Italy. In Moran 1991b, 25–50.

Efron, Noah J. 1998. Irenism and Natural Philosophy in Rudolfine Prague: The Case of David Gans. *Science in Context* 10(4): 627–49.

Ehrencron-Müller, H. 1924–35. *Forfatterlexikon omfattende Danmark, Norge og Island indtil 1814*. 12 vols. Copenhagen: H. Aschehoug.

Eisenstein, Elizabeth L. 1979. *The Printing Press as an Agent of Change*. 2 vols. New York: Cambridge University Press.

Elgenstierna, Gustaf. 1925–36. *Den introducerade svenska adelns ättartavlor*. 9 vols. Stockholm: P. A. Norstedt.

Ellehøj, Svend, ed. 1988. *Christian IVs verden*. Copenhagen: Nyt Nordisk Forlag Arnold Busck.

Ellehøj, Svend, Leif Grane, Knud Waaben, J. C. Melchior, Povl Johs. Jensen, Mogens Pihl, and Torben Wolff, eds. 1979–93. *Københavns universitet 1479–1979*. 14 vols. Copenhagen: G. E. C. Gad.

Engelstoft, Povl, and Svend Dahl, eds. 1933–44. *Dansk biografisk leksikon*. 27 vols. Copenhagen: J. H. Schultz. *See also* Bech 1979–84; Bricka 1887–1905.

Erlandson, Johanna, ed. 1996. *Tycho Brahe, stjärnornas herre*. Landskrona: Landskrona kommun.

Erslev, Kristian. 1879. *Danmark–Norges len og lensmænd 1513–1596*. Copenhagen: Jacob Erslev.

———. 1885. *Danmark–Norges len og lensmænd 1596–1660*. Copenhagen: Hoffensberg & Trap.

Evans, R. J. W. 1973. *Rudolf II and His World: A Study in Intellectual History 1576–1612*. Oxford: Clarendon Press.

———. 1984. Rantzau and Welser: Aspects of Later German Humanism. *History of European Ideas* 5(3): 257–72.

Fabritius, Albert. 1950. Brahe†. *Danmarks adels årbog*, pt. 2: 3–32. Copenhagen: J. H. Schultz.

———. 1961. Review of Carlquist 1948–85, vol. 8. *Personalhistorisk tidsskrift* 14th series, 3(3): 124–6.

Fichtner, Paul Sutter. 1976. Dynastic Marriage in Sixteenth-Century Habsburg Diplomacy and Statecraft: An Interdisciplinary Approach. *American Historical Review* 81(2): (April 1976): 243–65.

Figala, Karin. 1972. Tycho Brahes Elixier. *Annals of Science* 28(2): 139–76.

Fincke, Thomas. 1583. *Geometria rotundi*. Basel: Sebastian Henricpetri.

Findlen, Paula. 1989. The Museum: Its Classical Etymology and Renaissance Geneal-ogy. *Journal of the History of Collections* 1: 59–78.

———. 1991. The Economy of Scientific Exchange in Early Modern Italy. In Moran 1991b, 5–24.

———. 1994. *Possessing Nature: Museums, Collecting, and Scientific Culture in Early Modern Italy.* Berkeley: University of California Press.

Finlay, Robert. 1988. The Refashioning of Martin Guerre. *American Historical Review* 93: 553–71.

Flandrin, Jean-Louis. 1979. *Families in Former Times: Kinship, Household and Sexuality.* Trans. Richard Southern. Cambridge: Cambridge University Press.

Frandsen, Karl-Erik. 1988. 1536–ca. 1720. In *Den danske landbrugs historie.* Vol. 2, *1536–1810,* edited by Claus Bjørn, 2: 9–209. Copenhagen: Landbohistorisk selskab.

Frederiksen, Anders V. Kaare. 1976. *Familierekonstitution: En modelstudie over befolknings-forholdene i Sejerø sogn 1663–1813.* Copenhagen: Akademisk forlag.

Friis, F. R. 1871. *Tyge Brahe: En historisk fremstilling efter trykte og utrykte kilder.* Copen-hagen: Gyldendal.

———. 1889. Elias Olsen Morsing og hans observationer. Copenhagen: Martius Truelsen.

———. 1890–1901. *Bidrag til dansk kunsthistorie.* Copenhagen: Th. Linds efterfølger Hans Frandsen.

———. 1899–1900. Nogle meddelelser om Erik Lange den yngre til Engelsholm. *Samlinger til jydsk historie og topografi* 3d series, 2: 495–507.

———. 1902. *Nogle efterretninger om Tyge Brahe og hans familie.* Copenhagen: G. E. C. Gad.

———. 1904. *Peder Jakobsen Flemløs, Tyge Brahes første medhjælper, og hans observationer i Norge.* Copenhagen: G. E. C. Gad.

———. 1905. *Sofie Brahe Ottesdatter: En biografisk skildring.* Copenhagen: G. E. C. Gad.

Fussing, Hans H. [1942] 1973. *Herremand og fæstebonde: Studier i dansk landbrugshisto-rie omkring 1600.* Copenhagen: Nyt Nordisk Forlag–Arnold Busck.

Galison, Peter, and Bruce Hevly, eds. 1992. *Big Science: The Growth of Large-Scale Re-search.* Stanford, Calif.: Stanford University Press.

Galison, Peter, Bruce Hevly, and Rebecca Lowen. 1992. Controlling the Monster: Stan-ford and the Growth of Physics Research, 1935–1962. In Galison & Hevly 1992, 46–77.

Garstein, Oskar. 1953. *Cort Aslakssøn: Studier over dansk-norsk universitets- og lærdoms-historie omkring år 1600.* Oslo: Lutherstiftelsen.

Gascoigne, John. 1990. A Reappraisal of the Role of the Universities in the Scientific Revolution. In Lindberg & Westman 1990, 207–60.

Gassendi, Pierre. [1654]. *See* Norlind 1951.

———. 1655. *Tychonis Brahe, Equitis Dani, Astronomorum Coryphæi, Vita.* The Hague: Adrian Vlacq.

Geisberg, Max. 1974. *The German Single-Leaf Woodcut, 1500–1550.* 4 vols. Revised and edited by Walter L. Strauss. New York: Hacker Art Books.

Gillispie, Charles Coulston, ed. 1970–80. *Dictionary of Scientific Biography.* 16 vols. New York: Charles Scribner's Sons.

Gingerich, Owen. 1986. Copernicus's *De revolutionibus:* An Example of Renaissance Sci-entific Printing. In *Print and Culture in the Renaissance,* edited by Gerald P. Tyson and Sylvia S. Wagonheim, 55–77. Newark: University of Delaware Press.

———. 1989. Johannes Kepler. In Taton & Wilson 1989, 2A: 54–78.

Gingerich, Owen, and James R. Voelkel. 1998. Tycho Brahe's Copernican Campaign. *Journal for the History of Astronomy* 29: 1–34.

Gingerich, Owen, and Robert S. Westman. 1988. *The Wittich Connection: Conflict and Priority in Late Sixteenth-Century Cosmology. Transactions of the American Philosophical Society* 78(7). Philadelphia: American Philosophical Society.

Girouard, Mark. 1978. *Life in the English Country House.* New York: Penguin Books.

Gotfredsen, Edv. 1955. Tycho Brahes sidste sygdom og død. *Fund og forskning* 2: 3–8.

Goulding, Robert. 1995. Henry Savile and the Tychonic World System. *Journal of the Warburg and Courtauld Institutes,* 58: 152–79.

Greenblatt, Stephen. 1980. *Renaissance Self-fashioning: From More to Shakespeare.* Chicago: University of Chicago Press.

Gregersen, H. V. 1957. *Niels Heldvad (Nicolaus Helduaderus): En biografi.* Copenhagen: Historisk samfund for Sønderjylland.

Gunnarsson, Allan. 1993. Något om fruktträden i Renässansens trädgårdar og särskilt i Uraniborg. In Lundquist 1993, 41–64.

Haasbroek, N. D. 1968. *Gemma Frisius, Tycho Brahe and Snellius and Their Triangulations.* Delft: Rijkscommissie voor Geodesie.

Hall, A. R[upert]. 1983. *The Revolution in Science 1500–1750.* New York: Longman.

Hammond, N. G. L., and H. H. Scullard. 1970. *The Oxford Classical Dictionary.* 2d edition. Oxford: Clarendon Press.

Hanawalt, Barbara A. 1986. *The Ties That Bound: Peasant Families in Medieval England.* New York: Oxford University Press.

Hannaway, Owen. 1986. Laboratory Design and the Aim of Science: Andreas Libavius versus Tycho Brahe. *Isis* 77: 585–610.

Hansen, Svend Aage. 1964. *Adelsvældens grundlag.* Studier fra Københavns Universitets Økonomiske Institut, no. 6. Copenhagen: G. E. C. Gad.

Hareven, Tamara K. 1991. The History of the Family and the Complexity of Social Change. *American Historical Review* 96: 95–124.

Hartfelder, Karl. [1899] 1964. *Philipp Melanchthon als Preceptor Germaniae.* Berlin: A. Hofmann, reprinted Nieuwkoop: B. de Graaf.

Haskell, Francis. 1993. *History and Its Images: Art and the Interpretation of the Past.* New Haven: Yale University Press.

Hegelund, Peder. [1578] 1972. *Susanna.* Edited by Aage Jørgensen. Copenhagen: Munksgaard.

Heiberg, Johan Ludvig. 1846. Sophie Brahe: En hverdags-historie fra den 16de og 17de aarhundrede. In *Urania: Aarbog for 1846,* 171–210. Copenhagen: C. A. Reitzel.

Heiberg, Steffen. 1984. Samtidige portrætter af Frederik II. In G. Christensen et al. 1984, 183–204.

——— 1988. *Christian 4: Monarken, mennesket og myten.* Copenhagen: Gyldendal.

Helk, Vello. 1987. *Dansk–norske studierejser fra reformationen til enevælden 1536–1660.* Odense: Odense Universitetsforlag.

Hellman, C. Doris. 1944. *The Comet of 1577: Its Place in the History of Astronomy.* New York: Columbia University Press.

——— 1959. The New Star of 1572: Its Place in the History of Astronomy. *Actes du IXe Congrès International d'Histoire des Sciences,* 483–7.

Hellstedt, Jane. 1990. *Slott och borgherrar i Skåne.* Kristianstad: Rabén & Sjögren.

Herlihy, David. 1991. Family. *American Historical Review* 96: 1–16.

Hoffmann, Friedrich Lorenz. 1867. *Der gelehrte Buchhändler Georg Ludwig Frobenius in Hamburg.* Hamburg: Hoffmann & Campe.

Honnens de Lichtenberg, Hanne. 1985. John Gregor van der Schardt: Sculptor – and Architect. *Hafnia: Copenhagen Papers in the History of Art* 10: 147–64.

1991. *Johan Gregor van der Schardt: Bildhauer bei Kaiser Maximilian II., am dänischen Hof und bei Tycho Brahe.* Translated from Danish by Georg Albrecht Mai. Copenhagen: Museum Tusculanum.

1994. Tycho Brahe als Mäzen. In *Europa in Scandinavia: Kulturelle und soziale Dialoge in der frühen Neuzeit,* edited by Robert Bohn, 91–7. Frankfurt am Main.

Hounshell, David A. 1992. DuPont and the Management of Large-Scale Research and Development. In Galison & Hevly 1992, 236–61.

Hübner, Johann. 1737. *Genealogische Tabellen, nebst denen darzu gehörigen genealogischen Fragen, zur Erläuterung der politischen Historie.* 3 vols. Leipzig: Johann Friedrich Gleditschens seel. Sohn.

Ilsøe, Grethe. 1980. Den danske rigsråd 1570–88. In Jespersen 1980b, 9–33.

Jardine, Nicholas. 1984. *The Birth of History and Philosophy of Science: Kepler's "A Defense of Tycho against Ursus" with Essays on Its Provenance and Significance.* Cambridge: Cambridge University Press.

Jarrell, Richard A. 1989. The Contemporaries of Tycho Brahe. In Taton & Wilson 1989, 2A: 22–32.

Jensen, Frede P. 1978. *Bidrag til Frederik II's og Erik XIV's historie.* Copenhagen: Den danske historiske forening.

1982. *Danmarks konflikt med Sverige 1563–1570.* Copenhagen: Den danske historiske forening.

Jern, [Carl] Henrik. 1976. *Uraniborg: Herresäte och himlaborg.* Lund: Studentlitteratur.

1986. *Tycho Brahe-minnen på Ven.* Svenska Kulturminnen 3. Stockholm: Riksantikvarieämbetet.

Jespersen, Knud J. V. 1980a. Rekrutteringen til rigsrådet i Christian 4.s tid. In Jespersen 1980b, 35–122.

ed. 1980b. *Rigsråd, adel og administration 1570–1648.* Odense: Odense Universitetsforlag.

Kaae, Bue, ed. 1976. *Peder Hegelunds almanakoptegnelser 1565–1613.* 2 vols. Ribe: Historisk samfund for Ribe amt.

Kæmpe, Bent, and Claus Thykier. 1993. Tycho Brahe død af forgiftning? Bestemmelse af gifte i skæg og hår ved atomabsorptionsspektrometri. *Naturens verden* 425–34.

Kæmpe, B[ent]., C[laus]. Thykier, and N. A. Pedersen. 1993. The Cause of Death of Tycho Brahe in 1601. *Proceedings of the 31st TIAFT Congress 15–20 August 1993 in Leipzig,* 1–7.

Karker, Allan. 1964. Kong Frederik 2.s død. *Jyske samlinger* new series, 6(3): 253–81.

Kepler, Johannes. [1858–71] 1971. *Johannes Kepleri astronomi opera omnia.* Edited by Christian Frisch. 8 vols. Frankfurt–Erlangen: Heyder & Zimmer, reprinted Hildesheim: Gerstenberg.

1937–90. *Gesammelte Werke.* Edited by Walther von Dyck, Max Caspar, Franz Hammer, and Martha List. 20 vols. Munich: C. H. Beck.

Kettering, Sharon. 1986. *Patrons, Brokers, and Clients in Seventeenth-Century France.* New York: Oxford University Press.

1989. Patronage and Kinship in Early Modern France. *French Historical Studies* 16(2): 408–35.

Keuning, Johannes. 1956. The Van Langren Family. *Imago Mundi: Jahrbuch der alten Kartographie* 13: 101–9.

1973. *Willem Jansz. Blaeu: A Biography and History of His Work as a Cartographer and Publisher.* Amsterdam: Theatrum Orbis Terrarum.

Koeman, C. *Joan Blaeu and His Grand Atlas.* Amsterdam: Theatrum Orbis Terrarum.

Koestler, Arthur. 1960. *The Watershed: A Biography of Johannes Kepler.* Garden City, N.Y.: Anchor Books.

Kongsted, Ole. 1990. *Kronborg motetterne tilegnet Frederik II og Dronning Sophie 1582.* Edited by Jesper Düring Jørgensen and Erik Dal. Copenhagen: Det kongelige Bibliotek.

Kornerup, Bjørn. 1928–68. *Biskop Hans Poulsen Resen.* 2 vols. Copenhagen: G. E. C. Gad.

1959. Den lærde tidsrum 1536–1670. In *Den danske kirkes historie*, edited by Hal Koch and Bjørn Kornerup. 8 vols, 1950–66, 4: 11–379. Copenhagen: Gyldendal.

Krstović, Jelena O., and Marie Lazzari, eds. 1999. *Literature Criticism from 1400 to 1800.* Vol. 45, *Tycho Brahe, Nicolaus Copernicus, Galileo Galilei, Johannes Kepler.* Detroit: Gale Research.

Ladewig Petersen, Erling. 1980a. *Dansk social historie.* Vol. 3, *Fra standssamfund til rangssamfund 1500–1700.* Copenhagen: Gyldendal.

1980b. Rigsråd og adelsopposition 1588: En socialhistorisk studie. In Jespersen 1980b, 123–68.

[Langebek, Jacob, ed.] 1746. En skue-penge, med endeel mest utrykte efterretninger, om Tyge Brahe. *Danske magazin* 2: 161–372.

Laursen, L., ed. 1885–. *Kancelliets brevbøger.* Copenhagen: C. A. Reitzel. (The series began with the year 1561.)

Liliencron, R. von, F. X. von Wegele, and A. Bettelheim, eds. 1875–1912. *Allgemeine Deutsche Biographie.* 56 vols. Leipzig: Duncker & Humblot.

Lindberg, David C., and Robert S. Westman, eds. 1990. *Reappraisals of the Scientific Revolution.* New York: Cambridge University Press.

List, Martha. 1961. *Der handschriftliche Nachlaß der Astronomen Johannes Kepler und Tycho Brahe.* Munich: Bayerischen Akademie der Wissenschaften.

Lorenzen, Vilh. 1912. *Rantzauske borge og herresæder i 16 aarhundrede efter den Rantzauske tavle.* Copenhagen: Andr. Fred. Høst & søn.

Lundquist, Kjell. 1995. The plant material in the Renaissance garden of Tycho Brahe at Uranienborg (1581–1597) on the island of Ven – A reconstruction project. Abstract of paper presented at international symposium, Botanic Gardens: Past, Present, Future. Padua, Italy, 29–30 June 1995. Program and abstracts, 41.

1996. Kulturväxter i Tycho Brahes trädgårdar. In Erlandson 1996, 108–46.

1998. The plant material in the Renaissance garden of Tycho Brahe at Uraniborg (1581–1597) on the island of Ven – A restoration project in progress. *Museologica scientifica* 14(1) (supplement): 223–35.

ed. 1993. *Uraniborgs Renässansträdgård, Renässansens växtmaterial: Rapport från ett seminarium på Alnarp 19 feb 1991.* Stencil series 93: 7. Alnarp: Institutionen för Landskapsplanering, Sveriges Landbruksuniversitet.

DATE ORDER RECEIVED

Wednesday, December 13, 2023

The Maryland Book Bank

1794 Union Ave
Baltimore, MD 21211

```
*  5  9  4  3  0  9  *
```

Number of items: 1   Shipping Method: **STANDARD**

Your Order From:

Amazon

Locator Code	SKU	Title
	3-II-1-0111	On Tychos Island: ...

Customer Name: Thomas A. Anastasio

Order Date: 12/11/2023

Ship To: Thomas Anastasio
121 W MAIN ST
HYDE PARK, VT 05655-4421
USA

For Customer Service inquiries on your order, please contact us at:
customerservice@marylandbookbank.org

Our Guarantee: We appreciate your business and are committed to providing the best possible service. If you are unhappy with your purchase, please contact us immediately at customerservice@marylandbookbank.org and if we are unable to correct the problem to your satisfaction, we will refund your purchase. Thank you for your order!

Amazon Order Number: 112-4772273-7111455

LOCATION: TYPE:paperback	3.50
TITLE: On Tychos Island: Tycho Brahe, Science, and Culture in the Sixteenth Century	
ISBN/UPC No.: 9780521008846 INTERNAL SKU No.: 3-II-1-0111	
Description: Used - Very Good	
Amazon's Shipping and Handling	4.27
ITEM TOTAL - Amount Collected by Amazon	7.77

Maarbjerg, John P. 1995. *Scandinavia in the European World-Economy ca. 1570–1625: Some Local Evidence of Economic Integration.* New York: Peter Lang.

Martin, John. 1997. Inventing Sincerity, Refashioning Prudence: The Discovery of the Individual in Renaissance Europe. *American Historical Review* 102: 1308–42.

Mauss, Marcell. [1925] 1967. *The Gift: Forms and Functions of Exchange in Archaic Societies.* Translated by Ian Cunnison, introduction by E. E. Evans-Pritchard. New York: W. W. Norton.

Moesgaard, Kristian Peder. 1972a. Copernican Influence on Tycho Brahe. In Dobrzycki 1972, 31–55.

———. 1972b. How Copernicanism Took Root in Denmark and Norway. In Dobrzycki 1972, 117–51.

———. 1975. Tychonian Observations, Perfect Numbers, and the Date of Creation: Longomontanus's Solar and Precessional Theories. *Journal for the History of Astronomy* 6: 84–99.

———. 1977a. Cosmology in the Wake of Tycho Brahe's Astronomy. In *Cosmology, History, and Theology,* edited by Wolfgang Yourgrau and Allen D. Breck, 293–305. New York: Plenum Press.

———. 1977b. *Kosmos af Chronos: Træk af en udviklingshistorie for vort astronomiske verdensbillede.* Acta Jutlandica 47, Naturvidenskabelig serie 7. Aarhus: Acta Jutlandica.

Moesgaard, Kristian Peder, Kurt Møller Pedersen, and Bengt Strömgren. 1983. Astronomi. In Ellehøj et al. 1979–93, 12(1).

Møller Nicolaisen, N A. 1946. *Tycho Brahes papirmølle paa Hven: Udgravningen 1933–34 og forsøg til rekonstruktion.* Copenhagen: Gyldendal–Nordisk forlag.

Moran, Bruce T. 1981. German Prince-Practitioners: Aspects in the Development of Courtly Science, Technology, and Procedures in the Renaissance. *Technology and Culture* 22: 253–74.

———. 1982. Christoph Rothmann, the Copernican Theory, and Institutional and Technical Influence on the Criticism of Aristotelean Cosmology. *Sixteenth Century Journal* 13(3): 85–108.

———. 1991a. Patronage and Institutions: Courts, Universities, and Academies in Germany; an Overview: 1550–1750. In Moran 1991b, 169–83.

———. ed. 1991b. *Patronage and Institutions: Science, Technology, and Medicine at the European Court 1500–1750.* Rochester: Boydell Press.

Mortensen, Harald. 1931. Tycho Brahes bogejermærke. *Nordisk tidskrift för bok- och biblioteksväsen* 18: 141–5.

———. 1946. Portræter af Tycho Brahe. *Cassiopeia* 18: 53–76.

Needham, Joseph, and Wang Ling. 1959. *Science and Civilisation in China.* Vol. 3, *Mathematics and the Sciences of the Heavens and the Earth.* Cambridge: Cambridge University Press.

Nehamas, Alexander, and Paul Woodruff, trans. and eds. 1989. *Plato: "Symposium."* Indianapolis: Hackett.

Neher, André. 1986. *Jewish Thought and the Scientific Revolution of the Sixteenth Century: David Gans (1541–1613) and His Times.* Translated by David Maisel. New York: Oxford University Press, for Littman Library. *See also* Langermann 1990.

Nelson, John Charles. 1958. *Renaissance Theory of Love: The Context of Giordano Bruno's "Eroici furori."* New York: Columbia University Press.

Nielsen, Lauritz. 1946. *Tycho Brahes bogtrykkeri: En bibliografisk-boghistorisk undersøgelse.* Copenhagen: Vald. Pedersen.

Nordstrand, Ove K. 1961. *Danmarks ældste papirmøller og deres vandmærker.* Copenhagen: Egmont H. Petersen.

Norlind, Wilhelm. 1926. *Ur Tycho Brahes brevväxling.* Lund: C. W. K. Gleerup.

1944. Tycho Brahes världssystem: Hur det tillkom och utformades. *Cassiopeia* 16: 55–75.

1951. *Tycho Brahe: Mannen och Verket. Efter Gassendi översat med kommentar.* Lund: C. W. K. Gleerup.

1970. *Tycho Brahe: En levnadsteckning med nya bidrag belysande hans liv och verk.* Lund: C. W. K. Gleerup.

Norvin, William. 1937–40. *Københavns universitet i Reformationens og orthodoxiens tidsalder.* 2 vols. Copenhagen: Gyldendal–Nordisk forlag.

Nygren, Anders. 1953. *Agape and Eros.* Revised edition. Translated by Philip S. Watson. London: SPCK.

Nyström, Eiler. 1927. Tyge Brahes brud med fædrelandet. In *Festskrift til Kristian Erslev den 28. decbr. 1927 fra danske historikere,* 291–320. Copenhagen: H. Hagerup.

Ödman, Anders. 1996. Arkeologi i Tycho Brahe-lämningarna. In Erlandson 1996, 170–9.

Olsen, Gunnar. [1957] 1975. *Hovedgård og bondegård: Studier over stordriftens udvikling i Danmark i tiden 1525–1774.* Copenhagen: Landbohistorisk selskab.

Ovid. [1916] 1977. *Metamorphoses.* Translated by Frank Justus Miller. Loeb Classical Library. 3d edition, revised by G. P. Gould. 2 vols. Cambridge, Mass.: Harvard University Press.

Ozment, Steven. 1983. *When Fathers Ruled: Family Life in Reformation Europe.* Cambridge, Mass.: Harvard University Press.

Pade, Erling. 1976. *Milevognen og andre ældre opmålingssystemer.* Copenhagen: Høst & søn.

Pade, Marianne, and Minna Skafte Jensen, eds. 1988. *Renæssancen: Dansk–europæisk–globalt.* Renæssancestudier 2. Copenhagen: Museum Tusculanum.

Panofsky, W. K. H. 1992. SLAC and Big Science: Stanford University. In Galison & Hevly 1992, 129–46.

Peck, Linda Levy, ed. 1991. *The Mental World of the Jacobean Court.* New York: Cambridge University Press.

Pestre, Dominique, and John Krige. 1992. Some Thoughts on the Early History of CERN. In Galison & Hevly 1992, 78–99.

Pfeiffer, Rudolf. 1968–76. *History of Classical Scholarship.* 2 vols. Oxford: Clarendon.

Piø, Iørn, and Rita Pedersen. 1984. *Dronningens visebog.* Copenhagen: Foreningen Danmarks folkeminder.

Pogo, Alexander. 1935. Gemma Frisius, His Method of Determining Differences of Longitude by Transporting Timepieces (1530), and His Treatise on Triangulation (1533). *Isis* 22: 469–506.

Prickard, A. O., trans. 1916. The "Mundus Jovialis" of Simon Marius. *Observatory* 39: 367–81, 403–12, 443–52, 498–503.

Redondi, Pietro. 1987. *Galileo Heretic.* Princeton: Princeton University Press.

Resen, Peder Hansen. 1668. *Inscriptiones Haffnienses, Latinæ, Danicæ et Germanicæ, una cum Inscriptionibus Amagriensibus, Uraniburgicis et Stellæburgicis.* Copenhagen: Henrik Gøde.

Richter, Herman. 1939. Willem Jansz. Blaeu with Tycho Brahe on Hven, and His Map of the Island: Some New Facts. *Imago Mundi: Jahrbuch der alten Kartographie* 3: 53–60.

Robb, Nesca A. [1935] 1968. *Neoplatonism in the Italian Renaissance*. New York: Octagon Books.

Roberts, Michael. 1968. *The Early Vasas: A History of Sweden, 1523–1611*. Cambridge: Cambridge University Press.

Rørdam, Holger [Frederik]. 1863–74. *Kjøbenhavns universitets historie fra 1537–1621*. 4 vols. Copenhagen: Bianco Luno.

——— 1883–9. *Danske kirkelove*. 3 vols. Copenhagen: Selskabet for Danmarks kirkehistorie.

——— 1899–1901. Om Tyge Brahes præst paa Hven. *Kirkehistoriske samlinger* 4th series, 6: 614–17.

Rosen, Edward. 1986. *Three Imperial Mathematicians: Kepler Trapped between Tycho Brahe and Ursus*. New York: Abaris Books.

Sandys, John Edwin. 1921. *A History of Classical Scholarship*. 3d edition. 3 vols. Cambridge: Cambridge University Press.

Sarasohn, Lisa T. 1993. Nicolas-Claude Fabri de Peiresc and the Patronage of the New Science in the Seventeenth Century. *Isis* 84: 70–90.

Sarton, George. [1952–9] 1964–5. *A History of Science*. 2 vols. New York: John Wiley & Sons.

Schama, Simon. 1988. *The Embarrassment of Riches: An Interpretation of Dutch Culture in the Golden Age*. Berkeley: University of California Press, distributed by Random House.

Schepelern, H. D. 1971. *Museum Wormianum: Dets forudsætninger og tilblivelse*. Copenhagen: Wormianum.

——— 1988. Den lærde verden. In Ellehøj 1988, 276–301.

Schofield, Christine. 1989. The Tychonic and Semi-Tychonic World Systems. In Taton & Wilson 1989, 2A: 33–44.

Schwiebert, E. G. 1950. *Luther and His Times: The Reformation from a New Perspective*. St. Louis: Concordia.

Seidel, Robert. 1992. The Origins of the Lawrence Berkeley Laboratory. In Galison & Hevly 1992, 21–45.

Sello, G. 1896. *Des David Fabricius Karte von Ostfriesland und andere Fabriciana des Oldenburger Archivs*. Norden & Norderney: Herm. Braams.

Seters, W. H. van. 1969. Een Boheemse tak van het geslacht Gansneb genaamd Tengnagel tot de Camp. *Jaarboek van het Centraal Bureau voor Genealogie* 23(2): 41–65.

Shackelford, Jole. 1991. Paracelsianism and Patronage in Early Modern Denmark. In Moran 1991b, 85–109.

——— 1993. Tycho Brahe, Laboratory Design, and the Aim of Science: Reading Plans in Context. *Isis* 84: 211–30.

Skougaard, Mette, Helle Thordur Hansen, and Mona Rasmussen. 1984. *Bondens have: Bondehavernes udformning, dyrkning og anvendelse i 1800–årene*. Copenhagen: Nationalmuseet.

Skov, Hans, and E. Bastholm, eds. 1979. *Petrus Severinus og hans "Idea medicinæ philosophicæ": En dansk paracelsist*. Odense: Odense Universitetsforlag.

Skrubbeltrang, Fridlev. 1978. *Den danske landbosamfund 1500–1800*. Copenhagen: Den danske historiske forening.

Smith, Robert W. 1992. The Biggest Kind of Big Science: Astronomers and the Space Telescope. In Galison & Hevly 1992, 184–211.

Stevenson, Edward Luther. 1914. *Willem Janszoon Blaeu, 1571–1638: A Sketch of His Life and Work.* New York: Hispanic Society of America.

Stybe, Svend Erik. 1979. *Copenhagen University: 500 Years of Science and Scholarship.* Translated by Reginald Spink. Copenhagen: Royal Danish Ministry of Foreign Affairs.

Tamm, Ditlev. 1987. *Christian den fjerdes kanslere.* Copenhagen: G. E. C. Gad.

Taton, René, and Curtis Wilson, eds. 1989. *The General History of Astronomy.* Vol. 2, *Planetary Astronomy from the Renaissance to the Rise of Astrophysics,* Pt. A: *Tycho Brahe to Newton.* New York: Cambridge University Press.

[Thiset, A.] 1904. Mule† af Odense. *Danmarks adels årbog,* 322–33.

Thoren, Victor E. 1967. Tycho and Kepler on the Lunar Theory. *Publications of the Astronomical Society of the Pacific* 79: 483–9.

———. 1972. An "Unpublished" Version of Tycho Brahe's Lunar Theory. *Centaurus* 16: 203–30.

———. 1973. New Light on Tycho's Instruments. *Journal for the History of Astronomy* 4: 25–45.

———. 1985. Tycho Brahe as the Dean of a Renaissance Research Institute. *Religion, Science, and Worldview,* 275–95. Edited by Margaret J. Osler and Paul Lawrence Farber. Cambridge: Cambridge University Press.

———. 1989. Tycho Brahe. In Taton & Wilson 1989, 2A: 3–21.

———. 1990. With contributions by John R. Christianson. *The Lord of Uraniborg: A Biography of Tycho Brahe.* Cambridge & New York: Cambridge University Press.

Thykier, Claus. ed. 1990. *Dansk astronomi gennem firehundrede år.* 3 vols. Copenhagen: Rhodos.

Traweek, Sharon. 1992. Big Science and Colonialist Discourse: Building High-Energy Physics in Japan. In Galison & Hevly 1992, 100–28.

Trevor-Roper, Hugh. 1985. The Paracelsian Movement. In *Renaissance Essays,* 149–99. Chicago: University of Chicago Press.

Troels-Lund, Troels Frederik. 1910–12. *Historiske fortællinger: Tider og tanker.* 8 pts. in 4 vols. Copenhagen & Oslo: Gyldendal.

———. [1879–1901] 1968–9. *Dagligt liv i Norden i den sekstende århundrede.* Edited by Erik Kjærsgaard. 7 vols. Copenhagen: Gyldendal.

Vammen, Tine. 1977. Familiehistorisk forskning: En introduktion. *Fortid og nutid* 27(2): 163–93.

Van Helden, Albert. 1974. The Telescope in the Seventeenth Century. *Isis* 65: 38–58.

———. 1977. *The Invention of the Telescope.* Transactions of the American Philosophical Society, 67(pt. 4). Philadelphia: American Philosophical Society.

———. 1989. The Telescope and Cosmic Dimensions. In Taton & Wilson 1989, 2A: 106–18.

Ven. 1980. Stadsingenjörskontorets turistkarta över Ven i Landskrona, utgiven år 1980, skala 1:10,000.

Voelkel, James R. 2001. *The Composition of Kepler's "Astronomia nova."* Princeton: Princeton University Press.

Wad, Gustav Ludvig. 1893. *Breve til og fra Herluf Trolle og Birgitte Gjøe.* 2 vols. Copenhagen: Thaning & Appel.

Walker, Daniel Pickering. 1972. *The Ancient Theology: Studies in Christian Platonism from the Fifteenth to the Eighteenth Century.* Ithaca: Cornell University Press.

Warner, D[eborah]. J. 1971. The First Celestial Globe of Willem Janszoon Blaeu. *Imago Mundi: Jahrbuch der alten Kartographie* 25: 29–38.

1979. *The Sky Explored: Celestial Cartography 1500–1800.* New York: Alan R. Liss, and Amsterdam: Theatrum Orbis Terrarum.

Wattenberg, Diedrich. 1964. *David Fabricius: Der Astronom Ostfrieslands (1564–1617).* Vorträge und Schriften, 19. Berlin-Treptow: Archenhold Sternwarte.

Webster, Charles. 1975. *The Great Instauration: Science, Medicine, and Reform, 1626–1660.* London: Duckworth.

1982. *From Paracelsus to Newton: Magic and the Making of Modern Science.* Cambridge: Cambridge University Press.

Wegener, C. F. 1846. *Om Anders Sørensen Vedel, kongelig historiograph i Frederik IIs og Christian IVs dage.* Copenhagen: Bianco Luno.

Weistritz, Philander von der [pseudonym for C. G. Mengel]. 1756. *Lebensbeschreibung des berühmten und gelehrten dänischen Sternsehers Tycho v. Brahes. Aus der dänischen Sprache in die deutsche übersetzt.* 2 vols. Copenhagen & Leipzig: Friederich Christian Pelt.

Westergaard, P. B. C. 1930–3. *Danske portræter i kobberstik, litografi og træsnit: En beskrivende fortegnelse.* 2 vols. Copenhagen: P. Haase.

Westfall, Richard S. 1985. Science and Patronage: Galileo and the Telescope. *Isis* 76: 11–30.

Westman, Robert S. 1972. The Comet and the Cosmos: Kepler, Mästlin and the Copernican Hypothesis. In Dobrzycki 1972, 7–30.

1975a. The Melanchthon Circle, Rheticus, and the Wittenberg Interpretation of the Copernican Theory. *Isis* 66: 165–93.

1975b. Three Responses to the Copernican Theory: Johannes Praetorius, Tycho Brahe, and Michael Maestlin. In Westman 1975c, 285–345.

1980. The Astronomer's Role in the Sixteenth Century: A Preliminary Study. *History of Science* 18: 105–47.

1990. Proof, Poetics, and Patronage: Copernicus's Preface to *De Revolutionibus.* In Lindberg & Westman 1990, 167–205.

ed. 1975c. *The Copernican Achievement.* Berkeley: University of California Press.

Wittendorff, Alex. 1994. *Tyge Brahe.* Copenhagen: G. E. C. Gad.

Wolter, Hans Chr. 1982. *Adel og embede: Embedsfordeling og karriermobiliten hos den dansk–norske adel 1588–1660.* Copenhagen: Den danske historiske forening.

Yates, Frances A. [1964] 1969. *Giordano Bruno and the Hermetic Tradition.* New York: Vintage Books.

Yde-Andersen, David. 1952. To hodometre fra Renaissancen. *Fra National museets arbejdsmark* 72–8.

Zeeberg, Peter. 1987. Kemi og kærlighed: Naturvidenskab i Tycho Brahes latindigtning. In *Litteratur og lærdom: Dansk–svenske nylatindage april 1985,* edited by Marianne Alenius and Peter Zeeberg, 149–61. Renæssance studier 1. Copenhagen: Museum Tusculanum.

1988. Amor på Hven: Tycho Brahes digt til Erik Lange. In Pade & Skafte Jensen 1988, 161–81.

1991a. Adel og lærdom hos Tycho Brahe. *Latin og nationalsprog i Norden 1500–1800,* edited by Marianne Alenius, Birger Bergh, Ivan Boserup, Karsten Friis-Jensen, and

Minne Skafte Jensen, 21–31. Renæssancestudier 5. Copenhagen: Museum Tusculanum.

1991b. Science versus Secular Life: A Central Theme in the Latin Poems of Tycho Brahe. *Acta Conventus Neo-Latini Torontonensis*, 831–8. Binghamton: Medieval & Renaissance Texts and Studies.

1993. *Den praktiske muse: Tycho Brahes brug af latindigtningen.* Studier fra sprog- og oldtidsforskning udgivet af Den filologisk-historiske Samfund no. 321. Copenhagen: Museum Tusculanum.

1994a. Alchemy, Astrology, and Ovid – A Love Poem by Tycho Brahe. *Acta Conventus Neo-Latini Hafniensis*, 997–1007. Binghamton: Medieval & Renaissance Texts and Studies.

1994b. Neo-Latin Poetry in Its Social Context: Some Statistics and Some Examples from Sixteenth-Century Denmark. In *Mari Balticum – Mare Nostrum: Latin in the Countries of the Baltic Sea (1500–1800),* edited by Outi Merisalo and Raija Sarasti-Wilenius, 9–21. Helsinki: Academia Scientiarum Fennica.

1994c. *Tycho Brahes Urania Titani: Et digt om Sophie Brahe.* Renæssancestudier 7. Copenhagen: Museum Tusculanum.

1995. The Inscriptions at Tycho Brahe's Uraniborg. In *A History of Nordic Neo-Latin Literature,* edited by Minna Skafte Jensen, 251–66. Odense: Odense Universitetsforlag.

1996. En muse finder sit hjem: Tycho Brahe set med hans egne øjne. In Erlandson 1996, 67–78.

Zimmermann, T. C. Price. 1995. *Paolo Giovio: The Historian and the Crisis of Sixteenth Century Italy.* Princeton: Princeton University Press.

Zinner, Ernst. 1956. *Deutsche und niederländische astronomische Instrumente des 11.–18. Jahrhunderts.* Munich: C. H. Beck.

1967. *Deutsche und niederländische astronomische Instrumente des 11.–18. Jahrhunderts.* 2d edition. Munich: C. H. Beck.

INDEX

Note: Italic page numbers indicate illustrations; boldface ones, full biographical entries.

94072885R00206

Made in the USA
Lexington, KY
22 July 2018